Applying Remote Sensing and GIS for Spatial Analysis and Decision-Making

Mouhcine Batchi
University of Ibn Tofail, Morocco

Adil Moumane
University of Ibn Tofail, Morocco

IGI Global
Publishing Tomorrow's Research Today

Vice President of Editorial — Melissa Wagner
Managing Editor of Acquisitions — Mikaela Felty
Managing Editor of Book Development — Jocelynn Hessler
Production Manager — Mike Brehm
Cover Design — Phillip Shickler

Published in the United States of America by
IGI Global Scientific Publishing
701 East Chocolate Avenue
Hershey, PA, 17033, USA
Tel: 717-533-8845
Fax: 717-533-8661
E-mail: cust@igi-global.com
Website: https://www.igi-global.com

Library of Congress Cataloging-in-Publication Data

Names: Batchi, Mouhcine, 1981- editor. | Moumane, Adil, 1984- editor.
Title: Applying remote sensing and GIS for spatial analysis and decision-making / edited by Mouhcine Batchi, Adil Moumane.
Description: Hershey, PA : IGI Global Scientific Publishing, [2025] | Includes bibliographical references and index. | Summary: "The book aims to serve as a valuable resource for researchers, practitioners, students, and policymakers interested in understanding the diverse applications of remote sensing and GIS technologies for spatial analysis and decision-making across various domains"-- Provided by publisher.
Identifiers: LCCN 2024045718 (print) | LCCN 2024045719 (ebook) | ISBN 9798369364529 (hardcover) | ISBN 9798369364536 (paperback) | ISBN 9798369364543 (ebook)
Subjects: LCSH: Remote sensing--Case studies. | Geographic information systems--Case studies.
Classification: LCC G70.4 .A674 2025 (print) | LCC G70.4 (ebook) | DDC 621.36/78--dc23/eng/20241209
LC record available at https://lccn.loc.gov/2024045718
LC ebook record available at https://lccn.loc.gov/2024045719

British Cataloguing in Publication Data
A Cataloguing in Publication record for this book is available from the British Library.

This book is dedicated to the memory of Professor Imad Fenjiro, a remarkable scholar, educator, and mentor, whose profound impact on the field of geography and on the lives of his students and colleagues will never be forgotten. As a professor of geomorphology and modeling at our faculty, Professor Fenjiro was known not only for his vast knowledge but also for his passion in guiding and inspiring the next generation of geographers. His ability to bring complex concepts to life, particularly in the areas of geomorphological processes and spatial modeling, left an indelible mark on everyone who had the privilege of learning from him.

Beyond his academic contributions, Professor Fenjiro was a mentor who encouraged curiosity, fostered critical thinking, and exemplified a spirit of dedication to science and education. His commitment to advancing the study of geography and his tireless efforts in research and teaching have left an enduring legacy within our faculty and the broader academic community.

His sudden passing is a tremendous loss, but his influence endures through the many students he inspired and the work we continue in his memory. This book, rooted in the very subjects he was so passionate about, is a testament to his enduring impact on the field. We honor his contributions to geomorphology, spatial analysis, and geographic modeling, and we dedicate this volume to his memory with deep respect and gratitude.

May his legacy continue to inspire and guide future generations of scholars, and may he rest in peace.

Table of Contents

Detailed Table of Contents

With environmental concerns on the rise, the integration of modern technology with environmental sciences provides large scale and effective monitoring of the natural resources. This chapter aims to provide the basic understanding about Remote Sensing (RS) and Geographic Information Systems (GIS), thereby exploring, in depth, the applications of RS and GIS in environmental monitoring. It provides a comprehensive overview of the principles, methodologies, and applications of RS and GIS in environmental monitoring, and offers insights into the practical implementation of these technologies across diverse environmental contexts. Recognizing the interdisciplinary nature of environmental monitoring, the chapter explores the integration of technology with other scientific disciplines like ecology, hydrology, climatology, biodiversity conservation, and disaster management. The chapter also covers a case study on "Decadal Vegetation Cover Change Study in the Forest Near Sandstone Mines in Shankargarh, Prayagraj, India", which dives into the impact of sandstone mining on the vegetation cover of nearby forest.

Aeolian sediments cover roughly 6% of the Earth's surface, mainly in arid regions, accounting for 97% of their occurrence and spanning about 20% of the terrestrial surface. Sand drifts in these areas threaten fragile ecosystems by disrupting their stability. This study analyzes sand dune encroachment dynamics in the Tafilalet region of southeastern Morocco using geomatics techniques. Sand encroachment endangers human settlements, agricultural lands, and infrastructure. Employing remote sensing methods, including satellite imagery and the Fryberger method for estimating sand dune drifting, this research examines sand movement from 1984 to 2015. Results show a significant increase in sand-covered areas by approximately 44,872 hectares, driven by climatic conditions and human activities. The study highlights the severe risks to the Tafilalet oases and infrastructure, underscoring the need for continuous monitoring and sustainable management practices to mitigate sand encroachment impacts.

The Rif Mountains extends 31.375 km^2, is occupied in the north of Morocco, on the Mediterranean slopes, and the rainfall is relatively abundant in the western part and scarce in the eastern part, and irregular, with strong anthropic pressure. This has affected the use of forest and soil resources. This area has been subject to anthropogenic pressure which is steeped in history as an extensive agricultural- forestry-pastoral production system. This deterioration has resulted in intense erosive dynamics that can be seen in various forms, including, the areal etching, the gully and landslides. This dynamic erosion gradually and inevitably leads to soil degradation, causing serious problems in terms of loss soil and silting up of dams. To evaluate this dynamic, we have integrated of the Erosion Potential Method into a GIS & RS, for soil losses estimation allows to approach the severity of the erosive phenomenon, in order to identify as well as map the sensitive and high-risk areas of water erosion and the regions of intervention urgency in the Rif Mountains. Soil losses range from 17.254 to 0 t/ha/yr, the average rate of losses is 46 t/ha/ yr (Erosion in the plot). Potential total erosion in Rif Mountains in the order of 144.292.972 t/yr. Thus, the average amount delivered per unit area is 8.55 t/ha/yr (total of 26.825.625 t/yr).

Digital photogrammetry is a passive contactless, easy, and low-cost 3D scanning technique, that provides effective and powerful tools for 3D realistic reconstruction, as well as precise and high-resolution models based on Structure from Motion (SfM). Photogrammetry leverages data obtained by Unmanned Aerial Vehicles (UAVs), which have reached significant maturity in the sense that they offer a fast, affordable, and effective method for the acquisition of high-resolution geospatial information and image block (nadir and oblique) with high geometric and temporal resolutions. Data are processed by different available commercial or open-source photogrammetric software aiming at 3D reconstruction modelling. Then, the 3D-generated models can be further processed for quantity assessment by applying geometrical measurements. In addition, models through various geovisualization techniques can be used for realistic representations of the physical world. The current research investigates 3D model reconstruction by applying two photogram-metric software, Agisoft Metashape and 3DF Zephyr, in an example of an archaeological site in Greece focusing on the performance of each approach. Moreover, different distance and noise metrics are applied to compare and geometrically assess the 3D dense point clouds, meshes and Digital Surface Models (DSMs). Furthermore, traditional, and modern geovisualization forms of 3D models are discussed towards approaching the 3D models' representation holistically, aiming to provide guidelines to researchers in this field through a comprehensive workflow from data acquisition to manipulation and capabilities combined with other fields such as Computer Vision. Finally, open issues and future research directions are highlighted as a stimulus to further investigations.

This study integrates topographical, geographic, and infrastructure data into a GIS framework to investigate the complex dynamics of urban traffic congestion in M'diq. The research emphasizes common causes of congestion, including urban morphology, road network connectivity, public service distribution, and the impact of seasonal tourism. A composite heat map is utilized to identify heavily congested areas, such as Lalla Nezha and Nasr Streets. The study elaborates on the historical growth patterns, geographical difficulties, and randomness of the road networks, all of which exacerbate traffic issues. Oriented mainly toward the analysis of the causes of congestion, this research puts a spotlight on data-driven approaches to urban planning as a means to handle them. Findings feed into larger discussions of urban traffic management by underscoring the need for strategies in innovative planning, taking into account M'diq's unique topographical and infrastructural characteristics.

Throughout the history of Kenitra City, the urban policy concept has taken on two main dimensions. First, it is the mechanisms through which the former French colonial authorities deliberately perpetuated an ethnic and spatial divide between Europeans and Moroccan citizens, which was imposed by economic, social, and security motives. Second, it was a political choice that came after independence and remained confined between the sectoral approach to management and the urgent treatment of problems. These two dimensions have led to the fragmentation of the city and the absence of integration and cohesion between its various components. More than a century after the adoption of the urban policy in Morocco, the city of Kenitra is now at a unique turning point in its history. It is at a stage where it will soon decide, in the coming generations, whether the new development experiences will create a barrier and disintegration with the classical urban policy, or continue down a path that will not guarantee decent living prospects. For this reason, based on remote sensing techniques, this work seeks to determine the logic governing urban decision-making and reveal the flaws in urban planning and the city policy followed in Kenitra. This will be achieved by monitoring and analyzing the exacerbations of many urban, social, economic, and environmental problems over recent decades.

This chapter reviews research on the Google Earth Engine (GEE) platform and includes a case study in Vietnam. The literature highlights GEE's transformative role in geo-big data analytics, with its scalable and cost-effective processing of large datasets, despite some limitations in spatial functions and algorithms. The study employed GIS and remote sensing with GEE to generate land cover maps for 1985, 1995, 2005, 2015, and 2022, analyzing changes over this period. Using Landsat imagery and the random forest algorithm, the study achieved high classification accuracy, with overall accuracies above 96% and Kappa coefficients over 0.95. Findings show a shift from bare land and forest to increased crop areas and reduced forest cover due to socio-economic factors. Water bodies initially shrank but later expanded with irrigation development. The study provides insights for land management, while noting limitations like satellite resolution and weather conditions, and suggests future research should incorporate high-resolution data and additional factors affecting land cover changes.

Remote sensing (RS) and Geographic Information Systems (GIS) technologies are receiving significant attention due to their combined potential to revolutionize spatial analysis and enhance understanding of complex environmental and geographic phenomena. This integration has vast applications, spanning from urban planning to engineering technologies. Remote sensing and GIS play essential roles in detecting and mapping weed infestations in agricultural fields, facilitating precise weed management and reducing herbicide use. Advancements include high-resolution imagery and machine learning algorithms, which enhance detection accuracy and efficiency. This chapter reviews and discusses the framework of weed detection based on different techniques, modes of data collection, and categories of weed features, including spectral and spatial features of weeds. Additionally, it discusses the different challenges encountered in the detection of weed.

An artificial intelligence-based weed detection system is a computerized system designed to automatically identify and classify different types of weeds in agricultural fields. The system utilizes advanced computer vision techniques and machine learning algorithms to accurately detect and differentiate weeds from crops or other elements in the field. The weed detection system typically consists of hardware components such as cameras or drones that capture high-resolution images or videos of the agricultural area. These images are then analyzed by the artificial intelligence algorithms which have been trained on large datasets of weed images to recognize and distinguish various weed species. This paper explores the application of AI in weed detection and offers a promising solution for automating weed detection in crops. Furthermore, the work addresses the potential benefits of using automated weed detection systems such as reduced labor costs decreased herbicide usage, and improved environmental sustainability.

The integration of remote sensing technology in soil analysis represents a monumental leap in agricultural sciences. By harnessing the power of satellite imagery, scientists can now monitor and evaluate soil conditions across vast areas with unprecedented accuracy. This non-invasive technique allows for the continuous observation of soil health, aiding in the detection of changes over time due to natural or anthropogenic factors. The data collected from these observations are crucial for developing sustainable farming practices that not only increase crop yields but also protect the environment. Furthermore, the ability to analyze soil properties on such a scale is instrumental in managing land resources effectively, ensuring that the delicate balance between food production and ecological stewardship is maintained. As the global population continues to grow, the role of remote sensing in soil conservation and agricultural productivity becomes ever more essential, offering a beacon of hope for future generations.

Foreword

It is with great pride and enthusiasm that I introduce "Applying Remote Sensing and GIS for Spatial Analysis and Decision-Making," a comprehensive volume that reflects the remarkable strides in geospatial technologies and their application across various disciplines. This book, expertly edited by Dr. Batchi Mouhcine and Dr. Adil Moumane, stands as a testament to the growing importance of Remote Sensing and GIS in addressing the complex environmental, social, and economic challenges we face today.

The Faculty of Humanities and Social Sciences has long recognized the transformative potential of these technologies in advancing research and decision-making. As the doyen of our esteemed faculty, I am particularly pleased to see the diverse range of topics covered in this book, from specialized applications in geospatial analysis to the integration of artificial intelligence and machine learning in enhancing our understanding of the world.

Each chapter, meticulously researched and presented, offers valuable insights into the ways in which Remote Sensing and GIS can be leveraged to monitor environmental changes, manage natural resources, and support sustainable development. The inclusion of case studies from Morocco and beyond further enriches this work, providing practical examples of how these technologies can be applied in real-world scenarios.

I would also like to take this opportunity to express my gratitude to the Editorial Advisory Board— Fatima Ezzahra El Ghazali, Kumud Dubey and Abdessamad Elmotawakkil—for their dedication and valuable contributions to this project. Their hard work has been instrumental in ensuring the quality and cohesiveness of this book, and their expertise is deeply appreciated.

This book is not only a valuable resource for students, researchers, and professionals in the field of geospatial sciences, but it also serves as a guide for policymakers and practitioners seeking to harness the power of technology in their efforts to create a more sustainable and equitable future.

I congratulate all the contributors for their outstanding work and commend this book to all who wish to deepen their understanding of Remote Sensing and GIS. It is a significant contribution to the field and a reflection of the innovative spirit that drives our academic community.

Jamal Al Karkouri
Faculty of Humanities and Social Sciences, University Ibn Tofail, Kenitra, Morocco

Preface

In an era of unprecedented environmental challenges and growing demands on natural resources, the need for advanced tools in spatial analysis and decision-making has never been more critical. "Applying Remote Sensing and GIS for Spatial Analysis and Decision-Making" emerges as a comprehensive guide to these essential tools, focusing on their applications in diverse fields such as water resource management, land degradation, urban development, and environmental monitoring. This book explores how cutting-edge technologies like Remote Sensing (RS), Geographic Information Systems (GIS), and Artificial Intelligence (AI) can provide crucial insights for addressing global and local environmental issues.

The inspiration for this book stems from a recognition of the transformative role these technologies play in fostering informed decision-making across a wide array of sectors. The chapters included here represent the collective expertise of researchers and practitioners who have utilized RS and GIS to tackle real-world challenges. These contributions not only demonstrate the power of spatial technologies but also highlight their potential to foster sustainable solutions, particularly in the face of climate change, resource depletion, and land degradation.

Section I: GIS and Remote Sensing Applications for Water Resource Management delves into the intersection of water resources and spatial technologies. This section addresses the pressing concerns of water scarcity, groundwater contamination, and surface water dynamics, particularly in arid and semi-arid regions. Through case studies like the Akermoud Coastal Water Table and Rheraya Basin in Morocco, this section illustrates the use of GIS and Remote Sensing to evaluate seawater intrusion and optimize dam site selection. Chapters also explore how these tools can aid in monitoring the impacts of agricultural practices and climate change on water bodies, with examples from Morocco's Ben Mansour-Mnasra region and Thailand's tropical streamflow systems.

Chapter 1: This chapter introduces the fundamental concepts of Remote Sensing (RS) and Geographic Information Systems (GIS) in environmental monitoring. It highlights their interdisciplinary applications, covering principles, methodologies, and real-world use cases, including a case study on vegetation cover changes near sandstone mines in India.

Chapter 2: This study examines sand dune encroachment in southeastern Morocco using geomatics and remote sensing techniques. It details the environmental impacts of sand movement on ecosystems and infrastructure, emphasizing the need for sustainable management to mitigate these risks.

Section II: Land Degradation, Erosion, and Geomorphology focuses on the analysis of landscape dynamics using geomatics, with an emphasis on erosion, sand dune encroachment, and land degradation. As desertification and erosion continue to threaten agricultural productivity and biodiversity, the insights offered in this section are invaluable for conservationists and land managers alike. From the Rif Mountains of Morocco to the Tafilalet Region, this section offers a range of methodologies for assessing land vulnerability and environmental degradation through the application of GIS and Remote Sensing.

Chapter 3: Focused on the Rif Mountains in Morocco, this chapter uses GIS and RS techniques to evaluate the erosive impact of human activity on soil degradation. It identifies high-risk areas and proposes intervention strategies to address soil erosion and land loss.

Chapter 4: This chapter explores 3D model reconstruction techniques in archaeology, comparing the performance of two photogrammetric software tools. It discusses different visualization approaches and provides a comprehensive workflow for researchers working with 3D models.

Section III: Urban and Archaeological Studies using GIS and Remote Sensing highlights the use of these technologies in urban planning and heritage conservation. Urbanization, with its complex socio-economic and environmental implications, requires careful monitoring, as evidenced by chapters on traffic congestion in M'diq city and Kenitra's evolving urban policy. Moreover, this section touches on the preservation of archaeological sites through photogrammetric methods, showcasing the ability of GIS to bridge modern urban challenges with historical conservation.

Chapter 5: Using GIS, this study analyzes urban traffic congestion in M'diq, Morocco. It identifies key contributors to traffic issues and offers data-driven strategies for improving urban planning, public transit, and infrastructure design.

Chapter 6: This chapter reviews the urban development challenges in Kenitra, Morocco, using remote sensing. It analyzes decades of urban policy and planning decisions, proposing solutions to address social, economic, and environmental problems in the city.

Section IV: Advancements in Remote Sensing and GIS for Environmental Monitoring and Land Use Analysis captures the recent innovations in spatial analysis tools. From cloud-based platforms like Google Earth Engine to AI-driven analyses of Sentinel-2 imagery, this section explores the next frontier in environmental monitoring. The application of Remote Sensing in land cover change detection, weed detection, and environmental management underscores the versatility of these tools in addressing agricultural, ecological, and urban land use challenges.

Chapter 7: This chapter highlights the role of Google Earth Engine (GEE) in geo-big data analytics, presenting a case study from Vietnam. It examines land cover changes over time and offers insights into land management challenges and opportunities.

Chapter 8: This chapter reviews the use of RS and GIS technologies for weed detection in agriculture. It explores advancements in high-resolution imagery and machine learning algorithms that enhance precision weed management and reduce herbicide use.

Chapter 9: This chapter delves into AI-based weed detection systems, discussing how computer vision and machine learning algorithms can automate weed identification in agricultural fields, improving sustainability and reducing labor costs.

Chapter 10: Remote sensing in soil analysis is revolutionizing agricultural sciences by enabling large-scale, continuous monitoring of soil conditions. This chapter explores the role of satellite imagery in promoting sustainable farming practices and enhancing soil conservation efforts.

Throughout this book, we aim to present not only the technical expertise necessary for spatial analysis but also to underscore the importance of these methods in promoting sustainable decision-making. Each chapter has been crafted to guide readers through practical applications of GIS and Remote Sensing, backed by robust case studies and real-world data.

As we continue to witness rapid advancements in geospatial technologies, the potential to further refine our understanding of the Earth's processes grows. This book is intended to be a resource for students, researchers, and professionals seeking to deepen their knowledge of how GIS and Remote Sensing can be harnessed to solve pressing environmental challenges. We hope that these insights will inspire future applications of spatial analysis that contribute to a more sustainable and equitable world.

Editor

Acknowledgement

I would like to express my deep gratitude to **Jamal Al Karkouri**, Doyen of the Faculty of Humanities and Social Sciences at the University Ibn Tofail, Kenitra, for his unwavering encouragement and support throughout the development of this project. His belief in the significance of geospatial technologies for addressing pressing environmental and societal challenges has been a source of motivation and inspiration.

My heartfelt thanks go to the Editorial Board of this book: **Adil Moumane**, **Fatima Ezzahra El Ghazali, Kumud Dubey** and **Abdessamad Elmotawakkil**. Their dedication, insight, and valuable contributions have played an essential role in shaping the quality and coherence of this work. I deeply appreciate their collaborative spirit and commitment to ensuring that this book reflects the latest advancements and applications of Remote Sensing and GIS.

I also wish to thank all the reviewers who dedicated their time and expertise to ensuring the academic rigor and clarity of the chapters presented here. Their thoughtful feedback and constructive comments have been instrumental in refining the content and enhancing the overall quality of this publication.

Finally, my sincerest appreciation goes to all the authors who have contributed their research and expertise to this book. Their innovative work and passion for advancing the use of geospatial technologies are the foundation of this collective effort. I am honored to have collaborated with such a diverse and talented group of scholars and practitioners, whose contributions will undoubtedly advance the fields of spatial analysis and decision-making.

Thank you all for your commitment to this project and for making this book a reality.

Editor

Editorial Advisory Board

Chapter 1
Exploring the Earth From Afar:
Remote Sensing and GIS in Environmental Monitoring and Management

Kumud Dubey
https://orcid.org/0000-0001-6367-4617
ICFRE-ERC, Prayagraj, India

Vandita
https://orcid.org/0009-0006-4374-5522
ICFRE-ERC, Prayagraj, India & Forest Research Institute University, Dehradun, India

ABSTRACT

With environmental concerns on the rise, the integration of modern technology with environmental sciences provides large scale and effective monitoring of the natural resources. This chapter aims to provide the basic understanding about Remote Sensing (RS) and Geographic Information Systems (GIS), thereby exploring, in depth, the applications of RS and GIS in environmental monitoring. It provides a comprehensive overview of the principles, methodologies, and applications of RS and GIS in environmental monitoring, and offers insights into the practical implementation of these technologies across diverse environmental contexts. Recognizing the interdisciplinary nature of environmental monitoring, the chapter explores the integration of technology with other scientific disciplines like ecology, hydrology, climatology, biodiversity conservation, and disaster management. The chapter also

DOI: 10.4018/979-8-3693-6452-9.ch001

covers a case study on "Decadal Vegetation Cover Change Study in the Forest Near Sandstone Mines in Shankargarh, Prayagraj, India", which dives into the impact of sandstone mining on the vegetation cover of nearby forest.

I. INTRODUCTION

The monitoring and management of the environment is crucial for understanding the Earth's natural systems, identifying environmental changes, and implementing effective conservation measures. Monitoring helps track changes in ecosystems, assess the impacts of human activities, and make informed decisions to protect natural resources. The Remote Sensing (RS) and Geographic Information System (GIS) application in environmental monitoring has become indispensable, as, the integration of technology with environmental sciences provides large scale and effective monitoring of the natural resources. With the availability of data from multiple platforms and sensors, providing the information with a wide range of temporal, spatial, spectral and radiometric properties, remote sensing has become a vital tool for large scale environmental studies (Kakoti & Sarma, 2024). By providing accurate, timely, and comprehensive spatial data, RS and GIS empower researchers, policymakers, and environmental managers to make informed decisions aimed at promoting sustainability and resilience.

The process of gathering data about a phenomenon or an object without coming into direct contact with it is known as remote sensing. It utilizes sensors mounted on drones, aircrafts, satellites, or ground-based platforms to collect data. It measures the reflectance, emission, or scattering of electromagnetic energy to infer information about the Earth's features and properties. The remote sensing principle involves using sensors to detect and record electromagnetic radiation across different wavelengths. The data collected through RS can be in the form of images, spectral information, and other sensor readings, capturing a wide range of environmental parameters over large areas and extended periods. This ability to gather high-resolution data over broad spatial extents and in remote or inaccessible regions makes RS an invaluable tool for environmental monitoring. One of the most significant advantages of working with remote sensing satellite data is the accessibility to the information from several decades ago (Parra, 2022).

Geographic Information System, on the other hand, facilitates storing, analyzing, and visualizing spatial and geographic data. GIS integrates various types of data, including RS imagery, field observations, and statistical data, into a cohesive framework. This integration allows for complex spatial analyses, modelling, and visualization, facilitating a more thorough understanding of spatial patterns and relationships. GIS is not only a repository of spatial data but also a powerful analytical tool that supports decision-making processes in environmental management.

The unsustainability created by over usage of land and water resources need to be monitored for planning sustainable utilization of existing resources (Sultana *et al., 2023*). The integration of RS and GIS offers a sophisticated tool for understanding and managing our planet's resources and ecosystems. By analyzing satellite data over time, researchers can identify trends and patterns in the interdisciplinary field of environmental sciences. These insights are essential for not only developing sustainable practices, but also for mitigating negative environmental impacts. This chapter is aimed to provide researchers, practitioners, and policymakers with the knowledge and tools necessary to understand the remote sensing and GIS application for addressing environmental challenges.

I. i. Key Remote Sensing Tools for Environmental Monitoring:

Optical Satellite Imagery:

- **Multispectral Sensors**: Satellites like Moderate Resolution Imaging Spectroradiometer (MODIS), Sentinel-2, and Landsat collects data across multiple spectral bands, which may be used to monitor land cover changes, health of vegetation, and water quality.
- **High-Resolution Imagery**: High spatial resolution satellites (e.g., WorldView, SkySat, Spot 6, IKONOS, Pleiades 1A & 1B, GeoEye) provide detailed images for detecting small-scale changes.

Synthetic Aperture Radar (SAR):

- **All-Weather Monitoring**: SAR sensors, such as those on Sentinel-1, penetrate clouds and provide data regardless of weather conditions, making them ideal for monitoring soil moisture, deforestation, and urbanization.
- **Surface Deformation Analysis**: SAR data can be used to detect ground displacement and deformation, which are critical for monitoring landslides and subsidence.
- **Surface Roughness Detection**: SAR can also detect changes in surface roughness, which is indicative of urban development.

Thermal Imaging:

- **Temperature Monitoring**: Thermal sensors measure land and water surface temperatures, useful for monitoring thermal pollution, urban heat islands, and volcanic activity.
- **Evapo-transpiration Estimation**: Thermal data can be used to estimate evapo-transpiration rates, important for water balance studies.

LiDAR (Light Detection and Ranging):

- **3-Dimension Topographic Mapping**: LiDAR provides high-resolution elevation data, essential for flood modelling, forest structure analysis, and urban planning. It also provides high-resolution 3-D information about the urban landscape, useful for detailed urban analysis and modelling.

I. ii. GIS for Environmental Monitoring:

GIS integrates spatial data from various sources, including remote sensing, which may be utilized for analyzing, visualizing, and interpretation of changes in the features and phenomena of the Earth. It integrates remotely sensed data with other datasets like topographic maps, cadastral data, and socio-economic data for comprehensive analysis. Areas of change can be identified by overlaying maps from different period using multiple layers of spatial data. Some of the key techniques for environmental monitoring are:

- **Spatial analysis:** GIS enables spatial analysis by overlaying different layers of geographic data to identify patterns, relationships, and trends. It helps in assessing environmental impacts, planning land use, and identifying suitable locations for conservation efforts.
- **Mapping and Visualization:** GIS allows for the creation of maps and visualizations to communicate spatial information effectively. It can represent complex environmental data in a visually intuitive manner, facilitating decision-making and public outreach.
- **Image Differencing**: Detects changes between images of the same area taken at different times. The basic principle involves subtracting the pixel values of one image from the corresponding pixel values of another image to highlight changes. This method is useful for monitoring environmental changes, such as urban expansion, deforestation, land use and land cover (LULC) changes, and disaster impact assessment.

I. iii. Key Environmental Parameters Monitored Using Remote Sensing and GIS:

1. **Vegetation Health**: Assessed using indices like the Normalized Difference Vegetation Index (NDVI). It is an index to estimate vegetation health and coverage. High values of NDVI indicate dense and healthy vegetation, while lower values suggest sparse or stressed vegetation.

2. **Land Cover and Land Use**: Tracks changes in land cover types (e.g., forests, wetlands, urban areas) and land use practices. Normalized Difference Built-up Index (NDBI) is an index to identify and analyze built-up areas, such as urban infrastructure and developed land. High NDBI values indicate built-up or impervious surfaces, while low values correspond to non built-up areas such as fallow land, vegetation cover or water. NDBI is particularly useful for urban planning, monitoring urban sprawl, and assessing the impact of urbanization on the environment.
3. **Water Quality**: Monitors parameters such as chlorophyll concentration, surface temperature, suspended sediments in water bodies, and turbidity, which are indicators of water quality. This data is integrated with other spatial information to create comprehensive maps and models, enabling the identification of pollution sources and trends over time. This approach allows for efficient, large-scale monitoring of water quality, supporting effective management and conservation efforts.
4. **Soil Quality**: Productivity of the crops that grow on degraded soil is lower because the soil is less resilient (Mohamed, 2023). Various soil properties, such as nutrient content, organic matter, and erosion levels are assessed. Multispectral and hyperspectral sensors detect specific soil characteristics by analyzing the reflectance of different wavelengths. This data is integrated with other spatial information to produce comprehensive maps and models, enabling the analysis of soil quality trends and spatial variability. This approach aids in sustainable land management, precision agriculture, and environmental conservation by providing accurate and large-scale soil quality assessments.
5. **Atmospheric Conditions**: Atmospheric parameters such as humidity, pollutant concentrations and temperature are observed and analyzed. Multispectral and hyperspectral sensors provide information on the composition and behaviour of the atmosphere. Integrating this data with spatial and temporal information creates comprehensive models and maps, facilitating the tracking of weather patterns, air quality, and climate change. This combined approach supports environmental research, disaster preparedness, and policy making by delivering accurate and extensive atmospheric condition assessments.

II. APPLICATION OF RS AND GIS IN ENVIRONMENTAL MONITORING AND MANAGEMENT

1. Deforestation Monitoring

Deforestation, the large-scale removal of forest cover, poses a great deal of threat to biodiversity, climate stability, and ecosystem services. Deforestation monitoring helps detect loss of forest cover due to logging, agriculture, or urbanization; aiding in forest management and conservation efforts. Effective monitoring of deforestation is crucial for environmental conservation, climate change mitigation, and sustainable land management. Remote Sensing and GIS are essential technologies for tracking and analyzing deforestation patterns and impacts over time. (Solórzano *et al.* 2023) worked on deforestation detection with SAR and multispectral data using a spatio-temporal deep learning approach. They evaluated the potential of the U-Net 3D, a spatio-temporal deep learning algorithm, with multispectral and SAR data for detecting deforestation in a tropical rainforest of Southeast Mexico.

1. i. Importance of Deforestation Monitoring:

- **Biodiversity Conservation:** Forests are home to many kinds of species. Monitoring deforestation helps protect habitats and preserve biodiversity.
- **Climate Regulation:** Forests act as carbon sink by absorbing $CO_{2,}$ while deforestation releases stored carbon, exacerbating climate change.
- **Water Cycle Maintenance:** Forests play a vital role in the water cycle, influencing precipitation patterns and water availability.
- **Soil Conservation:** Forests prevent soil erosion by binding the soil with their root systems.
- **Sustainable Development:** Deforestation monitoring helps in management of forest resources, balancing economic benefits with ecological preservation.

1. ii. Applications of RS and GIS in Deforestation Monitoring:

- **Mapping and Visualization:** RS and GIS enable the creation of detailed maps showing the extent of forest cover, deforested areas, and changes over time. These visual representations help in understanding spatial patterns and trends in deforestation. (Júnior *et al.*, 2022) studied LULC dynamics in the Brazilian Caatinga Dry Tropical Forest using GEE platform, QGIS software and Landsat 8 satellite data. They examined the native vegetation loss in the Campo Maior Complex, from the year 2016 to 2020. For the analyses, they considered the temporal rainfall data, and trends and impacts of forest degradation vectors.

- **Change Detection:** By comparing temporal data, change in forest cover is detected. This helps identify areas of recent deforestation and monitor the effectiveness of reforestation efforts. (Islam *et al.*, 2021) studied decadal changes in forest cover in Nijhum Dwip National Park in Bangladesh using open source GIS tool (QGIS) and Landsat satellite data. They used maximum likelihood classification tool to classify the satellite images; and also evaluated the potential of SAVI (Soil Adjusted Vegetation Index) and NDVI based classification for detecting the changes in forest cover in comparison to supervised classification.
- **Forest Cover Analysis:** Data on various attributes, such as canopy density, species composition, and biomass, is used to assess the health and quality of forests.
- **Hotspot Identification:** Using RS & GIS deforestation hotspots can be identified, where the rate of forest loss is highest. This helps prioritize areas for conservation efforts and policy intervention.
- **Carbon Stock Estimation:** Data on forest biomass is used to estimate carbon stocks. It is integrated with other spatial information to model carbon sequestration and emissions from deforestation.
- **Biodiversity Conservation:** Deforestation data is overlaid with habitat maps to identify critical habitats and biodiversity hotspots that need protection, prioritizing them for conservation efforts. (Zhang *et al.*, 2021) used deep learning framework to study the deforestation changes with the help of remote sensing images. Their work aimed to help the department of forest manage forest resources more efficiently. For their research they used a U-Net+LSTM framework for detecting the changes in deforestation.
- **Disaster Management:** Forest fires and other natural disasters affecting forest are monitored. This helps in planning disaster response and assessing the impact on forest cover. Information on fire speed and direction is obtained by comparing the burnt area and active fires (Tichouiti, 2024). In their study they compared Normalized Burnt Ratio (NBR) and NDVI using Sentinel-2 imagery, to understand which index gives more precise results for characterization of burnt areas.
- **Climate Change Studies:** GIS integrates the data on land use changes and forest cover with climate models to study the impacts of deforestation on climate patterns and support mitigation strategies.
- **Community Engagement:** GIS tools can be used to share deforestation maps and data with local communities, helping them understand the impact of deforestation and engage in conservation efforts.

2. Urban Expansion Monitoring

Urban expansion refers to the growth of cities and towns as population increases and economic activities intensify. This phenomenon can result in significant environmental and socio-economic impacts, including habitat loss, increased pollution, and land use pattern changes. Monitoring urban expansion is crucial for sustainable urban planning, resource management, and environmental conservation. RS and GIS are powerful tools for tracking and analyzing urban growth.

2. i. Importance of Urban Expansion Monitoring

- **Environmental Protection**: Monitors the impact of urbanization on natural ecosystems and assists in mitigating negative effects.
- **Sustainable Development**: Ensures that urban growth is planned and managed in a way that balances economic development with environmental protection.
- **Resource Management**: Aids in the efficient allocation and management of the resources like energy, land, and water.
- **Disaster Management**: Assesses urban growth in relation to natural hazards, facilitating better preparedness and risk reduction strategies.

2. ii. Applications of RS and GIS in Urban Expansion Monitoring

- **Land Use and Land Cover Mapping:** A detailed maps of land use and land cover is created, showing expansion of urban area over time. This helps in understanding the patterns and trends of urbanization. (Wiatkowska *et al.*, 2021) studied spatio-temporal LULC changes in urban areas in Opole, Poland, using Sentinel-2 MSI, Landsat-5 TM, and Landsat-8 OLI data. They used ArcGIS and SNAP software for processing of images and LULC change analysis for the year 2000, 2005, 2010, 2015, and 2020. From the supervised classification of data, they obtained result showing increased built-up area, mainly at the cost of agricultural land.
- **Infrastructure Development Monitoring:** Using RS, the development of new infrastructure such as roads, bridges, and buildings is monitored. GIS integrates this information to assess how infrastructure development is driving urban expansion.
- **Green Space Monitoring:** Changes in green spaces within urban areas, such as parks and recreational areas is identified. This helps in planning for the preservation and expansion of these spaces to ensure a balanced urban environment.

- **Urban Sprawl and Urban Growth Modelling:** Urban sprawl, the uncontrolled expansion of urban areas into surrounding rural land is identified using RS & GIS. This helps in assessing the impact on the environment and planning for sustainable development. GIS integrates various data sets to model and predict future urban growth scenarios. This supports long-term planning and policy-making. (Mostafa *et al.*, 2021) monitored and forecasted urban expansion using machine learning techniques, in Gharbia, Egypt. Their study investigated the land use-land cover changes over the past 27 years and simulated the future dynamics. They used Landsat images to analyze the LULC changes between the year 1991 and 2018, and then, based on logistic regression-Markov chain model they predicted changes in LULC by the year 2048. They obtained results showing rapid urbanization corresponding to agricultural land diminution. Their results further revealed LULC dynamics in like manner confirming urban sprawl over agricultural land from the year 2018 to 2048.
- **Environmental Impact Assessment:** RS and GIS helps in assessing the environmental impacts of urban expansion, including changes in vegetation cover, soil erosion, and water bodies. This aids in planning mitigation strategies.
- **Heat Island Effect Analysis:** Urban areas often experience higher temperatures than its surrounding regions due to the heat island effect. Temperature data and spatial patterns are analyzed to identify hotspots and plan for cooling measures like green roofs and urban forests.
- **Disaster Risk Management:** Areas prone to natural disasters like floods and landslides are monitored, which helps in planning urban expansion to minimize risks. It also aids in post-disaster assessment and recovery planning.

3. Agriculture Monitoring

It involves the systematic observation and analysis of agricultural activities and crop conditions. This is essential for ensuring food security and optimizing resource use. Remote Sensing and GIS are tools facilitate tracking changes in cropland and pasture areas, assisting in agricultural planning and food security management.

3. i. Importance of Agriculture Monitoring

- **Food Security**: Monitoring crop health, growth, and yield helps in predicting and ensuring adequate food production.
- **Resource Management**: Efficient use of water, fertilizers, and pesticides can be achieved by monitoring crop health, nutrient status, and soil moisture.

- **Climate Adaptation**: Helps in assessing the impact of climate change on agriculture and developing adaptive strategies.
- **Disaster Management**: Early detection of droughts, floods, and pest infestations helps in mitigating their impacts on agriculture.
- **Sustainable Farming**: Promotes sustainable agricultural practices by monitoring and managing land use, crop rotation, and soil health.

3. ii. Key Remote Sensing Techniques:

Satellites like Landsat, Sentinel-2, and MODIS capture data in multiple spectral bands, which may be used to monitor crop health, growth stages, and soil conditions. High-resolution satellites like WorldView and IKONOS provides detailed images that can distinguish between different crop types and detect small-scale variations in crop conditions. Hyperspectral sensors capture data in hundreds of spectral bands, providing extensive information on crop health, nutrient status, and soil properties. SAR sensor's microwave penetrate through clouds and capture data irrespective of weather conditions, making them well-suited for monitoring crops in regions with consistent cloud cover. SAR data is also used for assessing soil moisture levels and crop structure, which are important for understanding crop health and yield potential. Thermal sensors measure land surface temperatures, which is used to estimate evapo-transpiration rates, helping in water management and drought assessment.

3. iii. Applications of RS and GIS in Agriculture Monitoring:

RS and GIS offer a range of applications that enhance the productivity, sustainability, and efficiency of agricultural practices. (Anyimah *et al.* 2021) worked on detecting stress areas in certain cocoa farms in Ghana, using RS and GIS. They used Sentinel-2 imagery to analyze monthly averaged NDVI of the selected cocoa farms and then created a model. This model served as the base to classify cocoa areas into healthy and environmentally stressed areas. Using the obtained results, they also developed a suitability map for spatial representation of the areas suitable for cocoa production. RS and GIS tools assist in agriculture monitoring through:

- **Crop Monitoring and Health Assessment:** Indices like NDVI is used to analyze crop health, vigour, and biomass. Stress caused by factors like drought, pests, or disease is detected and mapped which help manage affected areas for targeted interventions.
- **Soil Analysis and Management:** RS help in monitoring soil moisture levels, critical for irrigation planning and drought management. Using GIS, this data is integrated with other soil properties for comprehensive analysis. The

data on soil composition and fertility, aids in precision farming. Mapping this data helps in optimizing fertilizer application and improves soil management practices.

- **Irrigation Management:** Identifying water bodies and assessing water availability helps in planning and managing efficient irrigation systems. The data on crop water needs and soil moisture is utilized for developing precise irrigation schedules, reducing water wastage and improving crop yields.
- **Yield Prediction and Estimation:** Yield prediction model is prepared using data on crop growth stages. GIS helps visualize yield potential across different fields and regions. Accurate yield estimates support in harvest planning and market strategies.
- **Pest and Disease Management:** Using RS, early signs of pest infestation or disease outbreaks can be detected. Mapping these hotspots, enable timely and targeted interventions.
- **Land suitability and Crop Mapping:** RS provides data for classifying land cover types, including different crops. This data is mapped for land use planning and management. Historical RS data analyzed in GIS helps plan crop rotations and optimize land use for sustainable agriculture. (Debesa *et al.* 2020) studied remote sensing and GIS based physical land suitability of the Dabo Hana district in South-West Ethiopia for major cereal crops. To assess the suitability of the land, they used a multi-criteria evaluation approach. They acquired primary as well as secondary data for various physical land characteristics such as temperature, rainfall, slope, altitude, soil pH, soil texture, land use land cover, etc., and used it as input parameters for generating physical land suitability maps.
- **Precision Agriculture:** RS data guides Variable Rate Technology (VRT) applications, such as variable rate seeding, fertilization, and pesticide application. Management and mapping of these variable rates using GIS help in precise field management.
- **Disaster Management:** Detection of early signs of drought, and assessing its impact on agriculture, supports timely interventions and resource allocation. Monitoring of flood events, and mapping of affected agricultural areas aids in damage assessment and recovery planning.

4. Water Resource Management and Monitoring

Water resource management is the strategic planning and control of water usage to ensure quality and sustainable availability for various needs, including drinking, agriculture, industry, and ecosystem conservation. Effective management is critical in addressing challenges such as water scarcity, pollution, and climate change impacts. RS and GIS offer powerful tools for monitoring, analyzing, and managing water resources efficiently.

4. i. Importance of Water Resource Management

- **Sustainability**: Ensures that water resources are used efficiently and conserved for future generations.
- **Flood and Drought Mitigation**: Helps in predicting and managing natural water-related disasters.
- **Water Quality**: Monitors pollution levels to protect public health and ecosystem.
- **Agricultural Productivity**: Optimizes irrigation and water use in agriculture, which is essential for food security.
- **Urban Planning**: Supports the development of infrastructure for water supply and sanitation in urban areas.

4. ii. Remote Sensing in Water Resource Management:

Multispectral and high-resolution imagery are used to detect and monitor surface water bodies. SAR sensors are ideal for monitoring water bodies in regions with frequent cloud cover. SAR also detects changes in water extent and surface roughness, which are indicative of flooding or drought conditions. Thermal sensors measure land and water surface temperatures, useful for monitoring thermal pollution and understanding aquatic ecosystem health. Thermal data are used in estimation of evapo-transpiration rates, important for water balance studies. LiDAR provides high-resolution elevation data, which is utilized for flood modeling, watershed delineation, and infrastructure planning.

4. iii. Key Indicators Monitored Using Remote Sensing:

- **Surface Water Extent**: Monitoring changes in the size and shape of water bodies, such as lakes, rivers, and reservoirs.
- **Water Quality Indicators**: Assessing parameters like turbidity, chlorophyll concentration, and suspended sediments using spectral signatures.
- **Soil Moisture**: Estimating soil moisture levels, critical for agriculture and drought assessment.
- **Snow Cover**: Tracking the extent of snow, important for water availability in river basins fed by snowmelt.

4. iv. Applications of RS and GIS in Water Resource Monitoring & Management

- **Hydrological Monitoring:** RS provides detailed maps of water bodies like rivers, lakes, and reservoirs which helps in the management of these water resources.
- **Water Quality Monitoring:** Satellite data provides information about pollutants and water quality parameters such as chlorophyll concentration, turbidity, and algal blooms. This data is helpful in monitoring of water quality. RS also helps track sedimentation patterns in reservoirs and rivers. This is helpful in analyzing sediment transportation and deposition, crucial for maintaining water quality and storage capacity.
- **Watershed Management:** Delineation of watershed boundaries and integrating it with hydrological data helps analyze and manage watersheds effectively. (Kant *et al.*, 2023) conducted morphometric analysis for Beas River, Himachal Pradesh, using RS and GIS. They used digital elevation model from SRTM to extract various parameters and used ArcGIS software to carry out the modelling. The findings from such a study would prove beneficial in selecting sites, and designing structures for conserving water resource and for watershed management.
- **Groundwater Management:** Groundwater recharge areas and aquifer boundaries are identified, which is used in mapping and managing groundwater resources. RS tool also helps detect indicators of groundwater contamination. GIS integrates this data with well locations and land use information to assess and manage groundwater quality. (Sun *et al.*, 2022) in their research, aimed to conduct change detection studies on the desert edges of Wadi Asyuti, a tributary of Egypt's Nile basin, and to identify groundwater potential zones, using GIS techniques.
- **Irrigation Management:** Monitoring soil moisture and crop water requirements helps plan and optimize irrigation schedules to improve water use efficiency.
- **Climate Impact Assessment:** RS provides long-term data on climatic variables. GIS integrates this data to model and predict the impacts of climate change on water resources. (Marzouk *et al.*, 2021) assessed susceptibility of coastal areas of Al-Alamein New City, Egypt, to the effects of climate change. They used RS and GIS techniques for developing a model (Climate-Costal Model) to assess the impact of climate change and delineate the most vulnerable areas.
- **Urban Water Management:** RS data on land use changes and urban expansion is used to plan and manage urban water supply and drainage systems.

- **Wetland Monitoring:** Tracking changes in wetland areas is helpful in mapping and analyzing the impact on biodiversity and ecosystem services. (Faruque *et al.*, 2022) monitored changes in LULC by using RS & GIS techniques for human-induced mangrove forests, a wetland area in Bangladesh. They used satellite images of Landsat 5 and 8 for mapping of LULC of mangrove wetland and the areas adjacent to it. They evaluated change dynamics for the year 1990, 2000, 2010, and 2020. They also analysed NDVI and NDWI (Normalized Difference Water Index) to confirm changes in land features.
- **Disaster Monitoring:** Drought monitoring helps in detecting signs of drought and assesses its severity which helps in mapping drought-affected areas and plan mitigation measures. Detection and monitoring flood events in real-time and mapping of flood extents aids in disaster response and management.
- **Disaster Risk Reduction:** RS provides real-time data for early warning systems for floods, droughts, and other water-related disasters. This information is disseminated to relevant stakeholders for risk reduction. The impact of disasters can be assessed and mapped for planning recovery and mitigation.
- **Policy Development:** Development of policies and strategies for sustainable water resource management and decision making based evidence is aided by RS and GIS.

5. Air Quality Monitoring

Remote sensing and GIS tools are utilized for monitoring air quality parameters, like particulate matter, nitrogen dioxide, sulfur dioxide, carbon monoxide and methane concentrations. They enable the mapping of air pollution sources, dispersion patterns, and health impacts, supporting air quality management strategies and public health interventions.

5. i. Air Quality Monitoring Using Remote Sensing and GIS

Air quality monitoring is the process of measuring and analyzing the composition of the air to understand the presence and concentration of pollutants. Effective monitoring of air quality is essential for public health, environment protection, and regulatory compliance. Remote Sensing and GIS are essential tools that provide comprehensive and real-time data for monitoring and managing air quality. (Singh *et al.*, 2021) in their review paper, presented various sensors and tools available for air quality monitoring, assessment, and management.

5. ii. Importance of Air Quality Monitoring

- **Environmental Protection**: Helps in understanding the impact of air pollution on ecosystems and wildlife.
- **Regulatory Compliance**: Ensures adherence to air quality standards set by government agencies.
- **Climate Change Mitigation**: Monitors greenhouse gases and other pollutants that contribute to climate change.
- **Urban Planning**: Informs the development of policies and infrastructure to reduce air pollution in cities.
- **Public Health**: Protects people from harmful pollutants that can cause respiratory and cardiovascular diseases.

5. iii. Key Remote Sensing Techniques in Air Quality Monitoring

Satellite-Based Sensors:

- **Multispectral and Hyperspectral Sensors**: Satellites like Sentinel-5P, MODIS, and TROPOMI (Tropospheric Monitoring Instrument) measure atmospheric gases and aerosols in multiple spectral bands.
- **Geostationary Satellites**: Satellites such as GOES (Geostationary Operational Environmental Satellites) provide continuous monitoring of atmospheric conditions.

Aerosol Optical Depth (AOD):

- **Measurement of Particulate Matter**: AOD is a measure of the extinction of sunlight by aerosol particles in the atmosphere, used to estimate the concentration of particulate matter ($PM_{2.5}$ and PM_{10}).

Gas Concentration Retrieval:

- **Trace Gas Measurement**: Satellites like OMI (Ozone Monitoring Instrument) and GOME-2 (Global Ozone Monitoring Experiment) measure the concentration of trace gases such as ozone (O_3), nitrogen dioxide (NO_2), sulfur dioxide (SO_2), carbon monoxide (CO), and methane (CH_4).

LiDAR:

- **Vertical Profiling**: LiDAR systems provide vertical profiles of atmospheric pollutants, giving detailed information about the distribution of aerosols and gases in the troposphere and stratosphere.

5. iv. Key Indicators Monitored Using Remote Sensing

- **Particulate Matter ($PM_{2.5}$ and PM_{10})**: Measure of the concentration of fine particulate matter suspended in the air.
- **Ozone (O_3)**: Monitoring of tropospheric ozone, which is harmful for human health as well as vegetation.
- **Nitrogen Dioxide (NO_2)**: A significant pollutant from vehicle emissions and industrial activities.
- **Sulfur Dioxide (SO_2)**: A pollutant primarily emitted from the combustion of fossil fuel.
- **Carbon Monoxide (CO)**: A colorless, odorless gas resulting from incomplete combustion.
- **Methane (CH_4)**: A potent greenhouse gas contributing to climate change.

5. v. Applications of RS and GIS in Air Quality Monitoring

- **Urban Air Quality Management**: Monitors and manages air pollution in urban areas, supporting policies to reduce emissions from transportation, industry, and residential sources.
- **Industrial Emission Monitoring**: Tracks emissions from industrial facilities to ensure compliance with environmental regulations and implement pollution control measures.
- **Public Health Assessment**: Links air quality data with health statistics to evaluate the impact of air pollution on public health and develop targeted interventions. (Ahmad *et al.*, 2024) highlighted the importance of air quality monitoring and analysis to address human health risks. They visualized, monitored, and analyzed the distribution of air pollutants like $PM_{2.5}$, PM_{10}, sulphur dioxide, nitrogen dioxide, carbon monoxide, and ozone. Using the GIS tools they identified pollution hotspots and generated maps of spatio-temporal distribution of the pollutants.
- **Climate Change Research**: Monitors greenhouse gases and aerosols to study their effects on climate and evaluate mitigation strategies.
- **Emergency Response**: Provides real-time data on air quality during environmental emergencies, such as wildfires and industrial accidents, to protect public health and safety.
- **Environmental Impact Assessment**: Evaluates the impact of new developments and infrastructure projects on air quality, supporting sustainable planning and decision-making.

6. Disaster Monitoring and Management Using Remote Sensing and GIS

Disaster management involves the systematic process of preparing for, responding to, and recovering from disasters. Effective disaster management is essential to minimize the impact of natural and man-made disasters on human lives, property, and the environment. Remote sensing and GIS assist in disaster management and emergency response efforts by providing timely and accurate information about natural disasters, such as floods, wildfires, and earthquakes. They support disaster risk assessment, damage assessment, and evacuation planning, aiding in the mitigation and recovery process. (Hakim & Lee, 2020) reviewed the application of RS and GIS in detecting natural disasters, in Indonesia. They focused on Interferometric Synthetic Aperture Radar (InSAR) technique, susceptibility mapping and image classification. InSAR method is used for monitoring the disasters causing the deformation of the earth's surface, like land subsidence, earthquake, and volcanic activity. Image classification technique is helpful in monitoring pre and post disaster scenario.

6. i. Importance of Disaster Monitoring and Management

- **Risk Reduction**: Identifies and mitigates potential hazards to reduce disaster risk.
- **Preparedness**: Enhances the capacity of communities and authorities to respond effectively to disasters.
- **Response**: Facilitates timely and coordinated actions during a disaster to save lives and property.
- **Recovery**: Supports efficient recovery and reconstruction efforts to restore normalcy.
- **Resilience Building**: Strengthens the resilience of communities and infrastructure against future disasters.

6. ii. Remote Sensing in Disaster Monitoring and Management

Multispectral data is utilized to track changes in land cover, vegetation health and water bodies. High-resolution satellite data provide detailed images for detecting small-scale changes and damages caused by disasters. Regardless of the weather conditions, SAR sensors can gather radiations penetrating through clouds, making them ideal for monitoring floods, landslides, and earthquakes. SAR data can also be used to detect ground displacement and deformation, which are critical for earthquake and landslide monitoring. Thermal sensors measure land surface temperatures, which is used to detect and monitor wildfires and volcanic activities. LiDAR provides high-resolution elevation data, essential for flood modelling, landslide risk assessment, and infrastructure damage evaluation.

6. iii. Key Indicators Monitored Using Remote Sensing

- **Flood Extent and Depth**: Monitors the spatial extent and depth of floodwaters, helping in flood response and mitigation.
- **Vegetation Health**: Assesses the impact of droughts and wildfires on vegetation health using indices like NDVI.
- **Land Deformation**: Detects ground displacement and deformation caused by earthquakes and landslides.
- **Fire Hotspots**: Identifies active fire locations and monitors their spread.
- **Water Bodies**: Monitors changes in water bodies, crucial for drought and flood management.

6. iv. Applications of RS and GIS in Disaster Monitoring and Management

- **Flood Monitoring and Management**: Real-time monitoring of flood events, floodplain mapping, and risk assessment to inform emergency response and mitigation strategies. (Tillihal & Shukla, 2023) reviewed the importance and application of modern technological tools such as RS, GIS, machine learning, and artificial intelligence for flood disaster management, mitigation, adaptation, and awareness.
- **Earthquake Monitoring and Assessment**: Detecting ground deformation and assessing damage to infrastructure, supporting emergency response and recovery efforts. (Gitis & Derendyaev, 2023) presented an information technology for seismogenic field monitoring and analysis, as well as earthquake prediction, using RS data.
- **Wildfire Detection and Monitoring**: Identifying fire hotspots, monitoring fire spread, and assessing post-fire impacts on vegetation and ecosystems.
- **Drought Monitoring and Assessment**: Tracking soil moisture, vegetation health, and water body levels to predict and manage drought conditions.
- **Landslide Monitoring and Risk Assessment**: Detecting ground movements, assessing landslide risk, and planning mitigation measures to protect communities and infrastructure. (Arumugam et al., 2023) generated a Landslide Susceptibility Zonation (LSZ) map of Wayanad, India, for efficient and effective disaster management. They used DEM data for deriving topographical factors like elevation, slope, and aspects influencing landslide; and Landsat 8 imagery for LULC mapping. The map of LSZ was derived through

weighted overlay methodology by analyzing the climatic as well as geographical factors.

- **Volcanic Activity Monitoring**: Monitoring volcanic activity, detecting thermal anomalies, and assessing the impact of eruptions on nearby areas.
- **Infrastructure Damage Assessment**: Evaluating damage to buildings, roads, and other infrastructure post-disaster to support recovery and reconstruction efforts.

III. MONITORING OF BRAZILIAN AMAZON RAINFOREST USING REMOTE SENSING AND GIS: A CASE STUDY

With the advancement of technology over the past few decades, the utilization of satellite data and Geographic Information System tools for monitoring the Brazilian Amazon rainforest has significantly increased, resulting in improved results and understanding. This has opened up better opportunities for tracking large scale data and phenomena, thereby enhancing our understanding of the region's ecological dynamics.

(Floreano & Moraes, 2021) evaluated LULC changes, in the state of Rondonia, between 2009-2019. They used TerrSet 18.3 software to forecast LULC for the next ten years. They used Google Earth Engine (GEE) machine learning methods to classify images using Random Forest classifier. By comparing scenarios of one and three transitions, the Markov-CA deep learning algorithm was used to predict changes in LULC.

(Silva *et al.*, 2023) conducted analysis on fires and deforestation in the Brazilian Amazon from the year 2001 to 2020. Their study aimed to analyze the burned areas and deforestation, considering their impacts on Land Surface Temperature (LST) and rainfall variability. The following techniques and procedures based on GEE was used: (a) mapping the evolution of LULC; (b) utilising vegetation indices to analyse changes in vegetation cover; (c) mapping fires; (d) analysing rainfall and LST; and (e) utilising the geographically weighted regression method to analyse how land cover and climate affect hydrological processes.

(Crivelari-Costa *et al.,* 2023) estimated the changes in CO_2 emission and removal associated with LULC changes in the Brazilian Legal Amazon (BLA) from 2009-2019. The study analyzed carbon dynamics by using Gross Primary Productivity (GPP) and CO_2 flux models related to the LULC of the region. The data for concentration of CO_2 in the atmosphere was acquired using the Greenhouse gases Observing SATellite (GOSAT). The GPP and CO_2 flux variables were estimated using the data from MODIS sensor of Terra and Aqua satellite.

(Prudente *et al.*, 2022) used combination of optical and microwave satellite data for LULC mapping for more accurate classification in the area with cloud cover. They investigated the possible improvements when the SAR data was incorporated in the LULC classification process along with the optical data. They performed LULC mapping for the Roraima State, Brazil, for the year 2019 using MSI/Sentinel-2 and SAR/Sentinel-1. For their analysis, they used twenty-nine scenarios, involving a combination of different times of data acquisition, and optical and SAR based features. Their result determined that optical or SAR data when used individually were not enough to provide accurate LULC mapping.

(Oliveira *et al.*, 2020) studied deforestation incursion and fire-driven emissions of fine particulate pollutants in Apyterewa Indigenous Land, Brazilian Amazon. To install pastures or crop cultures in Brazilian Amazon, fire is used to remove natural vegetation. In their study they evaluated the relation between deforestation, LULC drivers and fire emissions in the study area. To perform LULC mapping, they used Geographic Object-Based Image Analysis (GEOBIA) besides official Brazilian deforestation data. They used Brazilian biomass burning emission model with fire radiative power (3BEM_FRP) for the estimation of emitted particulate matter ($PM_{2.5}$). The GEOBIA approach revealed advancement of deforestation, concurring with the official deforestation data, which was also associated with rise in $PM_{2.5}$ emissions from the fire.

IV. DECADAL VEGETATION COVER CHANGE STUDY IN THE FOREST NEAR SANDSTONE MINES IN SHANKARGARH, PRAYAGRAJ, INDIA: A CASE STUDY

Shankargarh is a town in the trans-Yamuna region of Prayagraj district. It is particularly renowned for its clusters of stone mines, which is an integral part of the local economy. The region is rich in deposits of sandstone, which is widely used as a construction and building material. The local economy primarily depends on mining and mining activities besides agriculture.

There are clusters of sandstone mines present in many parts of Shankargarh. A decadal vegetation cover change study was conducted for one of the mining sites and its surrounding forest area. The study was conducted for the year 2010 and 2020.

IV. i. Data & Methodology

Land cover classification was performed using Landsat 8 and Landsat 5 satellite imagery obtained from Earth Explore website (http://earthexplorer.usgs.gov). Landsat satellite provides multispectral data with 30 metres of spatial resolution.

Based on temporal relevance with the study, the data was acquired for the month of October, for both 2010 and 2020. Using ArcGIS software, maximum likelihood classification was performed for land cover analysis and change detection tool was used for detecting changes in vegetation cover between the two years.

Table 1. Specifications of the satellite imagery used for the study

Satellite	Sensor	Path/Row	Acquisition Date	Spatial Resolution
Landsat 5	TM*	143/043	October 23, 2010	30m
Landsat 8	OLI/TIRS**	143/042	October 18, 2020	30m

*TM: Thematic Mapper **OLI/TIRS: Operational Land Imager/Thermal Infrared Sensor

IV. ii. Result

Figure 1. Vegetation cover change in the study site of Shankargarh, Prayagraj, India, between the year 2010 and 2020

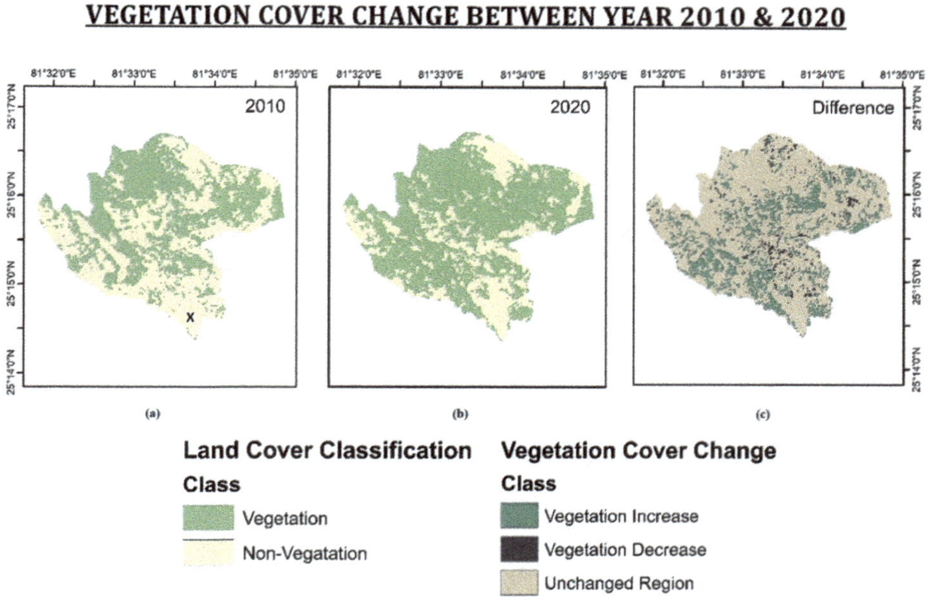

Figure 1(a) is the land cover classification map of the study site for the year 2010, where the mining site is marked by 'X', and the surrounding area is the forested region. Figure 1(b) is the land cover classification map of the study site for the year 2020. Figure 1 (c) is the map showing change in vegetation cover of the study site between the year 2010 and 2020.

IV. iii. Discussion & Conclusion

The study sites is spread in 11.63 km^2 of area. In 2010, 48% of this area was covered with forest, but this figure increased to 67% in 2020; although the region around the mining site (marked 'X') showed sparse vegetation, which grew denser away from the mining site. The total vegetation cover in the region increased by 2.19 km^2, although the increase was noted in the region farther away from the mines. This increase is possibly attributed to the fact that after a certain extent mining activities were stopped in this region. It may also be noted that change in vegetation cover depends on many factors, such as, precipitation, temperature, soil parameters, etc. So, to determine a definite cause for the results obtained, a detailed study is necessary including the other factors.

V. CHALLENGES AND FUTURE DIRECTIONS

V. i. Challenges:

- **Complexity of Environmental Processes:** Environmental processes are often complex and interconnected, involving multiple variables and spatial scales. Understanding and monitoring these processes require sophisticated analytical techniques and interdisciplinary approaches that integrate remote sensing data with other sources of information, such as ground-based measurements and modeling.
- **Data Availability and Quality:** Obtaining high quality and up-to-date remote sensing data can be challenging, especially for regions with limited coverage or frequent cloud cover. Also, most of the high-resolution data are not open source and are expensive.
- **Spatial and Temporal Resolution:** The spatial and temporal resolution of remote sensing data may not always meet the requirements of specific environmental monitoring applications.
- **Interference from Atmospheric Conditions:** Atmospheric conditions such as haze, aerosols, and cloud cover can affect the quality of remote sensing data by attenuating or distorting signals from the Earth's surface.

V. ii. Future Directions:

- **Advancements in Sensor Technology:** Advancements in sensor technology, including the development of next-generation satellites with improved spatial, spectral, and temporal resolution, will enhance the capabilities of remote sensing for environmental monitoring.
- **Big Data Analytics:** Better accessibility and understanding of machine learning algorithms, data mining techniques, and cloud computing platforms may help extract meaningful insights from massive datasets and support decision-making in environmental management.
- **Open Data Initiatives:** Open data initiatives, such as the availability of free and open access satellite imagery from government agencies.
- **Policy Integration and Decision Support Systems:** Integrating remote sensing and GIS data into decision support systems and policy frameworks can enhance the effectiveness of environmental management strategies.

VI. CONCLUSION

Remote Sensing and GIS have transformed how we monitor the environment, giving us powerful tools to observe, analyze, and manage natural and built spaces. This chapter highlighted how these technologies help in various areas like tracking land use and land cover (LULC) changes, monitoring forests, air quality, agriculture, water resources, urban growth, and managing disasters. By using high-resolution satellite images and advanced spatial analysis methods, RS and GIS allows us to detect environmental changes, understand patterns, and predict future trends. These capabilities help researchers, policymakers, and environmental managers make well informed decisions to promote sustainability and resilience.

As environmental challenges grow, the importance of modern technological tools in monitoring the environment will continue to rise. These technologies will drive innovations that improve our ability to understand, protect, and sustainably manage the environment. The future advances in remote sensing, geographic information system, machine learning, and artificial intelligence will provide greater precision, efficiency, and accessibility, helping us address the environmental challenges with more informed strategies. In conclusion, RS and GIS has emerged as an indispensable tool to observe, analyze, and interpret the Earth's surface remotely. While there are some challenges associated with monitoring and managing the environment using these technological tools, there are also opportunities for future advancements and innovations that can contribute to more effective and sustainable management of natural resources.

REFERENCES

AbdelRahman, M. A. (2023). An overview of land degradation, desertification and sustainable land management using GIS and remote sensing applications. *Rendiconti Lincei. Scienze Fisiche e Naturali*, 34(3), 767–808. DOI: 10.1007/s12210-023-01155-3

Ahmad, H. R., Mehmood, K., Bibi, S., Bashir, M. H., & Siddique, A. (2024). Revealing air quality dynamics: Leveraging remote sensing and GIS for active monitoring and risk evaluation. In *Geographical Information Science* (pp. 417–428). Elsevier., DOI: 10.1016/B978-0-443-13605-4.00021-7

Anyimah, F. O.Jnr, Osei, E. M.Jnr, & Nyamekye, C. (2021). Detection of stress areas in cocoa farms using GIS and remote sensing: A case study of Offinso Municipal & Offinso North district, Ghana. *Environmental Challenges*, 4, 100087. DOI: 10.1016/j.envc.2021.100087

Arumugam, T., Kinattinkara, S., Velusamy, S., Shanmugamoorthy, M., & Murugan, S. (2023). GIS based landslide susceptibility mapping and assessment using weighted overlay method in Wayanad: A part of Western Ghats, Kerala. *Urban Climate*, 49, 101508. DOI: 10.1016/j.uclim.2023.101508

Crivelari-Costa, P. M., Lima, M., La Scala Jr, N., Rossi, F. S., Della-Silva, J. L., Dalagnol, R., Teodoro, P. E., Teodoro, L. P. R., de Oliveira, G., de Oliveira Junior, J. F., & da Silva Junior, C. A. (2023). Changes in carbon dioxide balance associated with land use and land cover in Brazilian Legal Amazon based on remotely sensed imagery. *Remote Sensing (Basel)*, 15(11), 2780. DOI: 10.3390/rs15112780

da Silva, R. M., Lopes, A. G., & Santos, C. A. G. (2023). Deforestation and fires in the Brazilian Amazon from 2001 to 2020: Impacts on rainfall variability and land surface temperature. *Journal of Environmental Management*, 326, 116664. DOI: 10.1016/j.jenvman.2022.116664 PMID: 36370609

de Paula Sousa Júnior, V., Sparacino, J., de Espindola, G. M., & Sousa de Assis, R. J. (2022). Land-Use and Land-Cover Dynamics in the Brazilian Caatinga Dry Tropical Forest. *Conservation*, 2(4), 739–752. DOI: 10.3390/conservation2040048

Debesa, G., Gebre, S. L., Melese, A., Regassa, A., & Teka, S. (2020). GIS and remote sensing-based physical land suitability analysis for major cereal crops in Dabo Hana district, South-West Ethiopia. *Cogent Food & Agriculture*, 6(1), 1780100. DOI: 10.1080/23311932.2020.1780100

Faruque, M. J., Vekerdy, Z., Hasan, M. Y., Islam, K. Z., Young, B., Ahmed, M. T., Monir, M. U., Shovon, S. M., Kakon, J. F., & Kundu, P. (2022). Monitoring of land use and land cover changes by using remote sensing and GIS techniques at human-induced mangrove forests areas in Bangladesh. *Remote Sensing Applications: Society and Environment*, 25, 100699. DOI: 10.1016/j.rsase.2022.100699

Floreano, I. X., & de Moraes, L. A. F. (2021). Land use/land cover (LULC) analysis (2009–2019) with Google Earth Engine and 2030 prediction using Markov-CA in the Rondônia State, Brazil. *Environmental Monitoring and Assessment*, 193(4), 239. DOI: 10.1007/s10661-021-09016-y PMID: 33783626

Gitis, V., & Derendyaev, A. (2023). A Technology for Seismogenic Process Monitoring and Systematic Earthquake Forecasting. *Remote Sensing (Basel)*, 15(8), 2171. DOI: 10.3390/rs15082171

Hakim, W. L., & Lee, C. W. (2020). A review on remote sensing and GIS applications to monitor natural disasters in Indonesia. *Korean Journal of Remote Sensing, 36*(6_1), 1303-1322. DOI: 10.7780/kjrs.2020.36.6.1.3

Islam, M. R., Khan, M. N. I., Khan, M. Z., & Roy, B. (2021). A three decade assessment of forest cover changes in Nijhum dwip national park using remote sensing and GIS. *Environmental Challenges*, 4, 100162. DOI: 10.1016/j.envc.2021.100162

Kakoti, M., & Sarma, H. H. (2024). Applications of geospatial and remote sensing technologies in management of natural resources. *Indian Farmer*, 11(07), 230–235.

Kant, C., Kumar, G., & Meena, R. S. (2023). Modeling morphometric and geomorphological parameters of mountainous river basin for water resource management using remote sensing and GIS approach. *Modeling Earth Systems and Environment*, 9(2), 2151–2163. DOI: 10.1007/s40808-022-01614-0

Marzouk, M., Attia, K., & Azab, S. (2021). Assessment of coastal vulnerability to climate change impacts using GIS and remote sensing: A case study of Al-Alamein New City. *Journal of Cleaner Production*, 290, 125723. DOI: 10.1016/j.jclepro.2020.125723

Mostafa, E., Li, X., Sadek, M., & Dossou, J. F. (2021). Monitoring and forecasting of urban expansion using machine learning-based techniques and remotely sensed data: A case study of gharbia governorate, Egypt. *Remote Sensing (Basel)*, 13(22), 4498. DOI: 10.3390/rs13224498

Parra, L. (2022). Remote sensing and GIS in environmental monitoring. *Applied Sciences (Basel, Switzerland)*, 12(16), 8045. DOI: 10.3390/app12168045

Prudente, V. H. R., Skakun, S., Oldoni, L. V., Xaud, H. A., Xaud, M. R., Adami, M., & Sanches, I. D. A. (2022). Multisensor approach to land use and land cover mapping in Brazilian Amazon. *ISPRS Journal of Photogrammetry and Remote Sensing*, 189, 95–109. DOI: 10.1016/j.isprsjprs.2022.04.025

Singh, D., Dahiya, M., Kumar, R., & Nanda, C. (2021). Sensors and systems for air quality assessment monitoring and management: A review. *Journal of Environmental Management*, 289, 112510. DOI: 10.1016/j.jenvman.2021.112510 PMID: 33827002

Solórzano, J. V., Mas, J. F., Gallardo-Cruz, J. A., Gao, Y., & de Oca, A. F. M. (2023). Deforestation detection using a spatio-temporal deep learning approach with synthetic aperture radar and multispectral images. *ISPRS Journal of Photogrammetry and Remote Sensing*, 199, 87–101. DOI: 10.1016/j.isprsjprs.2023.03.017

Sultana, Q., Sultana, A., & Ara, Z. (2023). Assessment of the land use and landcover changes using remote sensing and GIS techniques. In *Water, Land, and Forest Susceptibility and Sustainability* (pp. 267-297). Elsevier. DOI: 10.1016/B978-0-323-91880-0.00022-2

Sun, T., Cheng, W., Abdelkareem, M., & Al-Arifi, N. (2022). Mapping prospective areas of water resources and monitoring land use/land cover changes in an arid region using remote sensing and GIS techniques. *Water (Basel)*, 14(15), 2435. DOI: 10.3390/w14152435

Tichouiti, K. (2024). NVDI-and NBR-based Comparison for Burnt Areas Characterization Using Sentinel-2 Imagery. In 2024 IEEE Mediterranean and Middle-East Geoscience and Remote Sensing Symposium (M2GARSS) (pp. 187-190). IEEE. DOI: 10.1109/M2GARSS57310.2024.10537321

Tillihal, S. B., & Shukla, A. K. (2023). River flood monitoring and management: a review. Advances in Water Resource Planning and Sustainability, 223-238. DOI: 10.1007/978-981-99-3660-1_13

Wiatkowska, B., Słodczyk, J., & Stokowska, A. (2021). Spatial-temporal land use and land cover changes in urban areas using remote sensing images and GIS analysis: The case study of Opole, Poland. *Geosciences*, 11(8), 312. DOI: 10.3390/geosciences11080312

Zhang, J., Wang, Z., Bai, L., Song, G., Tao, J., & Chen, L. (2021, July). Deforestation Detection Based on U-Net and LSTM in Optical Satellite Remote Sensing Images. In *2021 IEEE International Geoscience and Remote Sensing Symposium IGARSS* (pp. 3753-3756). IEEE. DOI: 10.1109/IGARSS47720.2021.9554689

Chapter 2

Geomatics-Based Analysis of Sand Dune Encroachment Dynamics in the Tafilalet Region, Morocco

Rachid Ouachoua
https://orcid.org/0000-0002-3194-7585
Ministry of Energy Transition and Sustainable Development, Morocco

Jamal Al Karkouri
Faculty of Human and Social Sciences, Ibn Tofail University, Morocco

Hamid Benssi
Faculty of Human and Social Sciences, Ibn Tofail University, Morocco

ABSTRACT

Aeolian sediments cover roughly 6% of the Earth's surface, mainly in arid regions, accounting for 97% of their occurrence and spanning about 20% of the terrestrial surface. Sand drifts in these areas threaten fragile ecosystems by disrupting their stability. This study analyzes sand dune encroachment dynamics in the Tafilalet region of southeastern Morocco using geomatics techniques. Sand encroachment endangers human settlements, agricultural lands, and infrastructure. Employing remote sensing methods, including satellite imagery and the Fryberger method for estimating sand dune drifting, this research examines sand movement from 1984 to 2015. Results show a significant increase in sand-covered areas by approximately 44,872 hectares, driven by climatic conditions and human activities. The study

DOI: 10.4018/979-8-3693-6452-9.ch002

highlights the severe risks to the Tafilalet oases and infrastructure, underscoring the need for continuous monitoring and sustainable management practices to mitigate sand encroachment impacts.

INTRODUCTION

In North Africa, arid lands represent more than 95% of the terrain area, except for the littoral zone of the Mediterranean Sea (Gherib et al., 2023). Arid lands are particularly vulnerable to soil degradation due to prolonged drought conditions and climatic variability, which can lead to severe ecosystem dysfunction and desertification (Ritsema et al., 2022). Manifestations of desertification in arid lands include soil salinization and sand encroachment, both of which significantly threaten agricultural productivity and infrastructure stability (Geist & Lambin, 2004; Cherlet et al., 2018). Mobile sand dunes are a major issue in these regions, encroaching on road networks, agricultural lands, and residential areas, exacerbating socio-economic challenges (Barchyn & Hugenholtz, 2012).

In terms of dune activity, recent studies have continued to identify active, dormant, and relict dunes, with active dunes posing the greatest hazard due to their lack of vegetation and abundant sand supply (Bourke et al., 2022; Thomas & Wiggs, 2021). Research on sand dunes has significantly advanced since the early 20th century, with modern techniques allowing for more precise measurements of wind velocities and sand flux (Swann & Walker, 2019; Baas, 2008). The problem addressed in this work is sand encroachment, manifesting as sandy accumulations at the edges of towns and within palm groves, leading to rural exodus. These sandy deposits sometimes cut off roads, further isolating communities and disrupting local economies (Issanova et al., 2023; Allahyari & Farjad, 2022).

One region of particular interest for sand dune monitoring is the Tafilalet region in southeastern Morocco, which is home to extensive and dynamic sand dune systems. The region's clear morphology and distinct dune forms, shaped by prevailing southwestern winds, provide an ideal setting for the application of remote sensing methodologies. In this region, the relentless advance of sand threatens human settlements and infrastructure, as exemplified by the late 1970s storm that devastated 60% of a village in the Tafilalet area.

Faced with this problem, which is likely to worsen in the future, the objective of this work is to analyze the spatio-temporal dynamics of sand encroachment with the overall aim of safeguarding the biodiversity of the oases of Tafilalet and protecting infrastructure such as roads, and irrigation canals. To do that it is essential to find effective ways to identify and monitor the dynamics and assess their environmental impact. Therefore, we employ remote sensing techniques alongside the Fryberger method—a widely recognized approach for estimating sand dune drifting rates—to investigate and quantify the spatial and temporal dynamics of sand dunes in the Tafilalet region (Fryberger, 1986). By combining satellite imagery analysis with ground truthing, we aim to provide comprehensive insights into the patterns of sand dune movement and their implications for desert landscape management and conservation.

MATERIALS AND METHODS

Study Area

The Tafilalet region, located in southeastern Morocco, subject of this study, covers an area of 17 031 km^2. is a vast and arid area that encompasses parts of the Drâa-Tafilalet region, mainly within the provinces of Errachidia and and Tinghir. Geographically, it stretches between approximately 30.6° to 32.0° North latitude and 3.6° to 5.0° West longitude. This region is flanked to the north by the rugged High Atlas Mountains and to the south by the expansive Sahara Desert, with its eastern border adjoining Algeria and its western periphery defined by the lush Ziz Valley (figure 1).

Figure 1. (A) Location of Morocco in the African continent, (B) Location of Tafilalet region in Morocco, (C) The Tafilalet region.

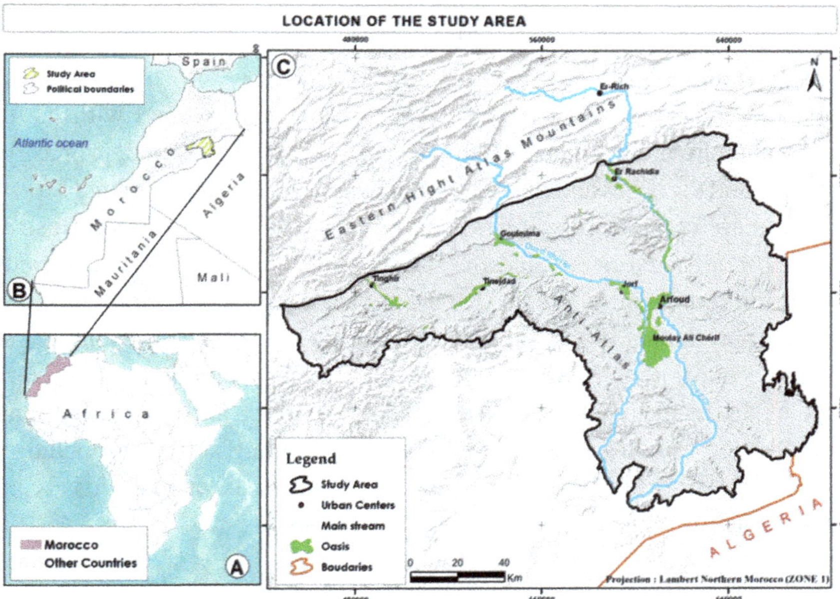

The region was selected for this study due to its unique environmental importance as home to Morocco's largest oases highly vulnerable to the effects of climate change, which exacerbate desertification and sand encroachment. The region's location between the High-Atlas Mountains and the Sahara Desert provides an ideal setting to study the interactions between deforestation, water erosion, and arid land processes. Understanding these dynamics in Tafilalet can inform sustainable management practices in similar arid regions globally.

The territory of Tafilalet, part of the Oases Biosphere Reserve of Southern Morocco, is recognized by UNESCO in November 2000 and has become an integral part of UNESCO's world program 'Man and the Biosphere (MAB). The region experiences an arid climate characterized by minimal annual rainfall, high evaporation rates, and significant temperature fluctuations, typical of desert environments. The landscape is dominated by flat plains, rocky plateaus, and the famous sand dunes, especially those near Merzouga, such as the Erg Chebbi dunes.

Environmental conditions in Tafilalet highlight its vulnerability to desertification and sand encroachment. The region's climate, coupled with human activities like overgrazing and deforestation, accelerates the degradation of its fragile ecosystem. Notably, the Tafilalet area features some of Morocco's most dynamic sand dunes, including the Erg Chebbi dunes, which can reach heights of up to 150 meters. These dunes exhibit various forms, including transverse, Barkhan, and linear dunes, shaped by predominant wind patterns from the southwest to the northeast. The sparse vegetation, primarily composed of drought-resistant shrubs and grasses, is mostly concentrated around the oases, which provide crucial support for agriculture, particularly the cultivation of date palms. The Tafilalet Oasis, nourished by the Ziz River, stands as a green haven amid the arid expanses, supporting the growth of crops and sustaining local communities.

The socio-economic fabric of Tafilalet is deeply intertwined with its unique geography and climate. Human settlements are sparsely distributed, primarily concentrated around the oases and major towns like Erfoud, Rissani, Jorf and Merzouga. These towns are connected by an expanding network of roads and infrastructure that cater to both local needs and the burgeoning tourism industry. Agriculture, centered around date palm farming, is a critical economic activity, with Tafilalet renowned for producing high-quality dates, especially the coveted Mejhoul variety. Tourism also plays a pivotal role, drawing visitors to the dramatic landscapes of the Erg Chebbi dunes for camel trekking, desert camping, and off-road adventures. Additionally, the region's economy benefits from mining activities, including the extraction of minerals like lead and barite, as well as the excavation of fossils, adding to its economic diversity.

The Tafilalet region's dynamic sand dunes make it an ideal site for studying sand dune activity and sand movement. The Erg Chebbi dunes, in particular, provide a valuable case for observing dune dynamics due to their size and mobility. Remote sensing techniques are especially suited for monitoring these changes, given the region's clear dune morphology and the logistical challenges of ground-based monitoring in such vast and remote areas. This study aims to analyze the spatio-temporal dynamics of sand dunes in Tafilalet, using remote sensing methodologies to gain insights into their movement patterns and impacts. By combining satellite imagery analysis with ground truthing and statistical modeling, the research seeks to inform effective strategies for managing and mitigating the effects of sand encroachment on local communities and infrastructure. The findings from this study will be crucial for understanding and preserving the delicate desert landscapes and the livelihoods they support in the Tafilalet region.

Data Used

To achieve a comprehensive understanding of sand dune dynamics and their impact over time, this study employs a combination of satellite imagery and wind data. The primary data sources include:

- Topographic and geological maps at 1/50.000 covering the study area;
- The soil map of Morocco at 1/1.500.000;
- Socio-economic data from the Census of Population and Housing of Morocco for 1994 and 2014.
- Wind Data from 6 hydro-meteorological stations (Errachidia, Erfoud, Jorf, Merroutcha, Taouz and Merzouga) for the period between 2000 and 2013 provided by the Guir-Ziz-Gheris Hydraulic Basin Agency.
- Landsat 5 Thematic Mapper (TM) imagery from 1984 and Landsat 8 Operational Land Imager (OLI) imagery from 2015, which are readily available and highly effective for monitoring sand encroachment over time. Landsat data offers consistent spatial resolution, making it well-suited for detecting changes in dune movement and landscape patterns.
- Sand Rose Graph software v3.0. This software is accessible for estimating sand drift potential (DP) by winds, resultant drift potential (RDP), and unidirectional Index (UDI), and for plotting sand roses;
- Exilis ENVI 5.3 for imagery analysis, and ArcGIS 10.7 for the spatial analysis and mapping.

The two time points (1984 and 2015) were chosen to provide a 31-year temporal span providing a valuable timeline for observing sand encroachment trends. Significant changes in sand encroachment can be linked to key climatic and socio-economic events in the region. Rising temperatures and decreasing precipitation (figure 8) have exacerbated aeolian processes and reduced vegetation cover, directly contributing to accelerated sand drift. Additionally, socio-economic shifts like population migration (figure 10) and changes in land use have also impacted sand encroachment. The abandonment of agricultural land due to rural exodus, for example, has reduced land management practices that historically helped stabilize dunes. By correlating these trends with major environmental and human-induced events, the study provides deeper insights into the drivers of sand encroachment over time.

Methodology

The methodology developed in this part combines the use of remote sensing and data layers (GIS) and the Fryberger method, including a large socio-economic database. The following diagram (figure 2) illustrates the methodology adopted in this work.

Figure 2. Workflow and the methodology adopted in this work.

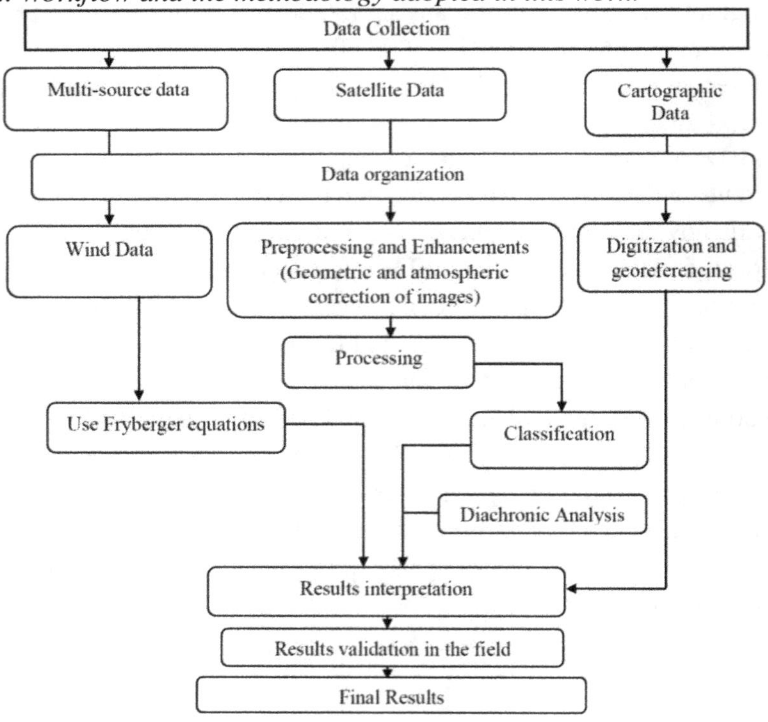

Image Pre-Processing

Image pre-processing involved correcting and enhancing satellite images to ensure accurate analysis under ENVI software environment. The steps included:

- Radiometric Correction: Adjusting the pixel values to correct for sensor noise and atmospheric conditions. In this step the Dark Object Subtraction (DOS) is used corrects for atmospheric scattering by subtracting the lowest reflectance value (often due to haze) from each band.
- Image Enhancement: Enhancing the visual appearance of the images to highlight specific features relevant to the study. Histogram Equalization (Linear Stretch) is used to spreads out the most frequent intensity values, improving contrast in areas of low dynamic range.

Sand Rose Analysis

To determine the effective direction of sand mass movement in Tafilalet, one must not consider the wind speed roses that cause particle drifting but rather the sand drift potential (DP), which is obtained by converting wind speed into wind force. However, strictly speaking, the effectiveness of winds depends not only on speed but also on wind turbulence (Chergui) (Benalla et al., 2003). The adopted approach is based on a theoretical calculation to determine the sand drift potential (DP).

Several methods and techniques exist for evaluating the DP value. However, after a bibliographic search on DPS calculation, it appears that the most widely used method in the field of sand encroachment is that developed by Fryberger (1979). Bullard (1997) used it to evaluate sand drift potential by wind. Livingstone (1989) used it to characterize changes in linear dune surfaces in the Namib Desert. Similarly, Callot and Oulehri (1996) applied it for the geodynamic study of eolian sands in the northwestern Sahara, and Wiggs et al. (1996) applied it to study wind energy variation in the southeastern Kalahari Desert and its implications for linear dune activity. Recent studies by Allahyari and Farjad (2022) and Issanova et al. (2023) have also utilized similar methods to analyze sand dune mobility and the impact of anthropogenic factors on sand encroachment dynamics.

The Fryberger method was selected to analyze the whole wind dataset as the following:

$$q = V^2(V - Vt) \tag{1}$$

where q is the amount of sand drift based on the vector units (VU), V represents the average wind velocity at a 10 m height (m/s), Vt is the impact threshold of wind velocity (m/s).

This formula allows us to calculate the potential sand transport coefficient for each wind speed and direction obtained from anemometer records. For a given direction, the observation frequency and each wind speed above the threshold speed should be multiplied by the entrainment potential using formula (2). To make this coefficient more practical, it should be divided by 100 (Fryberger, 1979) and to be compatible with the speed expressed in m/s, it should be divided by 12.5 (Callot, 1987). The sand transport coefficient Q can then be expressed by the new formula:

$$Q = \frac{q \times f}{12{,}5} \tag{2}$$

Other parameters of interest were calculated which are the resultant drift potential (RDP), the resultant drift direction (RDD), and the unit directional index (UDI) of wind estimated by vector analysis:

$$RDP = \sqrt{C^2 + D^2} \tag{3}$$

and

$$RDD = \tan^{-1}\left(\frac{C}{D}\right) \tag{4}$$

$$UDI = \left(\frac{RDP}{DP}\right) \tag{5}$$

$$C = \sum DP \times \sin\theta \; and \; D = \sum DP \times \cos\theta \tag{6}$$

Where θ is the angle in radians of orientation for each speed greater than the threshold speed, and Vt is assumed as 4.5 m/s based on studies in the same region (Benalla, 2003) and (Navas, Kabiri, Sadiki, & Faleh, 2013). The unit directional index (UDI), and index of directional variability (Fryberger S.G, 1979), is the ratio of the resultant drift potential (RDP) to the drift potential (DP) of sand. Lower UDI values indicate greater directional variability of wind, whereas higher UDI values mean that the wind comes from the same direction. Finally, C and D are sand drift potential in vertical and horizontal components.

RESULTS AND DISCUSSION

Sand Drift Potential (DPS)

The total DP expresses the overall sand drift potential caused by winds in 16 directions. Using DP, the wind regime of the region can be distinguished by referring to Fryberger S.G.'s classification of eolian environments in arid and desert regions (1979). Thus, anemometric data from selected sites were used to estimate DP. Analysis and processing revealed that the region belongs to areas with medium to high wind energy, with capacity values between 268.75 and 824.65 for the Merroutcha and Jorf sites, respectively (table 1).

Table 1. Classification of eolian environments using the Drift Potential and directional variability (Fryberger & Dean, 1979)

DP (vector units)	DP (m^3/m/ year)	Aeolian environment	RDP/ DP	Directional variability	Directional category (distribution probability)	Examples
<200	<17	Low	<0,3	High	Complex or obtuse bimodal	Star dune (Lancaster, 1983)
200-400	17-33	Medium	0,3-0,8	Medium	Obtuse or acute bimodal	Linear dunes (Fryberger, 1979)
>400	>33	High	>0,8	Low	Acute unimodal	Barchan (Fryberger, 1979)

Theoretically, according to the linear function of the relationship between DP and the amount of transported sand (figure 3), the values obtained for the sectors estimate an average drift of about 43 m^3/m/year under ultimate conditions. These quantities of moving sand per meter-wide section remain valid. According to Callot Y. (1987), in a medium-energy region, the amounts of displaced sand range from 20 to 30 tonnes.

Figure 3. Relationship between Drift Potential and the quantity of transported sand (Callot Y., 1987)

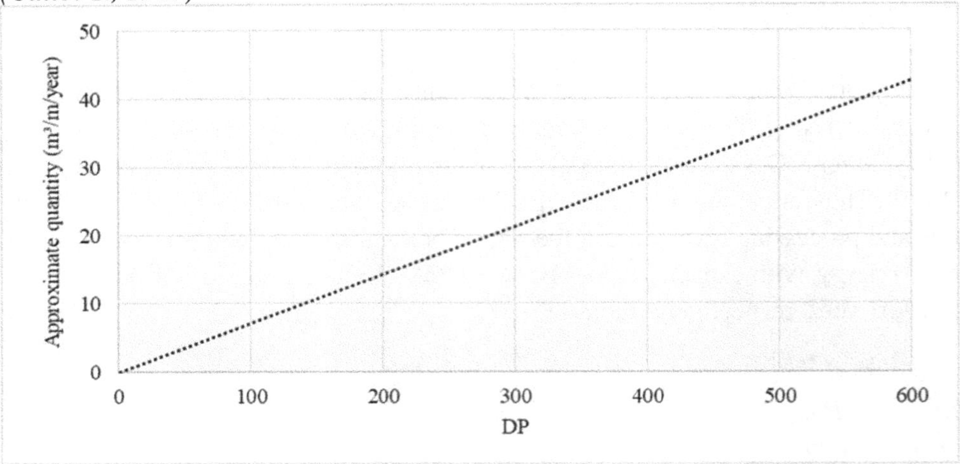

Direction and Module of the Resultant Drift Direction (RDD)

The Resultant Drift Direction (RDD) is determined by summing the sand Drift Potential (DP) vectors from 16 directional sectors. The RDD magnitude represents the net sand transport potential resulting from the combined influence of winds from all 16 directions. Additionally, potential energy coefficients were calculated, as presented in Table 2. Across all study sites, the predominant sand movement and potential drift during the period from 2000 to 2013 are directed towards the ENE. The findings are summarized in the following table:

Table 2. Average potential energy coefficients of sand displacement for the period (2000-2013)

Stations	RDD (°)	DP	RDP	RDP/ DP	Directional category	Theoretical quantity (m^3/m/year)
ERRACHIDIA	52 (ENE)	625.44	343.99	0.55	Obtuse or acute bimodal	44.41
ERFOUD	27 (NNE)	809.31	477.49	0.59	Obtuse or acute bimodal	57.46
JORF	43 (NE)	824.65	585.50	0.71	Obtuse or acute bimodal	58.55

continued on following page

Table 2. Continued

Stations	RDD (°)	DP	RDP	RDP/ DP	Directional category	Theoretical quantity (m^3/m/year)
MERROUTCHA	80 (ENE)	268.75	142.44	0.53	Obtuse or acute bimodal	19.08
TAOUZ	81 (ENE)	740.88	303.76	0.41	Obtuse or acute bimodal	52.60
MERZOUGA	341 (NNW)	305.35	79.39	0.26	Complex or obtuse bimodal	21.68

According to the table above, the sand migration coefficient is low in Merzouga compared to Jorf, where the maximum value in the zone is about 0,71. A low coefficient characterizes sand mobility without significant sand drift according to Fryberger S.G.'s classification (1979). When this coefficient approaches its maximum value, variability is low, and the wind tends to blow in a single direction, causing long-distance drift.

In reality, there are gaps concerning wind turbulence, which is not considered in the calculation. This phenomenon is crucial in triggering sand movements. Therefore, the resultant sand drift direction on the ground may differ from that determined by the method. One should be cautious about the results obtained from this quantification method, although it has proven valid in wind tunnel measurements and numerous studies (Callot Y., 1987).

Sand Drift Potential Roses and Migration Coefficient

The calculation for the sixteen directions of the sand transport coefficient allows obtaining the sand mobilization rose for each entrainment speed. The drift potential rose is obtained by summing the transport capacity coefficients for all speeds higher than the entrainment threshold. Thus, for each direction, one obtains a value representing the vectorial module of the sand transport capacity in the considered direction (figure 4). These results were obtained using the "SandRose Graph v3" software.

Figure 4. Potential sand drift roses for the period (2000-2013) for a Vt > 4,5 m/s, the resultant is marked in red.

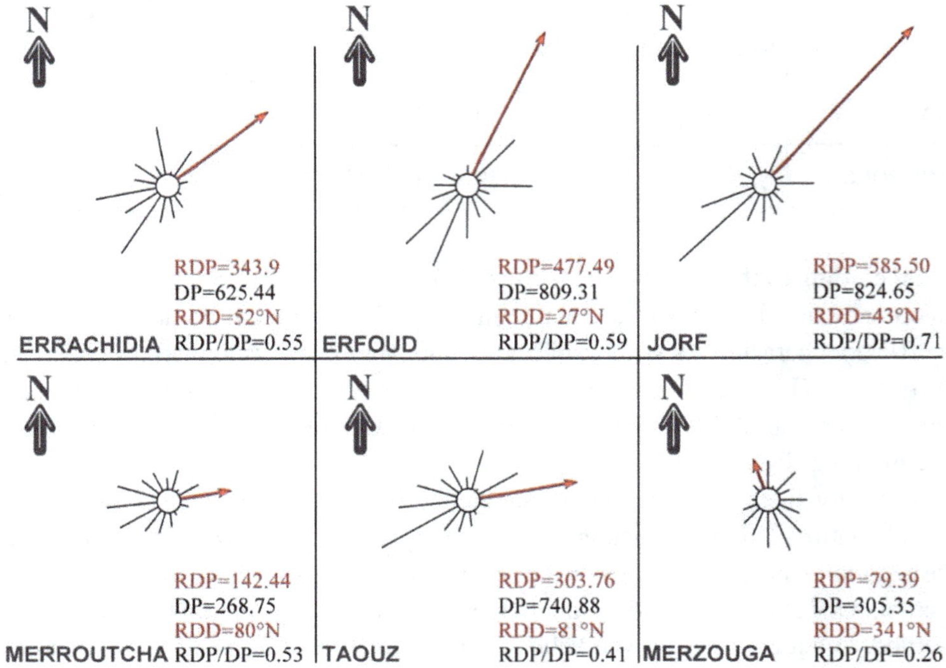

Sand drift roses provide a detailed representation of the current aeolian transport dynamics of sand particles. These visualizations are widely utilized by geomorphologists to analyze and compare dune morphologies in arid regions, specifically assessing both active and stabilized dune orientations. Variations between sand rose patterns and actual dune alignments may indicate historical shifts in aeolian circulation processes. The results generated through specialized software are typically presented as tables or graphs. Therefore, in our work mapping is essential for identifying current sand transport risks and for establishing priorities in the development of protective measures for vulnerable zones. Based on these analyses, a sand drift potential map (figure 5) was constructed to accurately reflect the identified patterns.

Figure 5. Sand drift potential mapping in the Tafilalaet region between 2000 and 2013

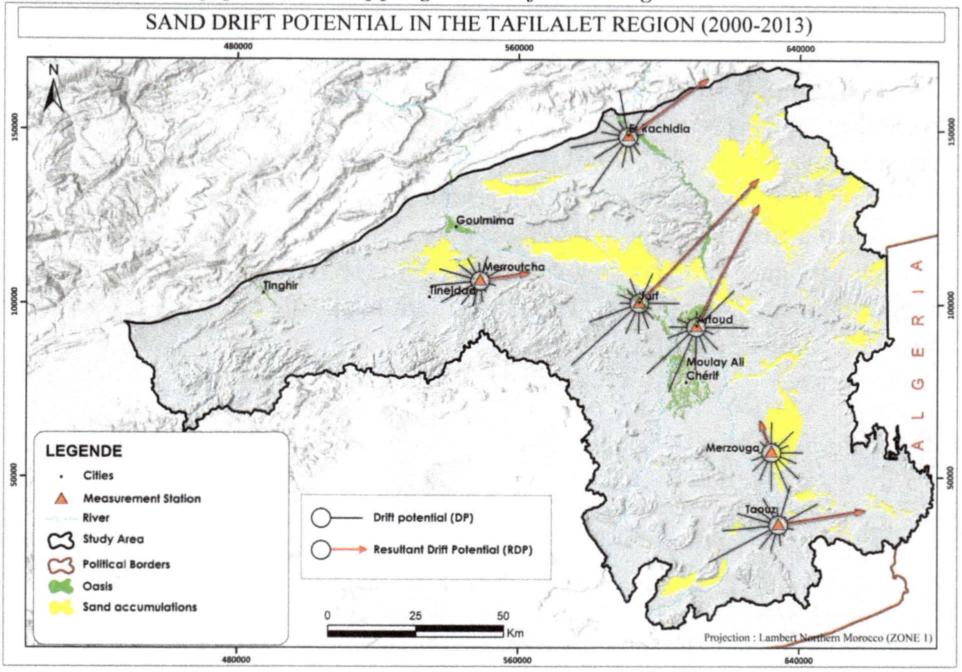

The SDP roses show the dominance of ENE and NE directions, with secondary opposite SW sector components. They are bimodal in all zones except for Merzouga, where the sand drift rose indicates potential migration in all directions, with a dominant direction towards the NW. In this zone, potential migration is weaker compared to other zones.

The Fryberger method is widely used for estimating sand drift potential, but it has limitations when applied to specific environments like the Tafilalet region. This method assumes stable wind conditions and minimal vegetation, which may not fully capture the variability in wind patterns, surface roughness, and vegetation cover in the region. In Tafilalet, seasonal wind shifts and sparse vegetation can influence sand transport, leading to potential inaccuracies. Validation through local field measurements and integrating satellite data can help address these limitations and improve model accuracy. Thus, we validated sand encroachment patterns using satellite monitoring, which allowed us to track and analyze spatial and temporal changes over time accurately. The findings and closely aligned with those obtained by Benalla (2003) in the same region, demonstrating the consistency and reliability of sand encroachment assessments in Tafilalet. This validation underscores the robustness of the combined methodology in capturing the extent and dynamics of sand encroachment, even in complex environments like Tafilalet.

Sand Encroachment Mapping

Before starting the classification procedure, we performed a statistical analysis of the separability of the different training plots for all channels (table 3). This pairwise separability analysis between classes was conducted using measures of transformed divergent separability (Richards J.A., 1999). According to Richards, separability values range between 0 and 2, indicating how statistically separated the pairs are. Values greater than 1.9 suggest that the class pairs are statistically well separated. However, for pairs with very low values (<1), it is recommended to either edit or combine them into a single class. The software automatically calculates these values after the creation of the ROIs.

Table 3. Transformed divergent separability values (Richards J.A., 1999) for the 2015 scene

	Anti-Atlas bare lands	Hamadas and stony plateaus	Sand accumulations	Alluvial lands	Vegetation	Water surface	Built-up areas
Anti-Atlas bare lands							
Hamadas and stony plateaus	1.86						
Sand accumulations	1.98	1.65					
Alluvial lands	1.81	1.94	1.96				
Vegetation	1.99	1.99	1.99	1.96			
Water surface	2.00	2.00	2.00	2.00	2.00		
Built-up areas	1.98	1.98	1.98	1.98	1.99	2.00	

It is important to note that this table only provides indications regarding the separability of the different classes in the initial nomenclature. The grouping of plots is based on the analysis of effective separability using the statistical file of these samples and the measurements of divergent separability transforms. Thus, even if samples do not physically represent the same land use, they may be confused due to their similar radiometric responses. In our study, training plots with a large standard deviation or a divergence separability transform value less than 1 are automatically replaced to minimize confusion between different land use classes. To address separability issues, a preliminary analysis of the spectral signature curves of each class was conducted to identify the most efficient classification channels (figure 6).

Figure 6. Spectral signature curves of the land elements classes for the 2015 scene

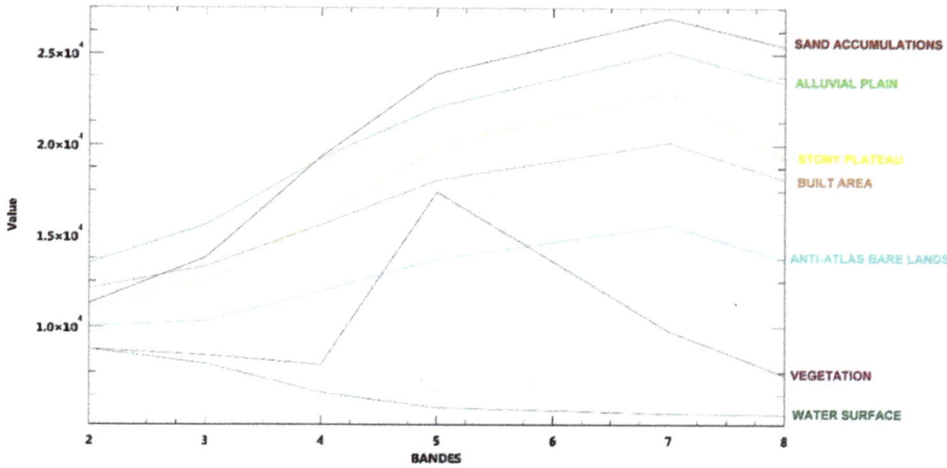

Finally, a confusion matrix was produced to obtain detailed information on the classification's accuracy and validate the selection of training plots. After several tests, the classification that most accurately reflected ground reality was selected. Validation was achieved by calculating cartographic precision, inventory precision, overall precision, and errors of omission and commission.

On average, 92.5% of the pixels were correctly classified on training sites according to the confusion matrix. The ecological variables defining the classes were detected on the image and were not confused. Given that the study area belongs to the desert climatic domain, a homogeneous landscape was not expected.

These results indicate a strong agreement between the classification results and field observations, demonstrating that satellite imagery is an important information source, primarily due to its spectral richness. Sampling is a tedious and restrictive task that heavily influences the results of supervised classification. A major drawback of these methods is the reliance on only three channels to visualize the color composition essential for sample selection.

Several methods were implemented, tested, and compared to determine which best reflected ground reality. In this work, after field reconnaissance, we classified our study area into five major classes.

- Hamadas and stony plateaus
- **Dunes and sandy soils (live sand);**
- Alluvial plain (Fields of agriculture and Oasis);
- Water surface;

- Built-up areas.

This classification into five major classes allowed us to clearly identify its spatio-temporal evolution of sand accumulations. The results obtained from the supervised maximum likelihood classifications were used to construct land cover and sand dunes mapping for the two scenes (1984 and 2015).

An important dynamic of sand drifting: Sandy accumulations are present almost totally on the depressions, basins and plateaus of the intermediate and southern zones of the basin, where deposition conditions are favored. Accumulation occurs when a natural or artificial obstacle forms traps to the sand carried by the wind and allows its accumulation, whether vegetation, an excavation, rough surface or slope may slow down the wind speed and cause the transported sand deposit. In our case, in addition to the developed areas (area of human activity), the most notable obstacle is the steep slopes of the slopes facing southwest, the opposite to the direction of the effective winds. The area covered by these accumulations increased from 99401.26 in 1984 ha to 144273.16 ha in 2015 (figure 7).

Figure 7. Evolution of sand accumulation between 1984 and 2015

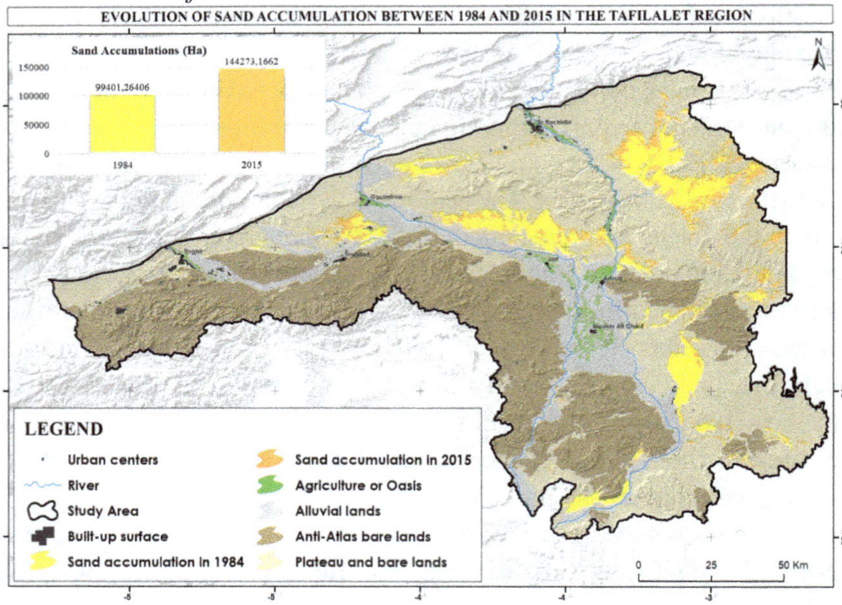

The study area experienced a significant increase in silting, with the sand-invaded area expanding by 44 871.9 ha from 1984 to 2015. This enormous evolution can be attributed to the severe droughts during the 1980s, accompanied by winds capable of transporting sand into bare lands. The alternation of wet and dry periods (figure 8) throughout 1984 to 2015 further contributed to the increase in silting, driving the geomorphological cycle characteristic of the region:

- **Water Erosion**: Destabilized bare slopes of the Atlasean unit and reliefs generate water erosion during relatively humid periods, leading to sediment deposition in low-lying areas. These alluvial deposits form the origin of the sands.
- **Wind Erosion**: Occurs during drought periods, leading to sand encroachment.

The Tafilalet region, increasingly impacted by climate change, faces significant challenges due to rising temperatures and decreasing precipitation (figure 8), both of which intensify sand encroachment. Higher temperatures lead to greater evaporation and reduced soil moisture, weakening the sparse vegetation that stabilizes the soil. This makes the land more susceptible to wind erosion and accelerates the movement of sand dunes. Simultaneously, decreased precipitation exacerbates arid conditions, further diminishing vegetation and water resources essential for sustaining local ecosystems and agriculture. The combined effects of higher temperatures and lower rainfall create a feedback loop, where the advancing sand further degrades the environment, threatening agricultural lands, infrastructure, and human settlements in this fragile region.

Figure 8. History of annual rainfall and mean annual temperature between 1983 and 2016 in Jorf region

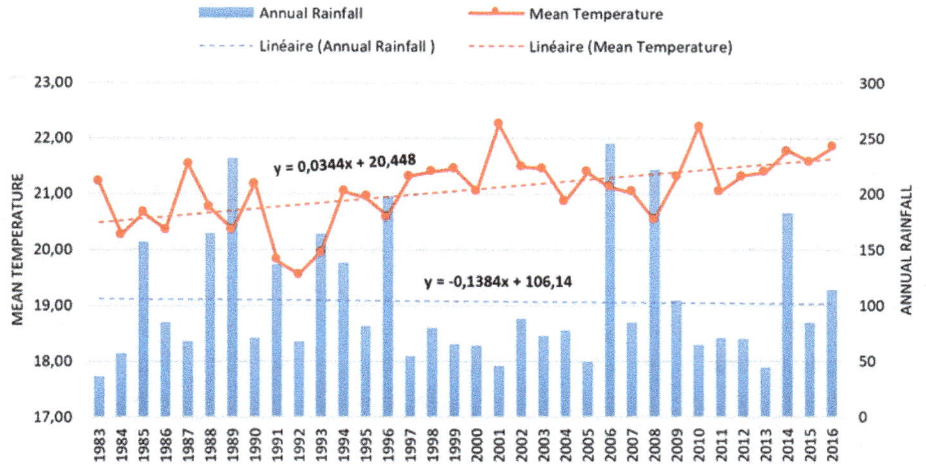

The increase in sand-covered areas is primarily due to the dynamic movement of accumulations driven by prevailing winds. These sandy accumulations represent a significant threat to palm groves, settlements, and hydro-agricultural infrastructure. Field observations have documented several instances of oases being completely invaded by sand, such as Tizgaghine. Figure (9) illustrates a diachronic analysis of agricultural fields, highlighting the extensive invasion that threatens the existence of this oasis.

Figure 9. Sand invasion of agricultural fields at Tizgagnine between 2011 and 2019 (from google earth imagery).

Roads, irrigation systems, and buildings are frequently buried or damaged by sands, leading to increased maintenance costs and, in some cases, the abandonment of settlements. The social impacts of sand encroachment are also significant. As traditional livelihoods become unsustainable and infrastructure deteriorates, many residents are forced to migrate to urban areas (figure 10), The figure illustrates a significant shift in population dynamics between 1994 and 2014, characterized by a positive population annual growth rate in urban areas and a correspondingly weak negative rate in rural communities. This trend suggests a substantial migration flow towards urban centers, primarily driven by factors such as sand encroachment and the consequent loss of agricultural lands in rural regions.

Jorf village, despite being a semi-urban center, experiences notably weak population growth, primarily due to its location in an area with the most significant sand encroachment dynamics in the region. The relentless advance of sand dunes has severely impacted the village, making it increasingly difficult for residents to maintain their livelihoods. This rural exodus exacerbates the depopulation of affected areas and leads to the erosion of cultural heritage due to the combined effects of sand encroachment and declining agricultural opportunities

Figure 10. Population Annual Growth Rate between 1994 and 2014 2014 for the Tafilalet communities (data source: Census of Population and Housing of Morocco 1994 and 2014)

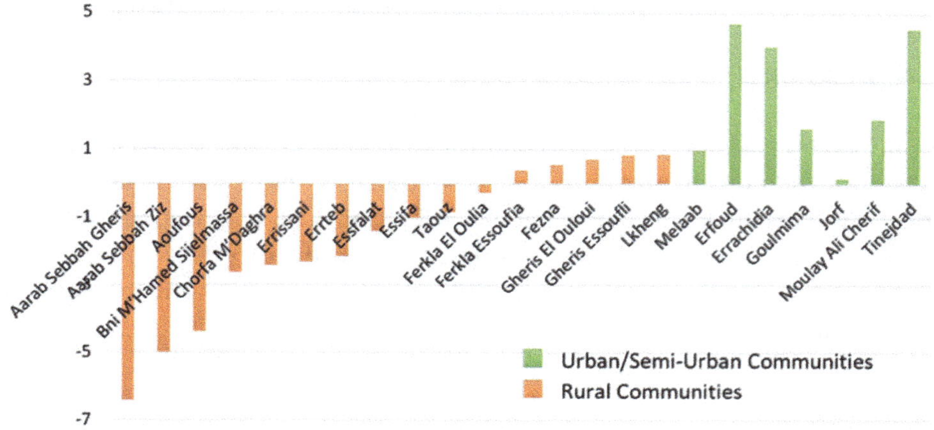

CONCLUSION

The comprehensive analysis of sand dune encroachment in the Tafilalet region reveals significant environmental and socio-economic impacts. The study quantified the spatial and temporal dynamics of sand dunes over a 31-year period using remote sensing techniques and the Fryberger method.

Key findings indicate that the mobility and encroachment of sand dunes in Tafilalet, particularly near the Jorf region, have increased notably, affecting critical infrastructure, agriculture, and human settlements. The sand drift potential (DP) values varied significantly across different sites, with the highest values observed in Jorf (up to 824,65) and the lowest in Merzouga (as low as 268.75). These values correlate with sand transport rates between 19.08 $m^3/m/year$ and 58.55 $m^3/m/year$.

Additionally, predominant sand movement towards the ENE direction was identified, posing a threat to oases and critical infrastructure such as water supply systems and transportation networks. The findings also indicate a notable increase in sand-invaded areas, driven by severe droughts and wind patterns that transport sand into previously unaffected regions. This increased sand coverage poses a substantial threat to palm groves, settlements, and hydro-agricultural infrastructure, leading to economic and ecological consequences.

To mitigate the effects of sand dune encroachment, several strategies are recommended such as continuous monitoring through ongoing remote sensing and ground truthing is essential to track sand movement and identify areas at high risk. Protective measures should be developed and maintained for vulnerable infrastructure, particularly in areas with high DP values. Promoting sustainable land management practices can help reduce the impact of desertification and preserve the delicate desert landscapes.

The involving of local communities in the development and implementation of conservation efforts is crucial for ensuring sustainable and effective mitigation strategies. Additionally, further research is needed to refine sand drift potential models and develop innovative technologies for sand dune stabilization. By focusing on these key areas, it is possible to mitigate the adverse effects of sand dune encroachment and support the resilience of the Tafilalet region's environment and local communities.

REFERENCES

Allahyari, A., & Farjad, B. (2022). Investigation of sand drift potential and sand dune mobility in Khartouran Erg, Sabzevar, NE Iran. *Aeolian Research*, 57, 100807. DOI: 10.1016/j.aeolia.2022.100807

Baas, A. C. W. (2008). Challenges in wind and water erosion research. *Geomorphology*, 100(3-4), 246–257. DOI: 10.1016/j.geomorph.2008.01.018

Bagnold, R. A. (1941). *The physics of blown sand and desert dunes*. Methuen.

Barchyn, T. E., & Hugenholtz, C. H. (2012). Predicting dune field state change with time-series vegetation data. *Earth Surface Processes and Landforms*, 37(8), 749–758. DOI: 10.1002/esp.3199

Benalla, M., Alem, E. M., Rognon, P., Desjardins, R., Hilali, A., & Khardi, A. (2003). Les dunes du Tafilalet (Maroc): Dynamique éolienne et ensablement des palmeraies. *Sécheresse, 14*(2), 73–83. ISSN 1147-7806.

Bourke, M. C., Fenton, L. K., & Balme, M. R. (2022). Dune Activity on Mars: Insights from Terrestrial Analog Studies. *Journal of Geophysical Research: Planets, 127*(4), e2021JE007108. https://doi.org/DOI: 10.1029/2021JE007108

Bullard, J. E. (1997). A note on the use of the "Fryberger method" for evaluating potential sand transport by wind. *Journal of Sedimentary Research*, 67(3), 499–501. DOI: 10.1306/D42685A9-2B26-11D7-8648000102C1865D

Callot, Y. (1987). *Geomorphology and paleoenvironments of the Saharan Atlas in the Western Grand Erg: Wind dynamics and Holocene paleolakes* (Doctoral thesis). Pierre and Marie Curie University.

Callot, Y., & Oulehri, T. (2015). Géodynamique des sables éoliens dans le Nord-Ouest saharien: Relations entre aérologie et géomorphologie. *Geodinamica Acta*, 9(1), 1–12. DOI: 10.1080/09853111.1996.11417259

Cherlet, M., Hutchinson, C., Reynolds, J., Hill, J., Sommer, S., & von Maltitz, G. (2018). *World atlas of desertification*. Publication Office of the European Union., DOI: 10.2760/06292

Fryberger, S. G. (1979). Dune form and wind regime. In McKee, E. D. (Ed.), U.S. Geological Survey Professional Paper: Vol. 1052. *A study of global sand sea* (pp. 137–169). Government Printing Office.

Fryberger, S. G. (1986). Dune form and wind regime. In McKee, M. E. (Ed.), U.S. Geological Survey Professional Paper: Vol. 1052. *Sand seas* (pp. 137–169).

Geist, H. J., & Lambin, E. F. (2004). Dynamic causal patterns of desertification. *Bioscience*, 54(9), 817–829. DOI: 10.1641/0006-3568(2004)054[0817:DCPOD]2.0.CO;2

Gherib, A., Ghazali, M. B., & Bouzid, S. (2023). Monitoring desertification in North Africa: Causes and solutions. *Environmental Science and Pollution Research International*, 30, 31395–31405. DOI: 10.1007/s11356-023-27883-9

Issanova, G., Tussupbekova, G., & Rau, S. (2023). Exploring the complexities of sand dune transformation: The role of anthropogenic degradation and climatic conditions. *Frontiers in Environmental Science*, 11, 1089321. DOI: 10.3389/fenvs.2023.1089321

Lancaster, N. (1983). Controls of dune morphology in the Namib sand sea. In *Developments in sedimentology* (Vol. 38, pp. 261–289). Elsevier., DOI: 10.1177/030913338901300105

Livingstone, I. (1989). Monitoring surface change on a Namib linear dune. *Earth Surface Processes and Landforms*, 14(4), 317–332. DOI: 10.1002/esp.3290140407

Ritsema, C. J., Albaladejo, J., Castaldi, F., & Keesstra, S. D. (2022). Climatic and land-use drivers of desertification. *Journal of Arid Environments*, 201, 104689. DOI: 10.1016/j.jaridenv.2022.104689

Thomas, D. S. G., & Wiggs, G. F. S. (2021). Aeolian Processes and Dune Systems: Advances in Understanding and Challenges Ahead. *Earth-Science Reviews*, 216, 103555. DOI: 10.1016/j.earscirev.2021.103555

Wiggs, G. F. S., Livingstone, I., Thomas, D. S. G., & Bullard, J. E. (1996). Airflow and roughness characteristics over partially vegetated linear dunes in the southwest Kalahari Desert. *Earth Surface Processes and Landforms*, 21(1), 19–34. DOI: 10.1002/(SICI)1096-9837(199601)21:1<19::AID-ESP508>3.0.CO;2-P

Chapter 3

Mapping of Land Degradation Using Revised Erosion Potential Method Into a GIS and RS: Case Study in Rif Mountains, Northern Morocco

Zahnoun Aman Allah
University of Ibn Tofail, Morocco

Al Karkouri Jamal
University of Ibn Tofail, Morocco

Batchi Mouhcine
https://orcid.org/0000-0002-0915-6592
University of Ibn Tofail, Morocco

ABSTRACT

The Rif Mountains extends 31.375 km², is occupied in the north of Morocco, on the Mediterranean slopes, and the rainfall is relatively abundant in the western part and scarce in the eastern part, and irregular, with strong anthropic pressure. This has affected the use of forest and soil resources. This area has been subject to anthropogenic pressure which is steeped in history as an extensive agricultural-forestry-pastoral production system. This deterioration has resulted in intense erosive dynamics that can be seen in various forms, including, the areal etching, the gully and landslides. This dynamic erosion gradually and inevitably leads to soil degradation, causing serious problems in terms of loss soil and silting up of dams. To evaluate

DOI: 10.4018/979-8-3693-6452-9.ch003

this dynamic, we have integrated of the Erosion Potential Method into a GIS & RS, for soil losses estimation allows to approach the severity of the erosive phenomenon, in order to identify as well as map the sensitive and high-risk areas of water erosion and the regions of intervention urgency in the Rif Mountains. Soil losses range from 17.254 to 0 t/ha/yr, the average rate of losses is 46 t/ha/ yr (Erosion in the plot). Potential total erosion in Rif Mountains in the order of 144.292.972 t/yr. Thus, the average amount delivered per unit area is 8.55 t/ha/yr (total of 26.825.625 t/yr).

1. INTRODUCTION

The development of human societies often comes at the expense of the natural resources (vegetation, water and soil). The unreasonable use these resources sometimes leads to the deterioration of the natural environment. The loss soil among the important aspects of this deterioration. It figures among the major processes of soil degradation worldwide, with the corollary of the soil loss productivity and environmental destruction.

The first scientific research on erosion took place in 1890 in Germany (Roose & Barthès, 2001). However, the systematic studies of erosion factors did not really take off until 1930 with Bennette in the United States (Roose E., et al., 2012). The erosion processes have been described since the 1940s and represented by the equations before the 1960s (Lane et al., 1997).

Water erosion, one of the major soil degradation process in the Mediterranean area, is one of the keys to desertification (UNEP, 1994), especially in the southern Mediterranean. This is caused by long periods of drought followed by severe storms with violent erosive rainfall, which falls on steep gradients with vulnerable soils.

In Morocco, the phenomenon of water erosion constitutes a serious problem for the managers of natural resources because the watersheds present very important soil losses cumulated to 100 million tons annually (Heusch, 1970), linked to different physical and anthropogenic factors. To this end, it loses the equivalent of 70 million m^3 of water through siltation of the dam reservoirs due to erosion, whether a decrease of 0.5% per year in the total capacity of the dams (Merzouk, 1992), and about 22 million hectares are threatened by erosion, of which more than 3 million would require urgent interventions. The Rif Mountains occupy the north of Morocco, which presents a great predisposition to the dynamics of water erosion and poses serious problems by inducing a loss of agricultural land causing siltation of the dams. In addition, it is characterized by abundant rains, very steep slopes, a large structural compartmentalization and extension of the substrates, an easily erodible and weakly resistant lithology (predominance of marl-limestone flyschs), an ancient and intense human presence as well as a much reduced vegetation cover (Salhi et al., 2019; Barathon, 1987).

For this study, consists of applying a model in a GIS as a decision-making tool to prevent the onset of water erosion and the development of runoff. In particular, it should enable land managers to analyses the information they need to intervene and correct the situation, which has been alarming through emergency measures in the Rif.

2. MATERIALS AND METHODS

2.1. Study Area

The Rif Mountains are characterised by an arched shape. Occupying the extreme North-West part of Morocco, between 34-36° north latitude and 2-7° west longitude. This is a mountain range with an area of 31.375 km^2, which encompasses all the mountains stretching in an arc from the Gibraltar Straits to the Low Moulouya. It is bounded to the north by the Mediterranean Sea, to the east by the Oued Moulouya, to the south by the prerif and to the west by the Atlantic Ocean (Fig. 1).

Administratively, the Rif Mountains covers three regions: Tangier - Tétouan - Al Hoceima, the northern part of the Fès-Meknès region and the western part of the Oriental region. It comprises 13 provinces, divided into 253 communes.

Morphologically, the Rif Mountains are dominated by more resistant terrains, particularly the Mesozoic calcareous or the sandstone-pelitic series (Masrour, 2012), an axial zone supporting the highest peaks of the chain (Jbel Tissouka 2122 m, Jbel Tidghine 2456 m) and a brutal Mediterranean fallout (El Gharbaoui, 1981; Sebbar, 2013; Suter, 1980).

From a physical point of view, the area is characterised by a number of specific features, including a highly compartmentalised geological structure, extensive substrata, rugged terrain, very steep slopes, easily erodible and not very resistant lithology (predominantly soft marl-limestone flysch and schist) and an extensive hydrographic network.

Since the 1990s, several studies on erosion involving different methods have been of interest: Sadiki A., 2005; Benmoussa et al., 1993; Daghmoumi R., 1994; Ait Fora A., 1995; Moukhchane M., 1998; Gauché E., 2002; Al karkouri J., 2003; Sadiki A., 2004; Mesrar et al., 2015; Manaouch et al., 2021.

Figure 1. Geographic location of the study area (U.S. Geological Survey, 2023)

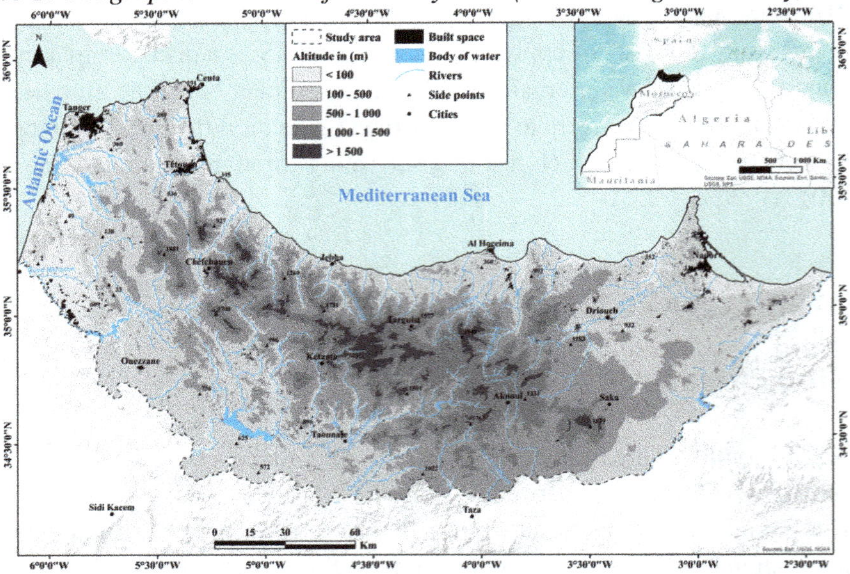

2.2. Erosion Potential Method

As there is no station measuring sediments in the study area, an empirical, semi-quantitative model was developed as part of the study in 1950s by Gavrilovic model. It was therefore used to determine the average annual deposit yield applied in the Balkan countries. Gavrilović, 1988 developed a method to determine the annual loss of soil caused by erosion. The EPM model is used to quantify the erosion of soils by water, and includes erosion mapping, annual yield of deposits, estimation of the quantity of deposits and classification of torrents. Currently, it is applied throughout the world, from:

Table 1.

Bulgaria	Pavlova-Traykova, 2024	Republic of Macedonia	Milevski, 2015
Croatia	Globevnik, 2003	Bosnia and Herzegovina	Tosic et al., 2013
Saudi Arabia	Azaiez, 2024	Slovenia	Globevnik, 2003
Italy	Fanetti & Vezzoli, 2007	Switzerland	Bemporad, 1997
Morocco	Zahnoun & Al Karkouri, 2020	Montenegro	Spalevic, 2013
		Iran	Solaimani, 2009

continued on following page

Table 1. Continued

Serbia	Dragičević, 2014	Iraq	Kareem et al., 2024
Algeria	Zeghmar et al., 2022	Chile	Kayimierski, 2013

The most frequently calculated method the results (Fig.2):

(i) Erosion coefficient Z,
(ii) Total annual soil loss W_a,
(iii) Actual sediment yield G_y.

The EPM model was not developed to examine the physical aspects of erosion and, as such, is particularly useful to zones with limited data or no published research on erosion. In this way, the method not only provides an estimation of the quantity of deposits production and transport, it provides an estimate of the intensity of erosion and indicates areas of potential threat to erosion, which differs from the USLE and PESERA which applies only to sheet erosion (Al Karkouri, 2003). The last is the total of the seven key coefficients of detached land based on Fig.1 (de Vente et al., 2007). These are:

Table 2.

(i) rainfall (P_a),	(ii) Temperature (T),
(iii) soil protection (X_a),	(iv) soil erodibility (Y),
(v) type and extent of erosion (δ),	(vi) π value (3.14),
(vii) slope (J_a),	

Table 3. Description coefficients of the EPM model (De Vente and Poesen 2005; Haghizadeh et al. 2009; Dragičević et al. 2018).

	Values
Soil erodibility (Y)	
Hard rock, erosion resistant	< 0.3
Rock with moderate erosion resistance	0.3–0.35
Weak rock, schistose, stabilised	0.35–0.4
rock with little resistance	0.4–0.45
Soils without erosion resistance	> 0.45

continued on following page

Table 3. Continued

Soil protection (X_a)		
Mixed and dense forest, Reforestation		< 0.2
Coniferous forest with little grove, scarce bushes, bushy prairie		0.2–0.4
Damaged forest and bushes, pasture		0.4–0.6
Damaged pasture and cultivated land		0.6–0.8
Areas without vegetal cover		0.8–1
Type and extent of erosion (δ)		
Erosion not apparent	0.1	< 0.2
Pellicle stripping	0.3	0.2 – 0.4
Dominant areolar stripping	0.5	0.4 – 0.6
Generalized areolar stripping	0.6	0.6 – 0.8
Individualized gully	0.7	
Localized gully	0.7	
Generalized gully (Bad land)	0.8	
Individualized mass movement	0.9	0.8 – 1
Localized mass movement	0.9	
Dominant mass movement	0.9	
Bank undercutting	0.9	
Associated processes	0.9	
Erosion coefficient (Z)		
Excessive erosion		>1
Severe erosion		0.7–1
Medium erosion		0.4–0.7
Slight erosion		0.2–0.4
Very slight erosion		0–0.2

This model's outputs are converted from m^3/km^2/year into t/ha/year by using the bulk density formula (Da) established by Blake in (1965):

Da = m0 / V

Where: m0: anhydrous mass of soil (g); V: volume of the cylinder (cm^3);

In 1985, Lazarević provided many modifications which were designed to determine the values assigned to the following parameters: coefficient of type and extent of erosion (δ) from 0.1 for low erosion to 1 for very high erosion, soil protection coefficient (Xa) less than 0.05 in dense mixed forest at 1 for bare soils and soil erosion resistance coefficient (Y) ranging from 0.2 for hard rocks to 2 in fine soils without erosion resistance. These modifications for the purpose of updating the classes of the intensity of potential erosion coefficient (Z) of less than 0.2 for very low erosion and more than 1 for very high erosion (Dragičević et al., 2016).

Figure 2. Description and equations of the EPM model

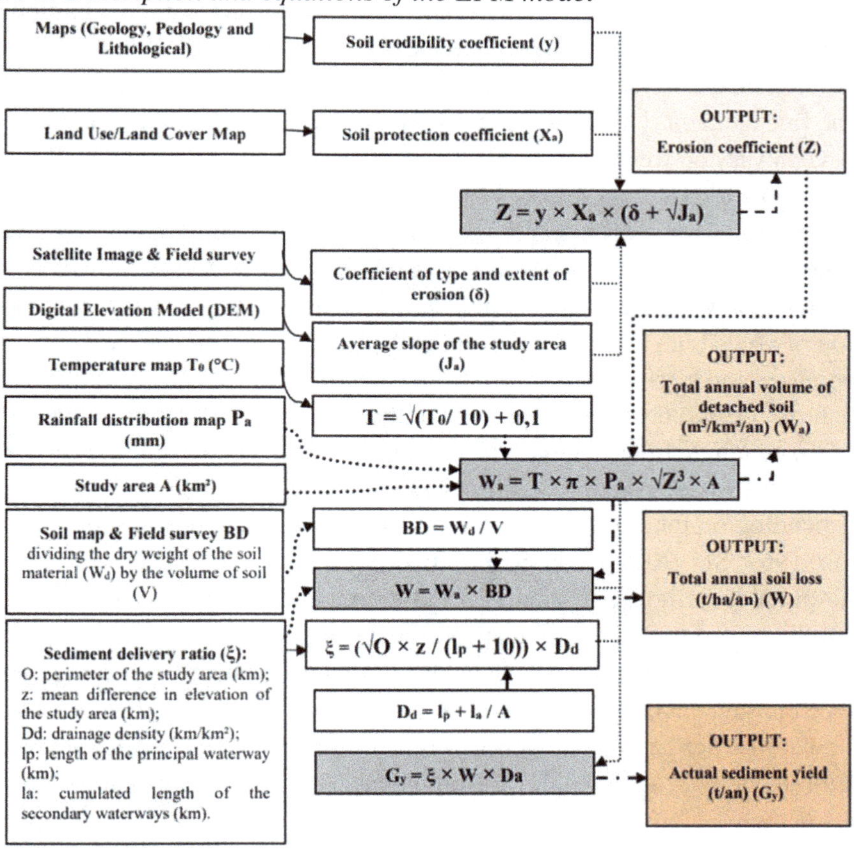

Soil erodibility coefficient (y) is the average response of the soil and the pedological profile to the erosive capacity of rainfall (Heusch, 1970). The Mediterranean soils are prone to deterioration when they are stripped and without a sustained supply of litter (Roose et al., 2012). Above all, is a key factor in soil erosion modelling (Le Bissonnais, 2005). Then, the types of soil become more sensitive to erosion with increasing the silt fraction (Wischmeier and Smith, 1978). To determine the vulnerability of the soil to erosion, we had recourse the formula used by RUSLE fraction (Wischmeier and Smith, 1978), which is written as follows:

$$Y=2.1*M^{1.14}*10^{-4}(12-MO)+3.25(B-2)+2.5(C-3)/100$$

Where, **M**: (% Fine sand +% silt)*(100-% Clay); **MO**: Percentage of organic matter; **B**: Soil permeability code (1 to 6); **C**: Profile permeability classes (1 to 4): 1 Very fine, 2. Fine, 3. Medium and coarse, 4. Very coarse

Further, the analyses were focused on hundred (100) samples taken from the Rif Mountains. A precise appreciation of the texture is made in the laboratory by a granulometric analysis, which gives the percentage distribution of each component. The results of the parameters involved in measuring the factor of soil erodibility and the granulometric analyses carried out on each type of soil, taking into consideration only the fine elements (fraction < 2 mm). The organic matter content of all the soil textures can vary between 0.4% and 4.2%, depending on soil moisture, soil use and the decomposition of these residues by microbes and other soil organisms.

The soil protection coefficient (X_a) it constitutes the most determining and dynamic parameter for Moroccan watersheds (Ait Brahim et al., 2002), as it influences on infiltration and evapotranspiration. In fact, the vegetation by its root part favors the creation of voids both at the surface and at depth, and thus the permeability of the soil increased and increases the infiltration (Ait-Fora, 1995). The works of Merzouk and Dahman (1998) on changes land use in the Tleta catchment area between 1976 and 1996, concluded that the natural vegetation was cleared with an average rate of 15 ha/year (Sossey Alaoui F., 2005). The coefficient (X_a) its value varies from 0 to 1 depending on the different types of surface cover (Adediji & Adepoju, 2010; Kalman, 1967; Shi, 2002; Salhi et al., 2019).

The climate acts through rainfall (P_a) and temperature (T) on erosion, but its action is primarily linked to the rainfall distribution during the year (Sadiki & Bouhlassa, 2004). One of the main problems which most often encountered is missing data. To fill in the missing data, we proceeded to interpolation from the surrounding stations and the reconstruction of homogeneous series by applying a statistical method of Thiessen (Lazarevic, 1985; Sadiki & Bouhlassa, 2004). It was calculated as follows:

$$CCLmoy = \sum ik \;=\; 1AiPi \,/\, A \; avec \; Pi = hi + hi + 1 \,/\, 2$$

With, $\mathbf{CCL_{moy}}$: Climate coefficients (Average rainfall in mm or average temperatures in °C); **A**: Total surface of the study area (km^2); $\mathbf{A_i}$: Surface between two isohyets (h_i and h_{i+1} (km^2); **K**: Total number of isohyets; $\mathbf{P_i}$: Average height between two isohyets (mm or °C).

The slope (J_a) of the surface affects erosion through its degree of inclination (Ait-Fora, 1995). The increase in the slope accentuates the runoff, which also becomes erosive and their energy exceeds that of rainfalls to be able to tear off and transport the solid particles (Ahamrouni, 1996). For this purpose, the knowledge of the slope system through their distribution and cartographic representation is an essential element for dealing with the erosion problems and for the precise description of topography, which also exert a direct influence on the geographical distribution of soils.

The values of type and extent of erosion coefficient (δ) and their development vary according to the size of the study area. Effectively, these values were used to determine the coefficient (δ) (Gavrilović, 2006). The type and extent of erosion coefficient classified based on a number of field visits into five classes, ranging from 0.1 to 1. However, the coefficient is identified the areas affected by the different forms of erosion; it is carried out from the field observations and the Google Earth satellite navigator, which permit to digitize the forms of erosion. The analysis of the map of erosion forms is multiplied because in most cases, the old forms have gained momentum, and the new forms have emerged.

3. RESULTS AND DISCUSSION:

Summarizing the analysis, it appears that the **Fig. 3.a** shows the distribution of Soil erodibility coefficient (Y) to erosion, which proceeds that the texture of the soils generally silty or silty-sandy increases the soil erodibility, particularly in the northern part of the Central-Western Rif, and that the rocky outcrops in the south of the Eastern Rif are mostly friable and very erodible and the soils are very degraded; these lands are poor in organic matter. The values of the (Y) vary from less than 0.2 to more than 0.45. Clearly, only 20% of the study area demonstrates a low erodibility, represented by the Brunified and Calcimagnesic class with Y less than 0.3. The class between 0.35 and 0.45 presents 76.7% of the total area predominance of the class of little evolved soils and raw mineral soils. Then, the rest of the studied surface, presents relatively very high values with 2.2% of the total area shows a strong with Y more than 0.45.

The Soil protection coefficient (X_a) is used to assess the effectiveness of different crop and land practices in terms of avoiding soil loss. Its value varies from 0 to 1 depending on the different types of surface cover (Adediji & Adepoju, 2010; Kalman, 1967; Shi Z.H., 2002). The **Fig. 3.b** permit to identify some information concerning the distribution of X_a values and their area. For this purpose, we note that the most encountered class is that of (< 0.6), it occupies more than half of the study area 87.7%, whether 27 515.2 km², while the values of X_a (> 0.6) occupy 12.3%.

The rainfall (P_a) plays an important role in erosion processes. However, the Rif Mountains record the most abundant rainfall in Morocco. Bearing in mind that the great extent and the diversity of topography make that the climate is very variable from one region to another (Ahamrouni, 1996). According to Al Karkouri, 2016, the most affected regions are not precisely known by the abundance of rainfall; they are the regions with a two-season climate regime, which are mostly affected by the soil erosion. Nevertheless, they are irregular in time-space, as well as very aggressive to the soil. The map of the spatial distribution of rainfall (**Fig. 3.c**) shows that the western Rif and the summits of the central (> 600 mm) are the most watered parts of the Rif Mountains (receiving more than 2000 mm/year and are concentrated in the south of the Tangier peninsula in the Bab Taza region, at the level of the high ridges of Bab Taza and Chrafate). The least watered regions essentially correspond to the low- and medium-altitude terrains of Ben Karrich, Bou Ahmed and Sebta.

Figure 3. Digital maps of the EPM model (a): Y coefficient; (b): X_a coefficient; (c): P_a coefficient; (d): T_o coefficient; (e): J_a coefficient; (f): δ coefficient; (g): D_d coefficient; (h): Watersheds of the Rif Mountains.

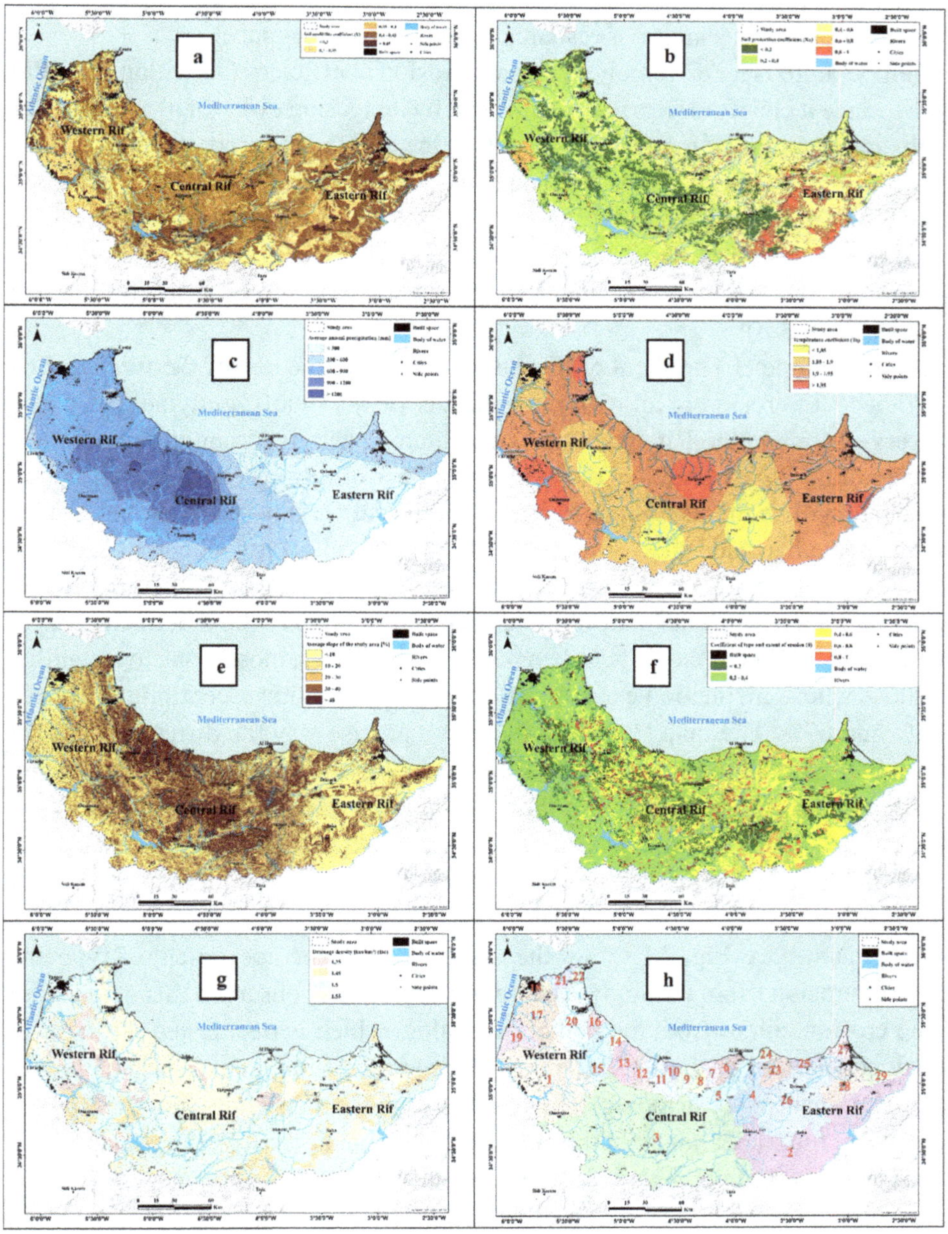

An analysis of farming and state management techniques for soil conservation in the Rif revealed a number of water and soil conservation practices. These techniques are classified as water use and mobilisation techniques, and soil fertility conservation techniques. The diversity of these water and soil conservation techniques reflects an awareness of the scarcity of resources and the risks they face.

Temperature is an extremely important erosion coefficient (T) within the EPM model, since it clearly affects water in the soil by increasing evapotranspiration. The results obtained in the map below show that the coefficient varies from 1.75°C to 2°C, distributed along a South-North and West-East gradient. Obviously, the maxima of > 1.95 are reached in the Eastern Rif, in part of the Central Rif the region of Targuest and the south of the Western Rif particularly in Ouezzane, whereas the minima of < 1.8 are encountered in the shaded area and descending from the altitudes (**Fig. 3.d**).

Average slope (J_a) gradient of the study area constitute one of the fundamental coefficients of erosion in all its forms (sliding, flows, runoff, gullying, etc.), they are very diverse in the Rif Mountains, which are generally strong despite the low altitude of the low mountain in the west. In the case of the presence of steep slopes, the terrain will be the seat of the frequency of the solifluxion sector, however, they are much localized and short (Maurer G., 1968), the runoff is concentrated and the gully is accentuated with the deposits downstream. The slopes of these mountain chains intervene by the absolute altitudes, the delimitation of the bioclimatic stages and the distribution of the soils. When mapping the slopes, most prone to instability are those whose inclination exceeds 10%, with a maximum susceptibility between 17.5% and 65%. The map (**Fig. 3.e**) show in a precise way the distribution of the slopes in the Rif Mountains. Indeed, the slopes (< 30%) are dominant, they represent more than 70% of the total surface, which are concentrated in the eastern and western parts of the mountains. But, the part with steep slopes (> 30%) are located in the middle of the study area, which coincides with the Central Rif and part of the Rif foreland.

The map above **Fig. 3.f** shows the distribution of the coefficient of type and extent of erosion (δ) in the study area. In this regard, it is observed that a good part of the erosion comes from the pellicle stripping, which is manifested the areas of low slope, particularly in the Eastern and Western Rif Mountains. This is followed by the dominant sheet stripping, which is developed in the Eastern and Central Rif, from where the tender nature of the marls and the relief declivity favor the extension of this type. Additionally, the mass movements affect the Central and Western Rif (especially the provinces of Al Hoceima, Chefchaouen and the regions of Aknoul), whence the preponderant areas of steep slopes and the presence of stocks of materials are easily mobilized.

The bulk density (**Fig. 3.g**) is a metric reflecting flow, the stability of the drainage network and the type of surface runoff (Franco-Plata et al., 2013). In this work, we have established the map of the distribution of the watersheds in the Rif, to address that the size of the northern basins is modeled by a dense network of small steep-sided and smaller valleys, are moreover more tectonically deformed in a soft and fragile rocky material (**Table 1**). On the other hand, the size of the southern basins are well characterized by the large synclines, installed at the limit of the Meso-Rif zone (**Fig. 3.h**). This indicates a great difference in the bulk density, the speed of sediment evacuation and the hydrological profile and that the sinking of the networks is very clear in the central part compared to the eastern or western part, which will lead to a high hydrographic density, the abundance of ravines, and finally a strong aggressiveness of the erosive action and an acceleration of the flow (Morphometric characteristics from 1 to 29 watersheds in the Rif Mountains).

Table 4. Morphometric characteristics of watersheds in Rif Mountains

N°	Watersheds	I_p	I_a	A	O	z
1	Loukous	1207.2	1132.5	3155.6	331	805
2	Moulouya	1855.9	1577.7	4099.5	560.3	1010
3	Sebou	3187.4	3230.5	9432.3	690.9	1240.5
4	Nekkour	314.2	360.7	932.6	194	978
5	Ghiss	270.6	325.8	842.1	174	1017
6	Isli-Rmod	113	104.1	336.2	116.7	571.5
7	Senada	86.5	74.2	246.4	81.3	737.5
8	Beni Boufrah	49.2	55.3	168	67.6	577.5
9	Feddal (Mestassa et Beni Gmil)	105.5	92.4	296.9	79.2	900
10	Fettouh-Takamout	50.6	33.7	139.4	56.6	685
11	Ouringa	177.3	173	578.9	132.1	1059
12	Amtar	124.6	82.1	311.2	103	1061
13	Tihaissass	187.5	196.3	644.7	114.3	1062.5
14	Bouahmed	31.9	44.3	138.6	59	910.5
15	Oued Laou	382.7	306.3	962	168.3	1073.5
16	Amsa-azla	71.7	54	205.7	78.9	807.5
17	Tahaddart (Mharher et El Hachef)	441.6	421.8	1120.3	169.2	488
18	Tangier(Mlaleh-Mghogha-Juifs-Zitoun-Boukhalef et Bougdour)	98.4	80.1	286.4	110.1	234
19	Gharifa	210.9	161.3	557	142.6	277.5
20	Martil	473.5	422.9	1307.3	190	898.5

continued on following page

Table 4. Continued

N°	Watersheds	I_p	I_a	A	O	z
21	Rmel-Ksar Esghir-Lmarse-Alienne	97.2	112.4	338.8	133.6	412.5
22	Smir-Negro-Fnideq	63	43.1	171.2	73.5	369.5
23	Boudinar	208.5	144.1	458.5	97.6	786
24	Trougout	44.6	39.7	128.1	55.5	332
25	Ait Said	79.4	33.6	155	77.2	459
26	Kert	1388.8	1352.2	2963.4	342.3	904.5
27	Beni Chikar	151	114.5	393.6	189.7	439
28	Nador	298.4	302.4	654.2	131.3	464.5
29	Arkmane-Cap de l'Eau	151.3	151.8	321.1	150	391

The erosion coefficient map (Z) gives an indication of the probability of erosion in the Rif mountains. The values of (Z) have been defined in terms of the erosion classes supplied by Gavrilović. In this regard, the values of these classes have been subdivided into five classes as shown in the **Fig. 4.a**. Each of the classes of this coefficient is classified in accordance with an erosion susceptibility code, which is assigned to it based on previous works and our knowledge of the study site. The analysis of the results shows that the major part of the study area is impacted by the very high erosion coefficients (> 1), it concerns 40.5% of the total area, whether 12738.6 Km^2.

The combination of the various coefficients that comprise the erosion Potential Method equation, combined with the digital values of the seven coefficients, produces the annual soil loss (W_a) at each point in the Rif Mountains. Every pixel on the map has a value that represents its potential for erosion. This permits to produce the map of average soil losses affected by water erosion. In order to better analyse and assess the results (**Fig. 4.b**), we opted for a grouping of the data by classes of soil losses. This grouping was inspired by previous studies with the exclusion of habitat areas and water bodies in the calculation of soil losses. The estimated global average 46 t/ha/year. The maximum and minimum losses by parcel are respectively 17.254 t/ha/year and 0 t/ha/year. The total annual losses of the Rif Mountains are 144.292.972 t/year. In comparison with other works, the results obtained are not far from those of (HCEFLCD), which found an average rate of soil losses of the order of 5.000 t/km^2/year, whether 156.875.000 t/year.

The erosion rate varies from area to area in the Rif Mountains, according to the impact of the various physical and anthropogenic forces that affect erosion dynamics. o facilitate reading of the map, the various units have been grouped into five classes (Fig. 4.b). Losses of more than 45 t/ha/year originate from 28.1% of the Rif

Mountain area. This shows the extent of the phenomenon. Thus, the lithological formations are clay, flysch, or marl soils.

Thus, this high sensitivity could be explained by its presence in areas that are severely devoid of vegetation cover and very steep slopes, in addition to the human factor, which results in a degradation of the natural vegetation against an invasion of agriculture and other human activities. The low to very low intensity classes (soil losses less than 15 t/ha/year) comes in the second position in terms of the surface area, covering the study area with 18.4% and 15.5% respectively. They are mainly located in the plains of the Atlantic and the Mediterranean facades or even river terraces. Although it is characterized by relatively high altitudes and steeper slopes, the development of a deciduous forest on fairly resistant lithological formations makes these lands less susceptible to erosion. In actuality, 35.6% of the total surface of the studied area generates losses between 15 and 45 t/ha/year, they are located in areas with relatively rugged relief with moderately high altitudes and medium to steep slopes.

The calculated value for the actual sediment yield (G_y) make it possible to estimate the quantities of sediment transported to the hydrographic network (outlet). Consequently, the erosion rates are not homogeneous in the study area. Some of these sediments are deposited upstream depending on the roughness of the topography and the distance of the sites from the production area. The analysis of the results given in map (**Fig. 4.c**), leads to the conclusion that the average quantities of soil delivered at the level of the Rif per surface unit and distributed in the watersheds from 1 to 29. The actual sediment yield for all the entire Rif Mountains amounts to 8.55 t/ha/year, whether a total of 26.825.625 t/year. The small northern basins (Fettouh-

Figure 4. Maps of the EPM model in the Rif Mountains (a): the erosion coefficient (Z); (b): soil losses (W_{sp}); (c): actual sediment yield (G_{sp}).

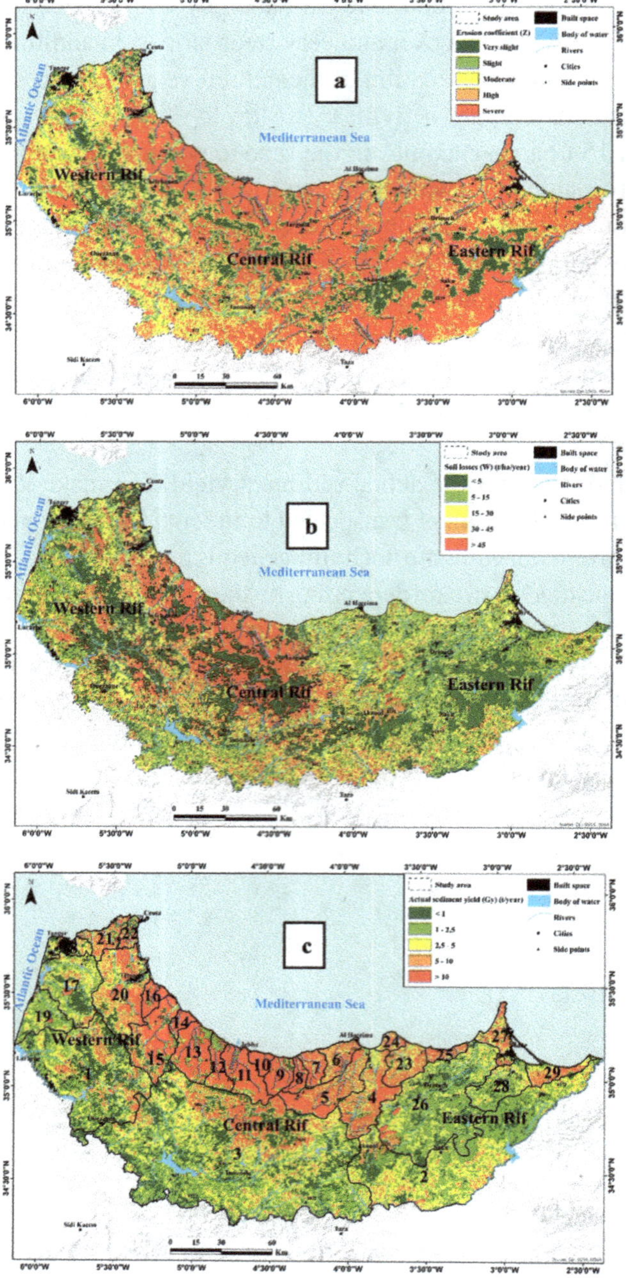

Takamout, Ouringa, Amtar and Bouahmed) are delivered in incredible quantities due to the inclination of the steep slopes and the nature of the lithology's which have been friable.

Evidently, in order to study the correlation between the erosion rate and the erosion coefficients (X_a, Y, T_o P_a, J_a et δ), one thousand (1000) random points were generated in the study area. The values of the erosion rate and the different coefficients were extracted at these points. Really, the Pearson correlation was established between the rate of erosion and each of its factors to know the most influential coefficients in the phenomenon of water erosion.

Table 5. Correlation between the soil losses and the erosion coefficients

Coefficients	W_a	Z	X_a	Y	T_o	P_a	δ	J_a
W_a	1	0,73	0,4	0,41	0,04	0,39	0,3	0,2
coefficient Z	0,73	1	0,58	0,36	0,01	0,05	0,43	0,07
coefficient X_a	0,4	0,58	1	0,04	0,02	0,27	0,25	-0,31
coefficient Y	0,41	0,36	0,04	1	0,11	0,3	0,14	0,15
coefficient T_o	0,04	0,01	0,02	0,11	1	0,02	-0,04	0,02
coefficient P_a	0,39	0,05	0,27	0,3	0,02	1	0,06	0,2
coefficient δ	0,3	0,43	0,25	0,14	-0,04	0,06	1	0,07
coefficient J_a	0,2	0,07	-0,31	0,15	0,02	0,2	0,07	1

The Table 2 shows a high correlation between soil loss (W_a) and Soil erodibility (Y), which appears to be the most important factor in erosion dynamics, With a correlation value of around 0.41 (23.5%), the soil protection parameter (X_a) controls 0.4 (23%) of losses, the rainfall (P_a) controls 0.39 (22.4%) of losses, the type and extent of erosion 0.3 (17.2%) and the slopes 0.2 (11.5%). The soil protection is more closely linked to erosion coefficients, with a correlation coefficient of 0.58, followed by the type and extent of erosion (correlation coefficient = 0.36). The soil erodibility is rather strongly linked to soil protection (correlation coefficient = 3). Thus, the slopes correlate well with the soil erodibility.

4. CONCLUSION

Using Gavrilović's model in the GIS, we were able to produce a synthetic map of the distribution of soil vulnerability, qualitative (Z) and quantitative (W_a) soil loss in the Rif. The research showed that mean soil loss was estimated at 46 t/ha/yr. The minimum and maximum losses per plot are respectively 0 t/ha/yr and 17.254 t/ha/yr.

The annual losses in the Rif around of 144.292.972 t/yr. The average actual sediment yield (G_y) of area for the Rif Mountains as a whole is 8.55 t/ha/yr. This is a total of 26.825.625 t/yr. The results of this model were analysed using GIS to determine the coefficients controlling water erosion, in order of importance: Soil erodibility, soil protection, rainfall, type and extend of erosion, slopes, and temperature.

In view of this situation, the local population and the state have long been implementing a variety of water and soil conservation techniques however remain insufficient, which differ from one area to another in the Rif Mountains, depending on the environmental dynamics.

From this chapter, we recommended:

- Increase the number of stations sampled to cope with climate variability in space, especially in the Rif mountains characterized by high microclimate variability;
- Accumulate climate data (height, intensity) over a long period;
- Maintaining crop rotation, which is beneficial for maintaining biological diversity and therefore helps to restore the level of organic matter that improves soil structure while protecting against run-off and soil erosion;
- Carrying out integrated and participative projects, as well as ongoing monitoring of forestry and agronomic infrastructures and developments.

At the end of this work, it is important to point out some development prospects, namely:

Study of cartographic modelling to enhance the traditional knowledge of the farmers. This work will make it possible to acquire developments adapted to micro-variations and likely to be integrated sustainably into farming systems, thus reducing investment costs;

Study on the costs of soil losses;

Study on the involvement of populations in anti-erosion projects.

REFERENCES

Adediji, A., & Adepoju, K. A. (2010). *Assessment of Revised Universal Soil Loss Equation (RUSLE) in Kastina Area, Kastina State of Nigeria using Remote Sensing (RS) and Geographic Information System (GIS) Environmental impact of metal recycling industry view project*. https://www.researchgate.net/publication/49593936

Ahamrouni, J. (1996). *Erosion hydrique dans le bassin versant de la Moulouya (Maroc Oriental) : Recherche des zones sources d'envasement de la retenue du barrage Mohamed V*. Université Cheikh Anta Diop, Faculté des Sciences et Techniques.

Ait Brahim, L., Chotin, P., Hinaj, S., Abdelouafi, A., El Adraoui, A., Nakcha, C., Dhont, D., Charroud, M., Soussey Alaoui, F., Amrhar, M., Bouaza, A., Tabyaoui, H., & Chaouni, A. (2002). Paleostress evolution in the Moroccan African margin from Triassic to Present. *Tectonophysics*, 357(1-4), 187–205. DOI: 10.1016/S0040-1951(02)00368-2

Ait-Fora, A. (1995). *Modélisation spatiale de l'érosion hydrique dans un bassin versant du Rif marocain : validation de l'approche géomatique par la sédimentologie, les traceurs radioactifs et la susceptibilité magnétique des sédiments*. Université de Sherbrooke.

Al Karkouri, J. (2003). *Dégradation du milieu naturel dans le bassin de Beni Boufrah (Rif central-Maroc) : analyse des facteurs et des processus, essai de quantification et modélisation spatiale*. Université Mohamed V, Faculté des Lettres.

Al Karkouri J (2016). *Les milieux montagneux marocains à l'épreuve du changement climatique (cas de la montagne rifaine)*. Hespéris-Tamuda LII (1) (2017): 237-267.

Amiri F, Shariff AR, Tabatabaie T (2012). Monitoring land suitability for mixed livestock grazing using Geographic Information System (GIS). *Application of Geographic Information Systems*, 241–266.

Azaiez, N. A. (2024). Insights into soil loss in the Jouah basin using EPM model and Caesium 137 as an isotopic tracer. *Egyptian Journal of Soil Science*, 64(1), 0. DOI: 10.21608/ejss.2023.245181.1681

Barathon, J. J. (1987). *Bassins et littoraux du rif oriental (maroc) : evolution morphoclimatique et tectonique depuis le neogene superieur*. Université de Poitiers.

Bemporad, G. A., Alterach, J., Amighetti, F. F., Peviani, M., & Saccardo, I. (1997). A distributed approach for sediment yield evaluation in Alpine regions.... *Journal of Hydrology (Amsterdam)*, 197(1-4), 370–392. DOI: 10.1016/0022-1694(95)02978-8

Berk, R. (2004). *Regression Analysis: A Constructive Critique*. SAGE Publications, Inc., DOI: 10.4135/9781483348834

Daghmoumi, R. (1994). Contribution of geology and GIS to the study of the watersheds of Aoudour (Central Rif) (Doctoral dissertation, Thesis 3rd cycle, Univ. Mohammed V, Fac. Sci. Rabat,(162p)).

De Ploey, J., Imeson, A., & Oldeman, L.-R. (1991). Soil erosion, soil degradation and climatic change. In Brouwer, F.-M., Thomas, A.-J., & Chadwick, M.-J. (Eds.), *Land use changes in Europe* (pp. 275–292). Kluwer Academic Publishers. DOI: 10.1007/978-94-011-3290-9_12

De Vente, J., Poesen, J., Arabkhedri, M., & Verstraeten, G. (2007). The sediment delivery problem revisited. *Progress in Physical Geography*, 31(2), 155–178. DOI: 10.1177/0309133307076485

Dragičević, N., Barbara, K., & Nevenka, O. (2016). A Review of the Gavrilović Method (Erosion Potential Method). *Application*, 68, 715–725. DOI: 10.14256/JCE.1602.2016

Dragičević, N., Karleuša, B., & Ožanić, N. (2018). Modification of erosion potential method using climate and land cover parameters. *Geomatics, Natural Hazards & Risk*, 9(1), 1085–1105. DOI: 10.1080/19475705.2018.1496483

Dragičević, N., Whyatt, D., Davies, G., Karleuša, B., & Ožanić, N. (2014). Erosion model sensitivity to land cover inputs: case study of the Dubračina catchment, Croatia. *In:Proc. 22nd Annual Conf. GISRUK 2014,* Glasgow, 340–348.

EarthExplorer (usgs.gov). 20/08/2023.

El Gharbaoui A (1981). *La terre et l'Homme dans la péninsule tingitane. Etude sur l'Homme et le milieu dans le Rif occidental.* Travaux de l'Institut Scientifique, n° 15.

Fanetti, D., & Vezzoli, L. (2007). Sediment input and evolution of lacustrine deltas: The Breggia and Greggio rivers case study (lake Como, Italy). *Quaternary International*, 173-174, 113–124. DOI: 10.1016/j.quaint.2007.02.008

Franco-Plata, R., Miranda-Vázquez, C., Solares-Hernández, H., Manzano-Solís, L. R., Bâ, K. M., & Expósito-Castillo, J. L. (2013). Implementing into GIS a Tool to Automate the Calculation of Physiographic Parameters of River Basins. *Open Journal of Modern Hydrology*, 03(02), 67–74. DOI: 10.4236/ojmh.2013.32009

Gavrilović S (1972). Inzenjering o bujic nimtokovima i eroziji. Beograd. Serbia. *Izgradnja (Special Issue).*

Gavrilović, Z., Stefanovic, M., Milojevic, M., & Bacevac, J. (2006). Erosion Potential Method, An Important Support for Integrate Water Ressource Management. *Presented atXXIII Conference of the Danubian Countries on the Hydrological Forecasting and Hydrological Bases of Water Management,* Bled, Slovenia.

Globevnik, L., Holjevic, D., Petkovsek, G., & Rubinic, J. (2003). Applicability of the Gavrilovic method in erosion calculation using spatial data manipulation techniques, erosion prediction in Ungauged Basins: Integrating methods and techniques. *IAHS Publ*, (279), 224–233.

Haghizadeh, A., Teang Shui, L., & Godarzi, E. (2009). Forecasting sediment with erosion potential method with emphasis on Land use changes at basin, electronic. *Journal of Geotechnical Engineering*, 14, 1–12.

Heusch B (1970). *L'érosion du pré-rif : une étude quantitative de l'érosion hydraulique dans les collines marneuses du pré-rif occidental.*

High Commission for Water and Forests and the Fight against Desertification — ERA-LEARN.

Joftic, L., Milliman, J., & Sestini, G. (1992). *Climate change and the Mediterranean, UNEPE.* Arnoid Pub.

Kalman, R. (1967). Essai d'évaluation pour le pré-Rif du facteur couverture végétale de la formule de Wischmeier de calcul de l'érosion. *Rapport Pour l'administration de La Forêt et d'eau, Rabat, 1*(12).

Kareem, I. J., Jasim, G. S., Ali, H. A., & Amin, R. M. (2024). Estimating the Extent of Water Erosion in Darbandikhan Lake Using a Model Gavrilović Method (EPM) (Erosion Potential Method). *International Journal of Religion*, 5(9), 358–369. DOI: 10.61707/neq6c157

Kayimierski, L. D., Irigoyen, M., Re, M., Menendey, A. N., Spalletti, P., & Brea, J. D. (2013). Impact of climate change on sediment yield from the upper Plata basin. *International Journal of River Basin Management*, 11, 1–11.

Lane, L. J., Hernandez, M., & Nichols, M. (1997). Processes controlling sediment yield from watersheds as functions of spatial scale. *Environmental Modelling & Software*, 12(4), 355–369. DOI: 10.1016/S1364-8152(97)00027-3

Lazarevic, R. (1985). The new method for erosion coefficient determination – Z. *Erosion Bulletin*, 13, 54–61.

Le Bissonnais, Y., Dubreuil, N., Daroussin, J., & Gorce, M. (2005). Modélisation et cartographie de l'aléa d'érosion des sols à l'échelle régionale. *Etude et Gestion Des Sols*, 11, 307–321.

Manaouch, M., Zouagui, A., & Fenjiro, I. (2021). A review of soil erosion modeling by R/USLE in Morocco: Achievements and limits. In *E3S Web of Conferences* (Vol. 234, p. 00067). EDP Sciences. DOI: 10.1051/e3sconf/202123400067

Masrour, A. (2012). *Le Jurassique moyen-supérieur du Rif externe et l'avant pays oriental: Stratigraphie, sédimentologie et modélisations paléogéographiques* (9783838181042nd ed.).

Maurer, G. (1968). *Les montagnes du Rif central. Étude géomorphologique*. Université Paris-Diderot.

Merzouk A, Bonn F, Badraoui M, Benkhaiti M, Tahri M (1992). Développement des SIG pour la mise en valeur agricole des terres arides et semi-arides : la problématique et les attentes. *Géo. Observateur. 2*, 77–84.

Mesrar, L., Akdim, M., Lakrim, M., Aroussi, O. E., Akhrif, I., & Jabrane, R. (2015). The physico-mechanical characteristics of the clays in and after doping with metal oxide Al2O3 in The region of Fez (Morocco). *Present Environment and Sustainable Development*, 9(1), 103–110. DOI: 10.1515/pesd-2015-0006

Milevski, I. (2015). General geomorphological characteristics of the Republic of Macedonia. *Geographical Reviews*, 48, 5–25.

Morgan, R. P. C. (1986). Soil erosion and conservation. *Rev. and Enl. Ed. of: Soil Erosion*, 256–295.

Moukhchane, M., Bouhiassa, S., & Chalouan, A. (1998). Cartographic and magnetic approach for the identification of sediment sources: case of the Nakhla watershed (Rif, Morocco). Science and Planetary Change/Drought, 9 (3), 227-232.

Pavlova-Traykova, E. (2024). ASSESSMENT SOIL EROSION RISK IN FOREST TERRITORIES IN SELSKA RIVER. Forest Science (0861-007X), 60.

Roose, E., & Barthès, B. (2001). Organic matter management for soil conservation and productivity restoration in Africa: a contribution from Francophone research. In *Managing Organic Matter in Tropical Soils: Scope and Limitations* (pp. 159–170). Springer Netherlands. https://doi.org/DOI: 10.1007/978-94-017-2172-1_15

Roose, E., Sabir, M., Arabi, M., Morsli, B., & Mazour, M. (2012). Soixante années de recherches en coopération sur l'érosion hydrique et la lutte antiérosive au Maghreb. *Physio-Géo*, 6(Volume 6), 43–69. DOI: 10.4000/physio-geo.2319

Sadiki, A., Bouhlassa, S., Auajjar, J., Faleh, A., & Macaire, JJ (2004). Use of a GIS for the assessment and mapping of erosion risks by the Universal Soil Loss Equation in the Eastern Rif (Morocco): case of the Boussouab wadi watershed. *Bulletin of the Scientific Institute, Rabat, Earth Sciences section, 26* (2004), 69-79.

Salhi, A., Benabdelouahab, T., El Hasnaoui, Y., El Moussaoui, M., El Mourabit, A., & Himi, M. (2019). Soil Erosion Assessment and Farmer's Perception in South Mediterranean Basins: A Moroccan Case Study. The 2nd Euro-Mediterranean Conference for Environmental Integration (EMCEI). Springer, Sousse, Tunisia.

Salhi, A., Benabdelouahab, T., Martin-Vide, J., Okacha, A., El Hasnaoui, Y., El Mousaoui, M., El Morabit, A., Himi, M., Benabdelouahab, S., Lebrini, Y., Boudhar, A., & Casas Ponsati, A. (2020c). Bridging the gap of perception is the only way to align soil protection actions. *The Science of the Total Environment*, 718, 137421. DOI: 10.1016/j.scitotenv.2020.137421 PMID: 32105933

Salhi, A., Martin-Vide, J., Benhamrouche, A., Benabdelouahab, S., Himi, M., Benabdelouahab, T., & Casas Ponsati, A. (2019). Rainfall distribution and trends of the daily precipitation concentration index in northern Morocco: A need for an adaptive environmental policy. *SN Applied Sciences*, 1(3), 277. DOI: 10.1007/s42452-019-0290-1

Sebbar A (2013). *Etude de la variabilité et de l'évolution de la pluviométrie au Maroc (1935-2005): Réactualisation de la carte des précipitations* [Université Hassan II]. https://doi.org/.DOI: 10.1684/sec.2009.0169

Shaban, A., & Khawlie, M. (1998). Geo-environmental assessment of riparian zones under extreme climatic events: a case study of representative rivers in Lebanon. Mediterranean rivers and riparian zones-processes and management symposium, Zaragosa, Spain, 2 1 September-2 October 1998, 25 p.

Shi, Z. H., Fang, N. F., Wu, F. Z., Wang, L., Yue, B. J., & Wu, G. L. (2002). Assessment of Erosion Risk with the Rusle and Gis in the Middle and Lower Reaches of Hanjiang River. *Proceeding of the 12th ISCO Conference,* Beijing, 73–78.

Solaimani, K., Modallaldoust, S., & Lotfi, S. (2009). Soil erosion prediction based on land use changes (A case in Neka wathershed). *American Journal of Agricultural and Biological Sciences*, 4(2), 97–104. DOI: 10.3844/ajabssp.2009.97.104

Sossey Alaoui F (2005). *Traitement et intégration des données satellitaires optiques et radar dans un sig en vue de l'élaboration de cartes de l'aléa lie aux instabilités de terrain dans la péninsule de Tanger (Rif septentrional, Maroc).* Universite Mohammed V, Faculté des Sciences.

Spalevic, V., Djurovic, N., Mijovic, S., Vukelic-Sutoska, M., & Curovic, M. (2013). Soil erosion intensity and runoff on the Djuricka River Basin (north of Montenegro). *The Malaysian Journal of Soil Science*, 17, 49–68.

Suter G (1980). Carte structurale de la chaîne rifaine au 1/500,000. *Notes et Mémoires Du Service Géologique Du Maroc*, 245.

Tangestani, M. H. (2006). Comparison of EPM and PSIAC models in GIS for erosion and sediment yield assessment in a semi arid environment. *Journal of Asian Earth Sciences*, 27(5), 585–597. DOI: 10.1016/j.jseaes.2005.06.002

Tošić, R., Kapović, M., Dragićević, S., & Lovrić, N. (2013). Assessment of soil erosion potential using RUSLE and GIS: A case study of Bosnia and Herzegovina. *Fresenius Environmental Bulletin*, 22(11a), 3415–3423.

Wischmeier, WH, Smith DD (1978). Predicting Rainfall Erosion Losses. A Guide to Conservation Planning. *The USDA Agricultural Handbook, 537.*

Zahnoun, A., & Al Karkouri, J. (2020). Estimation & Cartography the Water Erosion by Integration of the Gavrilovic "Epm" Model using a Gis in the Mediterranean Watershed: Oued Tleta (Western Rif, Morocco). *European Journal of Scientific Research*, ISSN 1450-216X / 1450-202X Vol. 155 No 3, 2020, pp.265 -278.

Zeghmar, A., Marouf, N., & Mokhtari, E. (2022). Assessment of soil erosion using the GIS-based erosion potential method in the Kebir Rhumel Watershed, North-east Algeria. *Journal of Water and Land Development*, •••, 52. DOI: 10.24425/jwld.2022.140383

Chapter 4
Comparison of Photogrammetric Reconstruction Methods:
The Case of an Archaeological Site With Two Software and Geovisualization Modelling Techniques

Kyriaki Tychola
https://orcid.org/0000-0002-2528-4295
MLV Research Group, Department of Informatics, Democritus University of Thrace, Kavala, Greece

Eleni Vrochidou
https://orcid.org/0000-0002-0148-8592
MLV Research Group, Department of Informatics, Democritus University of Thrace, Kavala, Greece

George A. Papakostas
https://orcid.org/0000-0001-5545-1499
MLV Research Group, Department of Informatics, Democritus University of Thrace, Kavala, Greece

ABSTRACT

Digital photogrammetry is a passive contactless, easy, and low-cost 3D scanning technique, that provides effective and powerful tools for 3D realistic reconstruction, as well as precise and high-resolution models based on Structure from Motion (SfM). Photogrammetry leverages data obtained by Unmanned Aerial Vehicles

DOI: 10.4018/979-8-3693-6452-9.ch004

(UAVs), which have reached significant maturity in the sense that they offer a fast, affordable, and effective method for the acquisition of high-resolution geospatial information and image block (nadir and oblique) with high geometric and temporal resolutions. Data are processed by different available commercial or open-source photogrammetric software aiming at 3D reconstruction modelling. Then, the 3D-generated models can be further processed for quantity assessment by applying geometrical measurements. In addition, models through various geovisualization techniques can be used for realistic representations of the physical world. The current research investigates 3D model reconstruction by applying two photogram-metric software, Agisoft Metashape and 3DF Zephyr, in an example of an archaeological site in Greece focusing on the performance of each approach. Moreover, different distance and noise metrics are applied to compare and geometrically assess the 3D dense point clouds, meshes and Digital Surface Models (DSMs). Furthermore, traditional, and modern geovisualization forms of 3D models are discussed towards approaching the 3D models' representation holistically, aiming to provide guidelines to researchers in this field through a comprehensive workflow from data acquisition to manipulation and capabilities combined with other fields such as Computer Vision. Finally, open issues and future research directions are highlighted as a stimulus to further investigations.

1. INTRODUCTION

In the last decades, 3D visualization representations have been increasingly applied to various fields. 2D data is completed by 3D and 4D, which enable improved approximations of reality. The latter opens new vistas (Haala & Cavegn, 2016; Kadobayashi, Kochi, Otani, & Furukawa, 2004). Nowadays, 3D representations are achieved by 3D reconstruction of surfaces (Ponto & Tredinnick, 2022; Tenze & Canessa, 2024). Photogrammetry has diverse applications in different scientific disciplines, such as in Archeology (Baltsavias, 1999; Gagliolo et al., 2018), providing powerful tools and techniques for accurate 3D reconstruction of photorealistic models (Hassani, 2015), allowing the exploration and navigation of virtual environments (Adami, Cerato, D'Annibale, Demetrescu, & Ferdani, 2014). Nevertheless, Photogrammetry has been utilized from the 19th century (Meydenbauer, 1867). The American Society for Photogrammetry and Remote Sensing (ASPRS), claimed that "Photogrammetry is the art, science and technology of obtaining reliable information about physical objects and the environment through processes of recording, measuring and interpreting images and patterns of electromagnetic radiant energy and other phenomena" (Grimm, 1980). It should be noted that the term derives from the combination of the words "photo" (light), "gram" (drawn), and "metry" (measuring) (Reulke & Döring, 2006).

Data is obtained in the simplest form - in terms of ease of utilization and low cost - by Unmanned Aerial Vehicles (UAVs), namely Drones, and provide large sets of nadir and oblique overlapping imagery with spatial and temporal information of the real world, while their processing is achieved by suitable methodologies or software (Nebel, Beege, Schneider, & Rey, 2020; Singla & Nand, 2023). A notable number of Open Source (OS) and commercial photogrammetric software such as Agisoft Metashape (Metashape, 2024), 3D Zephyr (3DFLOW, 2024), Pix4D (PIX4D, 2024), VisualSFM (Changchang Wu, 2024), MicMac (MicMac, 2024) etc. has been developed, providing tools and possibilities for 3D models' generation. However, all software is based on similar automatic basic principles.

3D model generation is achieved by algorithms for Structure-from-Motion (SfM) and Multiview Stereo (MVS) (Schöning & Heidemann, 2015) offering spectacular results, such as high-quality data and information with both low cost and reduced processing time. SfM was introduced in the computer vision field to automate the generation of 3D models from datasets that are unordered (Brown & Lowe, 2005; Ullman, 1979). It refers to a method according to the principles of typical photogrammetry, yet different since there is no need for 3D targets to estimate the relative camera's positions and orientation (Oliensis, 2000) and encompasses the utilization of image correspondence for image orientation. It should be noted here that to get everything into some known coordinate system, ground control points or direct georeferencing sensors are required (Agarwal et al., 2010).

Point cloud and Digital Surface Model (DSM) generated through SfM and dense image matching, exhibit comparable data quality and resolution to those obtained through airborne and terrestrial laser scanning (Light Detection and Ranging, LiDAR) (Fonstad, Dietrich, Courville, Jensen, & Carbonneau, 2013; Maurya & Kulkarni, 2021). A significant benefit of SfM is its simplicity in data acquisition and the capability to generate point clouds with no knowledge regarding the camera's calibration or position beforehand (Snavely, Seitz, & Szeliski, 2008). However, the quality of a photogrammetry 3D model reconstruction is influenced by the calibration of the camera. Furthermore, SfM is a specific name for the method utilized to generate a 3D object model from multiple angles (Setiyadi, Mukhtar, & Cahyadi, 2021).

The current literature provides many publications about 3D reconstruction modelling by using photogrammetric software. However, some authors focus on algorithms, such as the leverage and evaluation of SfM in the context of archaeological sites (F. Nex et al., 2015), or geomapping (Singh, Jain, & Mandla, 2014), while others combine and compare different photogrammetry tools for varying target objects, such as buildings, historical monuments (Luo, Pan, Zhang, & Guan, 2019; Murtiyoso, Grussenmeyer, Börlin, Vandermeerschen, & Freville, 2018), aerial views

(Eltner & Schneider, 2015), human figures and fabric (Wang et al., 2020; Zhang, Luo, Yang, & Yu, 2019). Another category of publications studies cultural heritage documentation by applying photogrammetric and machine vision methods consisting of real-time data (Gabara & Sawicki, 2023)or synthetic data (Marelli et al., 2023). Nevertheless, very few comparative studies have been performed considering 3D reconstruction model assessment (Kingsland, 2020).

To this end, this study aims to investigate the most recent developments in UAV image processing for 3D modelling applications through an explicit case study related to the 3D reconstruction of an archaeological site in Greece, using two photogrammetric software, Agisoft Metashape and 3DF Zephyr. This work includes the following contributions:

1. The comparison of data processing and model generation at every stage, highlighting merits and demerits in terms of the algorithms' implementation, qualitatively and quantitatively. The comparison is directed towards evaluating the level of photorealism of the software, including computational cost requirements.
2. The quantitative comparison between the two software includes the assessment of the following aspects: image quality, image matching, image alignment, sparse and dense point clouds, mesh and texture, digital surface model and orthomosaic model and processing time.
3. The qualitative comparison among the generated models from both software including the calculation and evaluation of distance and roughness, as well as the performance assessment of mesh and digital surface models.
4. The presentation of the point cloud model in modern geovisualization forms including: static 3D mapping visualization and animation, virtual (VR) and augmented reality (AR) application, holography, and web mapping and web building application.
5. Finally, this study extensively discusses the results, presents open issues and concludes with potential research directions.

The remaining of this paper is structured as follows. Section 2 reviews the literature on photogrammetric software and highlights the contribution of this work. Section 3 presents materials and methods, including data acquisition and the workflow of photogrammetric software. Section 4 describes data processing with a photogrammetric software analyzing each stage by focusing on their operating principles, while 3D re-constructed models are compared visually and quantitatively, also considering computational cost. In addition, point clouds are compared geometrically, and they are assessed. Moreover, this section focuses on 3D models' post-processing applying various geovisualization techniques providing 3D representations and interactivity.

Section 5 provides an extensive discussion on the research findings, highlighting advantages and limitations and open issues. Section 6 concludes the paper.

2. RELATED WORKS, MOTIVATION, AND CONTRIBUTION

Photogrammetry is considered a powerful instrument for 3D reconstructions, providing modern techniques based SfM and Multiview stereo algorithms (Maurya & Kulkarni, 2021). Photogrammetry software packages are competitive, and their development is continually offering many benefits due to their prospective of modeling based on images, involved in various fields, such as cultural heritage. The comparison of photogrammetric software has been broadly studied by multiple authors, from several diverse aspects, and through various case studies. Many works are referenced in the recent literature. In the study of Alidoost & Arefi (Alidoost & Arefi, 2017), the authors compare different photogrammetric software towards evaluating the performance of point clouds and DSMs, geometrically and visually, involving image acquisition, generation of point clouds, and assessment of accuracy. In the work of Pádua et al. (Pádua et al., 2018), the authors compare two photogrammetric software (Pix4Dmapper Pro and Agisoft Photoscan) and investigate their 3D modelling accuracy in terms of the ease of use of functionalities and the processing time of photorealistic 3D models. However, they are limited to small-scale cultural heritage sites and focus on the documentation of a monument using software. In the study of Dayal et al. (Dayal, Raghavendra, Pande, Tiwari, & Chauhan, 2017) the authors evaluate point clouds, emphasizing the geometric outcomes through photogrammetric software, such as Photoscan and VSfM, using obtained data by Terrestrial Laser Scanner (TLS).

In the work of Fassi et al. (Fassi, Fregonese, Ackermann, & De Troia, 2013) two methodologies are presented, referring to the application of laser scanner and photogrammetry techniques, including the execution time of the procedure and a proposed technique to improve the model's accuracy. Aguilera & Lahoz (Aguilera & Lahoz, 2006) authors investigate and compare image-based modelling and laser scanning techniques by applying data from digital color images acquired from a high-resolution camera, while the processing is based on image-based modelling software, focusing on the accuracy of 3D models and assessment of processing time. Bayram et al. (Bayram et al., 2015) in their work compare the Terrestrial Laser Scanning (TLS) method to typical (manual) photogrammetric and dense matching methods based on images, applying a new algorithm for dense point cloud generation using the PIXEL-PHOTO software. Moreover, they discuss the trustworthiness and the faced issues during the process of point cloud measurements. In the work presented by Adami et al. (Adami et al., 2014) the authors apply photogrammetric approaches, such as dense image matching and spherical photogrammetry, using data obtained from laser scanners, to evaluate the operational results and explore the virtual models' utilization.

In the study of Gagliolo et al. (Gagliolo et al., 2018), photogrammetric software workflows are studied using three datasets of the same object under different conditions, emphasizing the quality of the dense point clouds. Burns & Delparte (Burns & Delparte, 2017) compare 3D reconstruction models at various spatial scales using three photogrammetric software, such as SfM software packages, Pix4Dmapper Pro 3.1 and Agisoft Photoscan Pro, focusing on the evaluation of the alignment of images, reprojections and spatial errors. In their work, Niederheiser et al. (Niederheiser et al., 2016), five different sensors of commercial cameras with varying resolutions, size, and prices are considered, and four different software packages are used for their comparison in terms of easiness in the workflow, similarity, visual appeal, and quality of the 3D point cloud. Haala & Cavegn (Haala & Cavegn, 2016) use two photogrammetric approaches to evaluate 3D point cloud and DSM, employing two different benchmark datasets; the first contains airborne nadir and oblique imagery, and the second consists of several image datasets in various scales and parameters, such as number of images and camera positions, based on the scene's complexity. In the work of Haala & Cavegn (Haala & Cavegn, 2016), as well as in the work of Nikolov & Madsen (Nikolov & Madsen, 2016), the authors apply both commercial and open-source photogrammetric software focusing on the utilization and the characteristics of algorithms at the dense image-matching stage, aiming to geometrically compare the dense point clouds. They use eight datasets with convergent and regular (parallel axes) images in varying scales, including scenes captured from aerial and terrestrial means. In the study of Remondino et al. (Remondino, Spera, Nocerino, Menna, & Nex, 2014), the authors use different scenarios to compare the performance of six photogrammetric software (Context Capture, Photoscan, Pix4D, Reality Capture, Memento, and 3DF Zephyr) using images under different lighting conditions and different camera positions to evaluate the reconstructed model from each software by employing distance and roughness metrics.

Undoubtedly, all aforementioned works are substantial and contribute to the body of knowledge. Most of these works compare and analyze the performances of photogrammetric software for 3D reconstruction models and estimate the 3D models geometrically. However, none of them provides a holistic perspective, that gathers, presents, and analyzes in detail the workflow of the 3D reconstruction model, comparing the 3D point clouds geometrically, and describing further post-processing techniques. The lack of a complete investigation on 3D reconstruction encouraged us to implement a holistic investigation using case study experiments towards presenting all parameters and factors that are important both to 3D modelling reconstruction and the post-processing of data. Therefore, the contributions and innovations of the current study lie in the conducted experiments using two photogrammetric software packages emphasizing the 3D reconstruction workflow, 3D point cloud, mesh, and DSM assessment, both qualitatively and quantitatively. In addition, we provide modern geovisualization techniques for 3D representations, such as 3D modelling, VR/AR, web applications, and holograms. Thus, the distinctive contribution of this work is a targeted and comprehensive analysis of 3D reconstruction models and geovisualization representation forms, also including advantages, limitations, and open issues. Table 1 summarized the primal features of the related works mentioned above, regarding 3D reconstruction modeling through photogrammetric software comparison, with the aim to highlight comparatively the contributions of the current review article (Proposed) compared to the previous ones and therefore to establish the novelty of the proposed work by comparing it with comparable previous work of the literature. This sound literature review in tabulated form is considered useful to perform meta-analysis of available works and establish the ground for the proposed work.

Table 1. Comparative table of features of the present work (Ours) versus related works from the literature.

Ref., Year	3D reconstruction models through Photogrammetry comparison							
	Data	Comparison	Assessment			Geovisualization	Highlights	
	UAVs	Photogrammetric software	Point cloud	Mesh	DSM	VR/AR, Web mapping, web apps, hologram	Pros/ Cons	Open issues
(Alidoost & Arefi, 2017), 2017	✓	✓	✓	×	✓	×	×	×
(Pádua et al., 2018), 2018	✓	✓	×	×	×	×	×	×
(Dayal et al., 2017), 2017	×	✓	✓	✓	×	×	×	×
(Fassi et al., 2013), 2013	×	✓	×	×	✓	×	×	×
(Aguilera & Lahoz, 2006), 2006	×	✓	✓	×	✓	×	×	×
(Bayram et al., 2015), 2015	×	✓	×	×	×	×	×	×
(Adami et al., 2014), 2014	×	✓	×	✓	×	×	×	×

continued on following page

Table 1. Continued

Ref., Year	3D reconstruction models through Photogrammetry comparison							
	Data	Comparison	Assessment			Geovisualization	Highlights	
	UAVs	Photogrammetric software	Point cloud	Mesh	DSM	VR/AR, Web mapping, web apps, hologram	Pros/ Cons	Open issues
(Gagliolo et al., 2018), 2018	✓	✓	✓	×	×	×	×	×
(Burns & Delparte, 2017), 2017	×	✓	×	×	×	×	×	×
(Niederheiser et al., 2016), 2016	×	✓	✓	×	×	×	×	×
(Haala & Cavegn, 2016), 2016	partially	✓	×	×	×	×	×	×
(Nikolov & Madsen, 2016), 2016	×	✓	✓	×	×	×	×	×
(Remondino et al., 2014), 2014	partially	✓	✓	×	×	×	×	×
Proposed	✓	✓	✓	✓	✓	✓	✓	✓

3. MATERIALS AND METHODS

The proposed methodology, illustrated in Figure 1, consists of five steps: (a) Data acquisition; (b) Data processing by applying two different full automatic software, Agisoft Metashape and 3D Flow Zephyr; (c) Acquisition of a dense point cloud, a DSM and an Orhomosaic as results; (d) Comparison of the workflow performance of the software at every stage of model generation, including the assessment of the 3D dense point cloud and meshes calculating the roughness and the distance, DSM elevation using Cloud Compare software and QGIS software package (Milenkovic & Ressl, 2013; Sankaranarayanan, Samet, & Varshney, 2007); (e) Geovisualization techniques include 2D/3D maps, animation, virtual/augmented reality and web applications aiming to valorize the 3D model representation towards better understanding of reality and improving interactivity.

Figure 1. Flowchart of the five-stages proposed method.

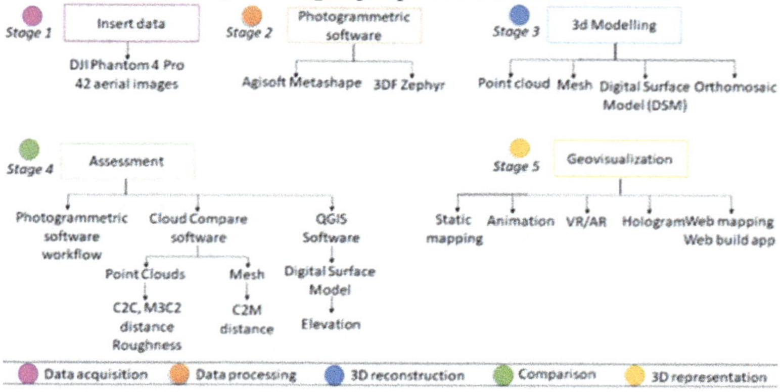

3.1. Study Area

The area under study is an archaeological site in Platania (Figure 2), sited approximately 30 km north-east from Drama, which is a city in eastern Macedonia, Greece (N 41011′48.4″, E 024023′42.0″; elevation: 257 m a.s.l.; Hellenic Grid, EGSA'87) and it is deemed of an Acropolis. The location is a segment of a collection of seven fortifications that exist in the broader region, having visual contact between them. Excavations disclosed that the regions was occupied from pre-historic until the Byzantine period (Vasileiadis, Tsoukala, & Kostopoulos, 2019).

Figure 2. Location of the archaeological site in Drama, Greece.

3.2. Data Acquisition

Data acquisition includes the flight of a drone and the gathering of images. In this work, it was used a DJI Phantom 4 Pro with four propellers. The settings of the mounted camera are: resolution of 20 Mpx, sensor width of 13.2 mm, sensor height of 8.8 mm, 24 mm focal length, pixel size 2.4 microns sensor. Moreover, the drone possesses 3-axis stabilization (pitch, roll, yaw), controllable range of −90° to +30° and range of angular vibration of ± 0.02° (Papadopoulou, Vasilakos, Zouros, & Soulakellis, 2021). The flight mission was not specific, due to the steep slopes of the region. Nevertheless, the region was covered adequately. The total number of images was 100, however, after a manual visual inspection, 42 images were considered suitable for processing. Figure 3 shows the flight plan mission and Table 2 summarizes the information of aerial images.

Figure 3. The flight plan.

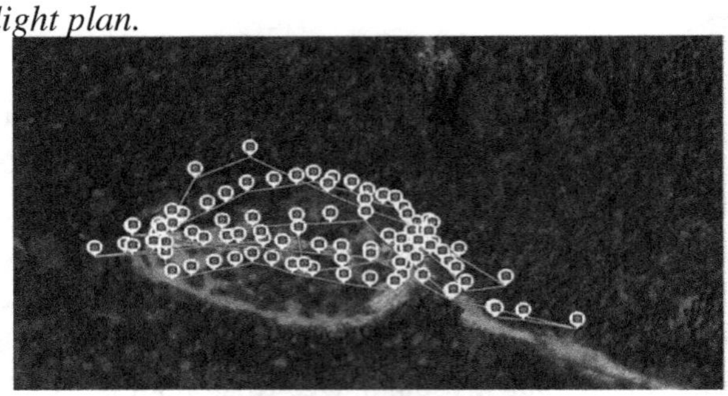

Table 2. Information of the aerial images obtained by the drone.

Date	Number of images	Dimension	Flying altitude	FC6310
22/3/2019	100/42	5472 x 3078	28.6m	8.8mm

Furthermore, Ground Control Points (GCPs) and their coordinates were collected for the accurate position recording of a point. Specifically, GCPs are points with known global or local geodetic coordinates and their distribution should follow the geometry of the region to avoid large deviations and spatial errors (Oniga, Breaban, & Statescu, 2018). Information obtained is longitude, latitude, and altitude, with an error to the order of 10^(-2) m. It is noted that a checkpoint should be displayed in, at least, three images. In this work, 18 GCPs were distributed in the study area, while 17 of them were used in the study for the orientation of the model and the accuracy of the Georeference. One GCP was excluded since it did not provide accurate spatial references and measurements within the captured imagery and did not contribute to final results' accuracy, displaying a big error, therefore it is not visualized in Figure 4. From the 17 GCPs, 14 were used as control points (numbers 2, 3, 5, 6, 7, 8, 9, 10, 11, 12, 14, 16, 17and 18 of Figure 4) and 3 as check-points (numbers 4, 13 and 15 of Figure 4) for the assessment of the spatial error. GCPs are targets using black marker on stone and they were measured by using the GPS. Figure 4 shows the distribution of GCPs in the under-study area.

Figure 4. The distribution of GCPs in the under-study area; GCPs are considered in the examined area. Points 2, 3, 5, 6, 7, 8, 9, 10, 11, 12, 14, 16, 17 were used as control points, while points 4, 13 and 15 were used as checkpoints.

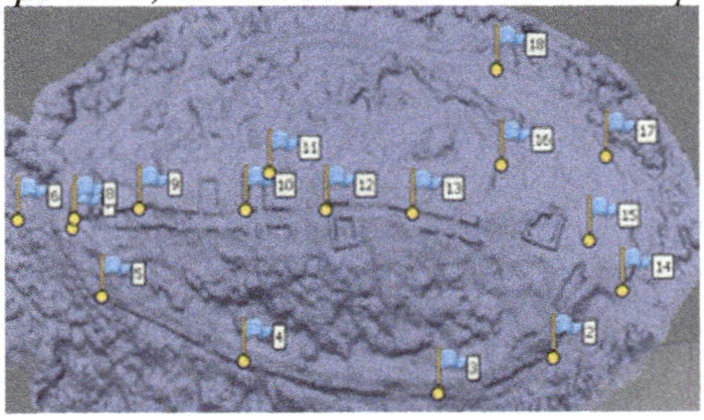

3.3. Data Processing

Obtaining records and geometric documents of the real world, for 3D photo-realistic model reconstructions, initially requires fieldwork and data acquisition, i.e., oblique and vertical overlapping images acquired by UAVs so that the region under study to be covered sufficiently, as well as the establishment of the GCPs. Then, data is inserted into the photogrammetric software and the automatic processing begins. In this sub-section, the workflow process of Agisoft Metashape and 3DF Zephyr software packages is described. In general, the 3D reconstruction of a model using photogrammetric software follows standard steps. However, there are some significant differences. Figure 5 shows the workflow of Agisoft Metashape and 3DF Zephyr towards comparing the data process between them.

Initially, the images are aligned and oriented (Kaminsky, Snavely, Seitz, & Szeliski, 2009) and they are checked qualitatively using an index, which indicates any kind of distortion that results from the combination of the following: the correlation loss, the distortion of brightness and the contrast (Zhou Wang & Bovik, 2002). Then the software packages detect homologous points among images and match them using the Scale-Invariant Feature Transform (SIFT) (Eskicioglu & Fisher, 1995) and 3DF Samantha algorithm (Hansman & Ring, 2019) in Agisoft Metashape and 3DF Zephyr, respectively.

In the next step, the SfM algorithm is applied, and images are georeferenced using GCPs towards integrating the model in the desired reference system. A seven-parameter linear similarity transform is used to solve the Georeference: three translation, three rotation and one scale parameter. Then, the camera's parameters are improved for better accuracy, applying a Bundle adjustment algorithm. In the next stage, a sparse point cloud is created as a result of the SfM algorithm, using Nearest Neighbor techniques and Random Sample Consensus (Taejung Kim & Yong-Jo Im, 2003). Sparse point clouds consist of a large collection of points and has low resolution. Specifically, vectors are created that compose the keypoints in many images. The cloud detects the overlapping pairs of images i.e., the pairs which have enough valid homologous points (tie points) representing their positions. Subsequently, a denser point cloud is generated, via the MVS algorithm which is increased at least by twice its original size (Seitz, Haas, Noack, & Wieprecht, 2018) while the quantity of collected coordinates per unit area refers to the density of point clouds (Alidoost & Arefi, 2017). Especially, in 3DF Zephyr, dense point cloud generation is achieved by (MVS) 3DF Stasia providing accuracy from a set of 2D images as every single pixel of the input images is used. After that, a mesh is created i.e., the compacted surface of the dense point cloud. The interpolation methods of Triangulated Irregular Network (TIN) is applied, and subsequently it acquires texture by using information from the original images (Xu, Zhang, & Xu, 2016). In 3DF Zephyr, the mesh is generated using the (MVS) 3DF Sasha algorithm applying the Poisson approach (Kazhdan & Hoppe, 2013) allowing to get sharp edges on the 3D model. In addition, the program sets the most suitable color for each pixel by using the 3DF Tessa algorithm automatically, to deal with the model color balance problem due to the images captured in different light conditions (Hansman & Ring, 2019). Details about both software can be found in their homepages, in (Metashape, 2024) for Agisoft, and in (3DFLOW, 2024) for 3DF Zephyr. The algorithms, however, used in both software are the developers' intellectual property, and they are not accessible for refinement and analysis; therefore, limited details are provided in terms of similarities/differences of the integrated algorithms between the two software used in this work.

Figure 5. Flowchart of the workflow comparison of the procedure for creating a photo-based 3D model in Agisoft Metashape and 3DF Zephyr software packages.

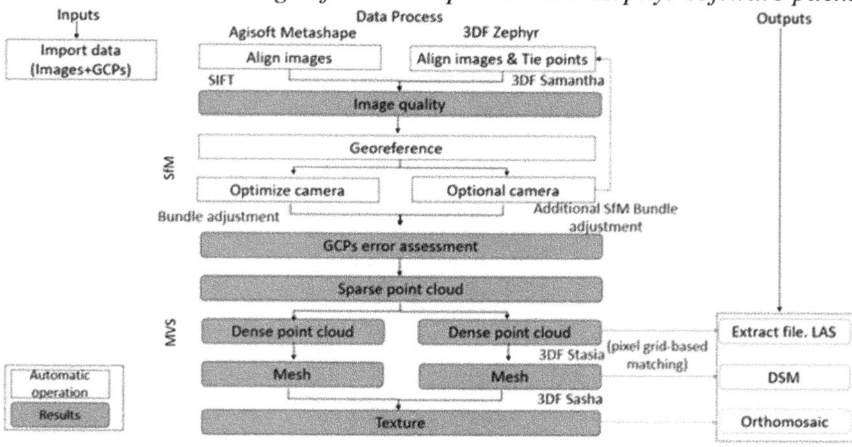

The final products of 3D reconstruction are the Orthomosaic, the DSM and the *.Las file, which are derived from texture, mesh and dense point cloud, respectively, and they can be used for further analysis and processing (Bulgakov, Sayfeddine, Bock, & Fares, 2020). An Orthomosaic, also known as an orthophoto is a photo-grammetrically orthorectified image mosaicked from an image collection, where distortion of geometry, referring to relief displacement errors caused by perspective effects has been removed from images, while the images have been color-balanced towards producing a continuous mosaic dataset. Hence, the scale is united, and it provides distances, area, and coordinates' measurements. A DSM correspond to the topography of the surface of the earth incorporating natural and artificial features of the environment and provides the elevations of regions. A*.LAS file (LASer) is the standard data file in binary format that is obtained from the dense point cloud, and it is useful to exchange and archive point cloud data. Its structure can be divided into (a) Variable Length Records, (b) Public Header Block, (c) Point Data Records, and (d) Extended Variable Length Records.

4. EXPERIMENTS AND RESULTS

In this section, software workflow assessments are described focusing on the visual qualitative and quantitative comparison at every stage of 3D reconstructions, including the specific features of software packages and the operating procedure, as well as the geometrical precision of the dense point cloud. In addition, final various geovisualization techniques are presented. To be fair, all results and experiments were conducted on the same computer (Windows 64-bit, RAM 8 GB, CPU Intel(R) Core (TM) i7-8565U CPU @ 1.80GHz, GPU(s) Radeon (TM) 520 Hainan). Table 3 shows the features of the software used in this work.

Table 3. Comparison of Agisoft Metashape and 3DF Zephyr requirements and characteristics.

Software	Operating System	Scalability	Photogrammetry type	Data source
Agisoft Metashape	Linux, macOS Microsoft Windows	Multiple images, cluster distributed processing	Aerial, close-range, UAV, satellite	Images, video
3DF Zephyr	Microsoft Windows	Multiple images	Aerial, close-range, UAV	Images, video, laser scan

4.1. Quantitative Comparison

4.1.1. Image Quality

Image quality assessment is achieved from the calculation of 3D image information. Agisoft software uses an index and the images having an index value less than 0.8 are rejected. In 3DF Zephyr software a value for each image and a corresponding score is automatically calculated. The greater the value, the greater the image quality. The calculations are based on the best focus and the high density of key-points. However, it is noted that a high value does not necessarily indicate a properly captured image; yet, a low value certainly indicates that the image must be rejected. Figure 6 shows the outcomes of the image quality assessment.

Figure 6. (a) Image quality index of Agisoft Metashape; (b) Image quality index of 3DF Zephyr.

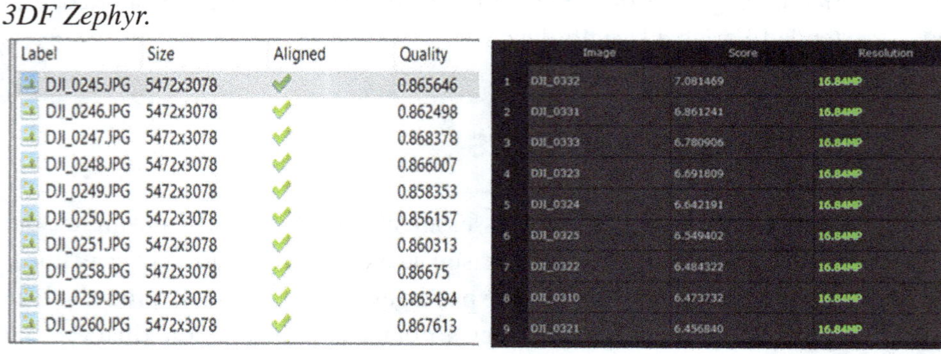

4.1.2. Image Matching

Feature detection and matching are achieved by the SfM algorithm (Francesco Nex & Remondino, 2014). In Agisoft the matches are represented with blue and red lines, which indicate the correct and the false cases, respectively. In addition, a table provides information about the number of valid and invalid keypoints in each image. In 3DF Zephyr the lines are multicolored. The orange color indicates the correct matches, while the rest of colors indicate the false, resulting usually due to the lack of texture and opacity. Moreover, information, e.g., the number of matches and the average Ground Sampling Distance (GSD) are provided. GSD refers to the distance linking two successive pixel centers evaluated on the ground. The greater this value, the poorer the image spatial resolution, resulting in less visible image details (Lee & Sull, 2019). Figure 7 presents the outcomes of the matches. In Agisoft Metashape the GSD is 0.0057 m and in 3DFZephyr 0.0085 m. Hence, the images' alignment in Agisoft Metashape excels over 3DF Zephyr resulting in a bigger spatial resolution.

Figure 7. Image matching outcomes in (a) Agisoft Metashape and (b) 3DF Zephyr.

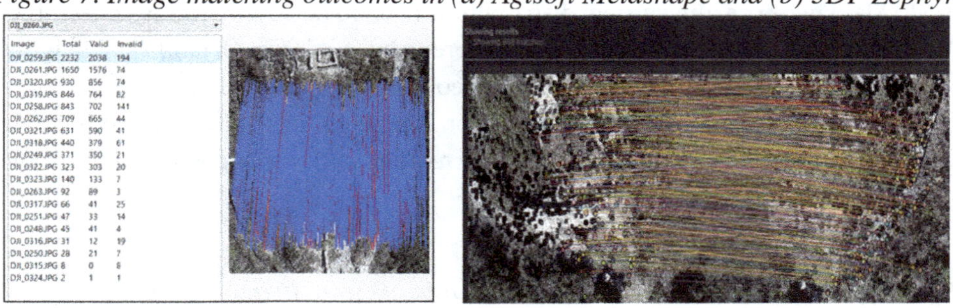

(a) **(b)**

4.1.3. Align Images and Ground Control Points (GCPs)

To align images, the objects must be captured in at least three images, utilized for the computation of 3D location and the sparse point cloud formation (Westoby, Brasington, Glasser, Hambrey, & Reynolds, 2012). A united set of images is created, linked to a given geodetic reference system, while the external orientation of all images is calculated using the collinearity. Figure 8 shows image aligning where the orientation and location of the camera are defined after the camera's calibration; 3DF Zephyr uses the Samantha algorithm and provides an online database to store the camera calibration parameters for future leverage, in contrast to the Bundle adjustment algorithm used in Agisoft Metashape.

Georeference is the processing of the images associated with the geodetic reference system through Affine transformation. Four at least photoconstants are required to estimate the Root Mean Square Error (RMSE) (Padró, Muñoz, Planas, & Pons, 2019). In this work, 17 GCPs are located in the study area, as already mentioned. Both programs used 13 GPSs as control points and 3 as checkpoints. The total RMS in Agisoft Metashape was 0.031m and 0.19 m from control and checkpoints, respectively, while in 3DF Zephyr the error was 0.08 m and 0.13 m, from control and checkpoints, respectively. According to the literature, accuracy is considered optimal, when an error less than 0.05m is reported (Cramer, Stallmann, & Haala, 2000). The latter does not occur in this case study, since greater RMS errors are reported from both software. The large values of RMSs indicate that a generated model utilizing the selected GCPs exhibits a poorer DSM regarding its reliability. The influence of GPSs on the DSMs accuracy is undoubtable. Research reveals that better results can be obtained by following a uniform distribution of the GPSs in the horizontal as well as in the vertical plane. Moreover, for GCPs of low-lying areas, the accuracy must not be assumed in the high-lying areas and vice-versa (Long et al., 2023). Therefore, the vertical distributions need to be strictly followed for high-undulating terrains such as in our case, where there is a large variability in heights.

It should be noted that the chosen archaeological site is sited in a remote mountainous area with steep slopes and elevation gradients. In addition, there is dense wild vegetation, ancient walls, as well as irregular stones that create shadows and blockages.

Figure 8. (a) Orientation and location of cameras from image aligning in Agisoft Metashape; (b) Orientation and location of cameras from image aligning in 3DF Zephyr.

(a) **(b)**

Finally, the used low-cost and lightweight UAV, the four-propeller DJI Phantom 4 Pro, does not meet strict in-build accuracy requirements and the utilization of GCPs towards achieving a sure level of accuracy is necessary for preventing any biases related to focal length during the self-camera calibration process. The authors acknowledge that DSM accuracy depends on multiple factors, such as the ones mentioned above as well as the UAVs flight height. However, these factors were not considered in our analysis. This work aims to cover the literature gap of the qualitative and quantitative comparison between the extracted models from the two software, including aspects not previously considered within the same research, such as the presentation of the extracted point cloud model in modern geovisualization forms of static 3D mapping visualization and animation, VR/AR application, holography, and web mapping and web building application, and to extensively discuss the results and present open issues and potential future research directions.

Figure 9 shows the distribution of photos in Agisoft Metashape and 3DF Zephyr. In addition, in 3DF Zephyr the distances among GCPs are calculated automatically.

4.1.4. Sparse and Dense Point Cloud

In Agisoft Metashape the sparse point cloud is generated using the SIFT algorithm, while 3DF Zephyr uses the 3D Samantha algorithm. Visually, the cloud in Agisoft Metashape appears denser in contrast to the 3DF Zephyr cloud, which has holes as depicted in Figure 10. Nevertheless, the number of points in 3DF Zephyr is bigger, providing 113.469 points, compared to the Agisoft Metashape providing 47.588. It is noted that the dense point cloud generation is based on images. In addition, the selected filter method for the depth of each point plays a vital role in the improvement of model's quality. In both Agisoft and 3DF Zephyr, Aggressive filtering was applied to increase the model accuracy due to the steep slopes of the region. In this case, the dense point cloud in Agisoft Metashape consists of 17.018.480 points in contrast to the cloud in 3DF Zephyr (12.445.780). Figure 10 shows the sparse and dense point clouds reconstructed by Agisoft Metashape and 3DF Zephyr. Although both software packages generate different sparse clouds in terms of the points' number, their dense point clouds do not seem to differ qualitatively. However, the dense point cloud from Agisoft Metashape, is more complete than that of 3DF Zephyr, providing more details from the surrounding area.

Figure 9. Distribution of GCPs and measurement error in: (a) Agisoft Metashape. Points 4, 13, and 15 were utilized as check points, while the rest as control points. Excluded point 1 is also visualized. The different ellipse colors represent the Z error while the ellipse shape represents the X, Y error; (b) 3DF Zephyr.

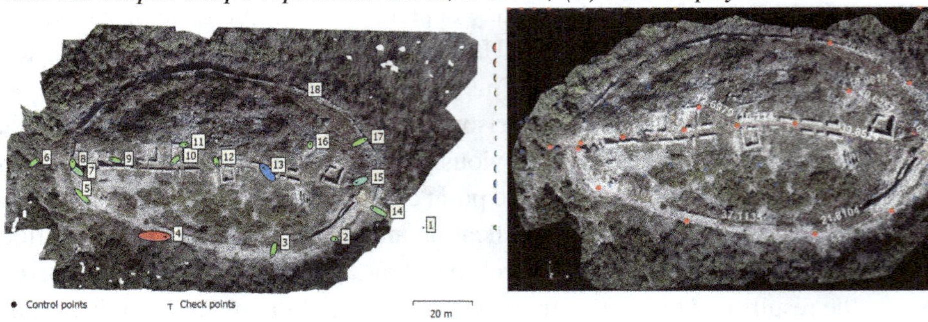

(a) **(b)**

Figure 10. (a) Sparse point cloud from Agisoft Metashape; (b) Dense point cloud in Agisoft Metashape; (c) Sparse point cloud in 3DF Zephyr; (d) Dense point cloud in 3DF Zephyr.

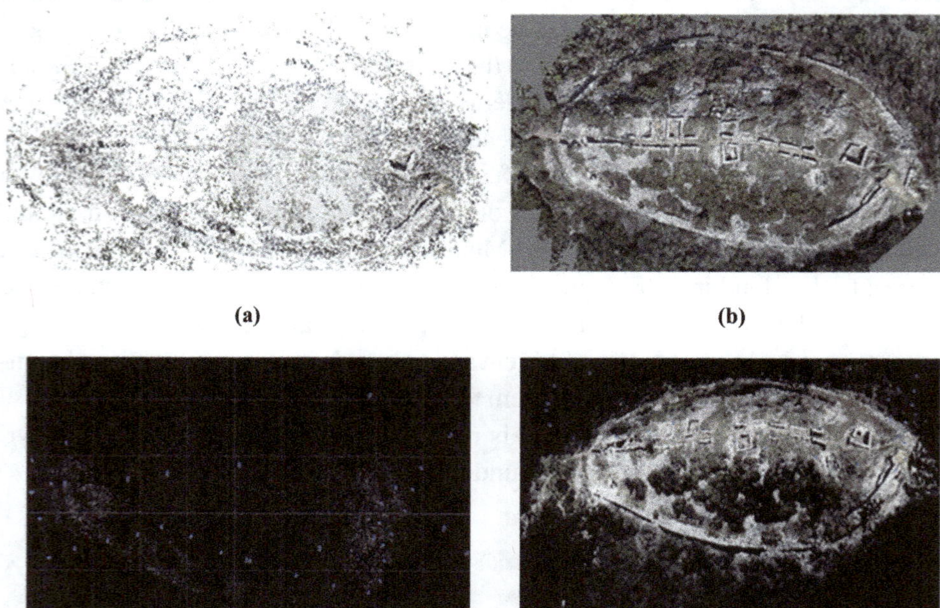

(a) **(b)**

(c) **(d)**

4.1.5. Mesh and Texture

Triangular meshes obtained by Agisoft Metashape and 3DF Zephyr contain 54.685 and 76.820 triangles, respectively. Figure 11 shows the meshes obtained by two software packages. In addition, Agisoft Metashape provides three different mesh forms (solid, edges, color) in contrast to 3DF Zephyr, and a Photoconsistency technique-based optimization embedded filter. This filter smooths out the excessive geometry that the model tends to have. Figure 12 shows the mesh quality after the application of the filter. In 3DF Zephyr this filter can be manually applied many times, considering the previous improvement results, thus, leading to a growth in all details of the reconstructed surface, with no compromises in accuracy or alterations of the shape of the original scene. Furthermore, area and volume information are provided. After the filter implementation, the number of triangular surfaces, the area and the volume are significantly increased, while after the second application of the filter in 3DF Zephyr, the area and volume of the region are negligibly changed. In Agisoft Metashape the number of triangles was increased from 54.685 to 1.084.488, while in 3DF Zephyr from 76.820 to 664.127. Regarding the area, in Agisoft Metashape and 3DF Zephyr from 17.795 m^2 and 16.123 m^2, respectively, they were increased to 18.162 m^2 and 22.004 m^2. It is understood that between two meshes, initially, Agisoft Metashape achieves better outcomes in terms of more triangles compared to 3DF Zephyr; however, the difference in the surfaces' number is not great. It is notable that after the Photoconcistency technique's implementation, the number of triangles was increased in Agisoft Metashape almost twenty times. In 3DF Zephyr this Photoconsistency can be repeatedly applied to increase the triangle, and consequently the quality. Furthermore, the number of triangles, the area and the volume in 3DF Zephyr were recalculated using the filter twice. The outcomes are included in Table 4.

Figure 11. Mesh and the forms of mesh in (a) Agisoft Metashape; (b) in 3DF Zephyr.

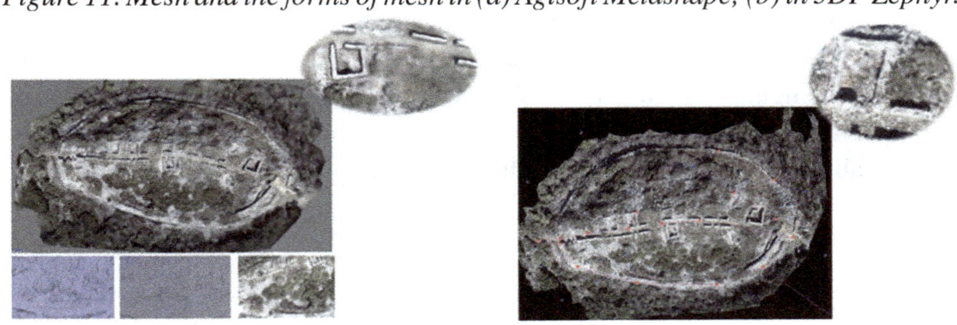

(a) **(b)**

Figure 12. Mesh quality after the filter implementation in (a) Agisoft Metashape; (b) in 3DF Zephyr.

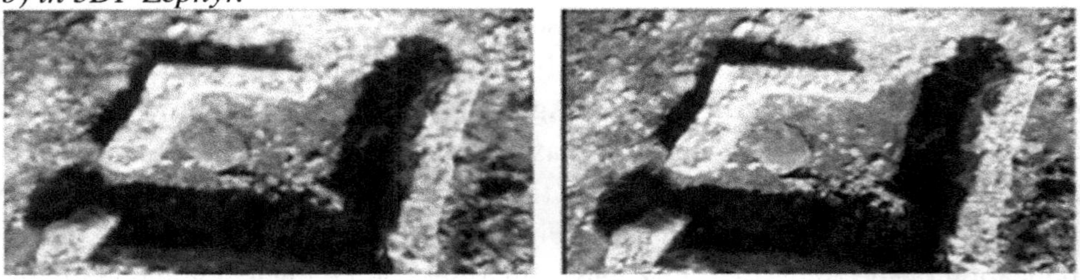

(a) **(b)**

Table 4. Outcomes from the Photoconsistency based optimization filter application in 3DF Zephyr.

Description	Triangle count	Area (m²)	Volume (m³)	Filter
First application	664127	22004	235438	Photoconsistency based optimation
Second application	8754819	24722	234631	Photoconsistency based optimation (x2)

Then, the mesh obtained texture through filtering techniques, i.e., by employing hole filling and ghosting, since the model had gaps because of the intense vegetation in the area under study. It is noted that the ghosting filter exists only in Agisoft Metashape. At the blending level, the mosaic was selected since it combines low frequency of overlapping images to avoid weighted average and high frequency problems which are responsible for the details received from a single picture (usually the picture with the best resolution). In 3DF Zephyr, during the procedure of obtaining texture, an algorithm was applied (multitexture) to texture color balance allowing the automatic selection and adjustment of the utmost appropriate color for every pixel. This feature is useful when photos have been obtained under different conditions of lighting. Figure 13 presents the texture of models and examples of their details. Evaluating the two objects visually, it is observed that they do not differ dramatically. However, the model obtained by 3DF Zephyr presents better cohesion.

Figure 13. Mesh texture obtained by (a) Agisoft Metashape and (b) by 3DF Zephyr.

(a) **(b)**

Figure 14 shows a chart of the reconstruction information of point clouds, meshes, and textures for both software packages.

4.1.6. Digital Surface Model (DSM) and Orhomosaic

DSM depends on the dense point cloud quality. Note that a set of options are available in both software to define the orthophoto resolution by changing the GSD and other parameters or can also set the resolution by specifying the image size. However, currently, DEM that rely on the dense cloud can be created in the default resolution only, for both software. In Agisoft Metashape the pixel dimension is defined automatically, and each pixel obtains a value based on the domination value of each point from the total of points which belong to the corresponding cell. The cell sizes are 0.029 in Agisoft Metashape and 0.03 in 3DF Zephyr, while in Agisoft Metashape there is a resolution of 2.92 cm/pix and point density of 0.117 point/cm^2 and a resolution of 2.63 cm/pix and point density of 0.115 point/cm^2 in 3DF Zephyr. The errors' distribution is normal and no outliers or other systematic errors in both DSMs are noticed. The Orthomosaic characteristics are the cell size, which is 0.007 and 0.003, in Agisoft Metashape and 3DF Zephyr, respectively. Figure 15 shows the DSM and Orthomosaic for both software packages.

Figure 14. Number of points of reconstructed models in Agisoft Metashape and 3DF Zephyr.

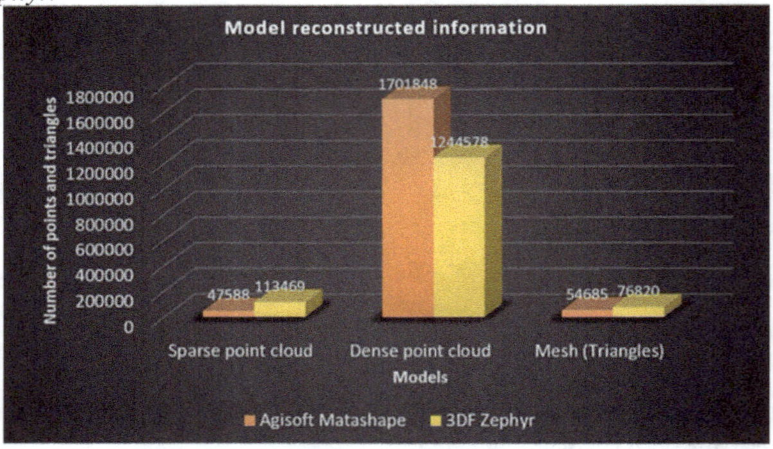

Figure 15. (a) Orthomosaic obtained by Agisoft Metashape; (b) DSM obtained by Agisoft Metashape; (c) Orthomosaic obtained by 3DF Zephyr; (d) DSM obtained by 3DF Zephyr. Red cycles indicate the holes in Agisoft Metashape Orthomosaic.

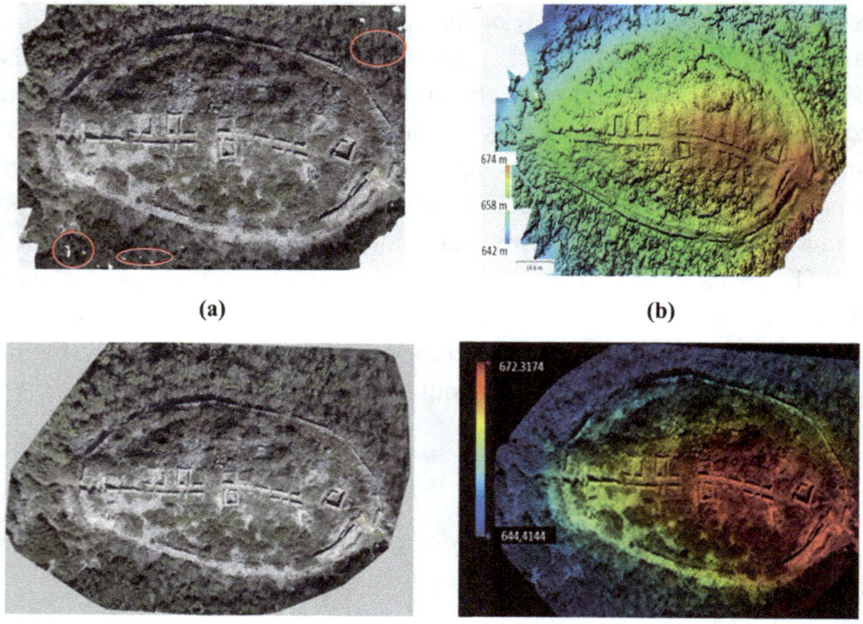

(a) (b)

(c) (d)

4.1.7. Processing Time

In this subsection, the execution time of the 3D reconstruction processing is assessed. In Figure 16 the processing time of 3D model reconstruction between Agisoft Metashape and 3DF Zephyr is compared; 3DF Zephyr seems to spend more time for generating the sparse point cloud, mesh, and texture. Especially, for the sparse point cloud generation it needs much more time compared to the Agisoft Metashape. This is justified by the fact that for generating a sparse point cloud, the procedures' calculations are merged in 3DF Zephyr, in contrast to the Agisoft Metashape. Figure 17 shows the average percentage of the work processing in Agisoft Metashape and 3DF Zephyr. It is notable the great difference regarding image aligning; 3DF Zephyr requires 12% of the whole processing time in contrast to Agisoft Metashape which requires just 1%. However, for the Orthomosaic reconstruction, Agisoft Metashape spends twice as much time compared to 3DF Zephyr, as it is illustrated in the pie chart of Figure 16.

Figure 16. Processing time comparison of the 3D reconstruction workflow in Agisoft Metashape and 3DF Zephyr.

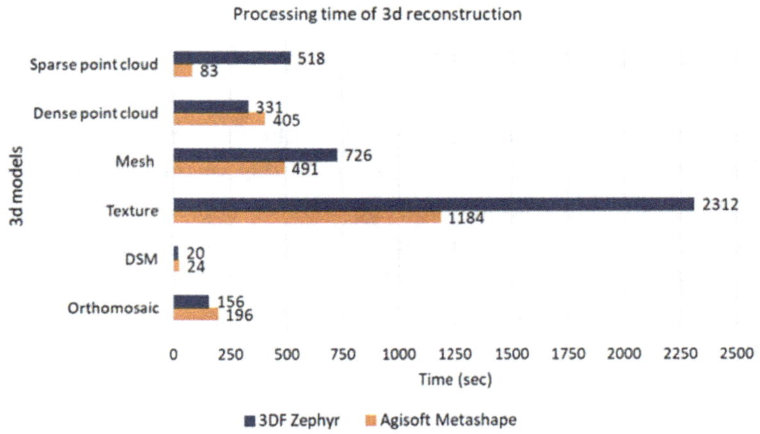

Figure 17. Pie chart indicating the average percentage of processing time for each step of the workflow in: (a) Agisoft Metashape; (b) 3DF Zephyr.

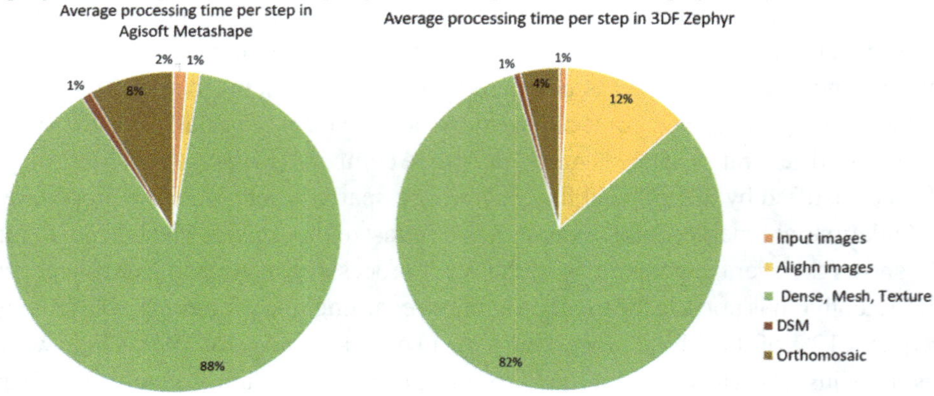

(a) **(b)**

4.2. Qualitative Comparison

4.2.1. Point Cloud Performance Assessment

Regarding the point cloud assessment, Cloud compare software (CloudCompare, 2024) was used. The data was inserted in the form of *.LAS. files, which were extracted by the dense point cloud and could be depicted in three dimensions, as shown in Figure 18. The distribution of points in Agisoft Metashape is visually denser in contrast to 3DF Zephyr, including 17k and 12k points, respectively. The higher density of the Agisoft Metashape point cloud may be credited to the used parameters.

Figure 18. Visual representation of point cloud obtained by (a) Agisoft Metashape and (b) 3DF Zephyr, in Cloud software.

(a) **(b)**

4.2.2. Distances Calculation

Point cloud geometry assessment plays an important role since it brings out details of the model's structure and surface. Therefore, a point cloud comparison obtained by two different software packages is useful for a better understanding of the scene. We consider that point clouds contain noise calculations including outcomes before and after noise removal. Initially, Cloud to Cloud (C2C) distance was calculated. This is a typical way to detect changes and characteristics of point clouds through the nearest neighbor distance, e.g., by computing distances between two points. For every single point of the associated cloud, the algorithm looks for the nearest point in the reference cloud and calculates the Euclidean distance between them (Ahmad Fuad, Yusoff, Ismail, & Majid, 2018; Shen, Lindenbergh, & Wang, 2016). Considering that the 3D models contain noise mainly due to the lighting conditions, they were compared by setting the Least Square Plane (LSP) parameter, which was more resistant to noise. Then the outliers were removed, and the models were recalculated. Figure 19 shows the results of the C2C distance before and after the noise removal, visualized as a pseudo color heatmap, while Figure 21 shows the corresponding Gauss histogram. The mean of the distance between the two models is 0.0378 m and the standard deviation (SD) is 0.0451 mm before noise removal, while after noise removal the mean is 0.0336 m, and the SD is 0.0310 mm. The corresponding RMSs are 0.0589 m and 0.0457 m. It is clear that between the two models, there are minor differences in excluding the most remote points. The value of standard deviation indicates if the point cloud optimally denotes the plane's surface; 3DF Zephyr has a higher degree of planarity as shown in Figure 20. The latter depends on the properties of the UAV. However, in general, the generated point cloud depends on the different algorithms that are utilized to correspond features and resolve the epipolar geometry or other photogrammetric parameters.

Figure 19. C2C distance between two point clouds (a) before and (b) after noise removal. Green color indicates distance above the ground truth, red color indicates distance below the ground truth and blue color indicates where the surfaces overlap. Comparison of both point clouds reveals no systematic errors or discrepancies. However, it can be observed a distribution of discrepancies and correspondences of vegetated areas that are not visually detectable.

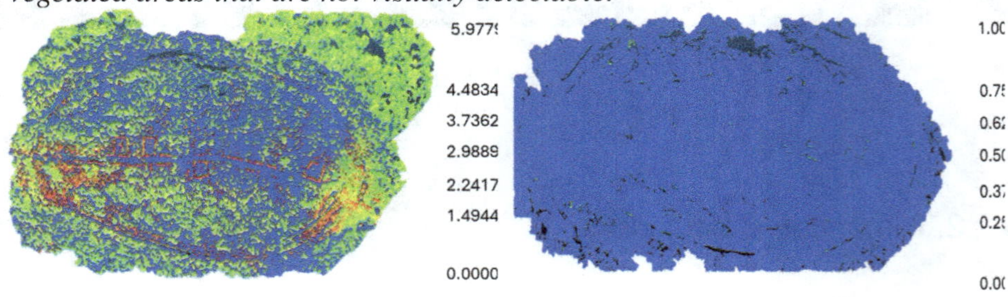

(a) **(b)**

In addition, after the removal of the outliers, the point clouds were recalculated based on Euclidean point-plane distance, which is appropriate for regions with large curvature (Antova, 2019), as illustrated in Figure 21. The mean distance and standard deviation are 0.0590 m and 0.0013 mm, respectively, while the maximum residual is 0.18 m. The histogram shows a strong peak between 0.03 m - 0.04 m interval range. Statistical values are within the expected range.

Figure 20. Gauss distribution of C2C distance between two clouds (a) before and (b) after noise removal.

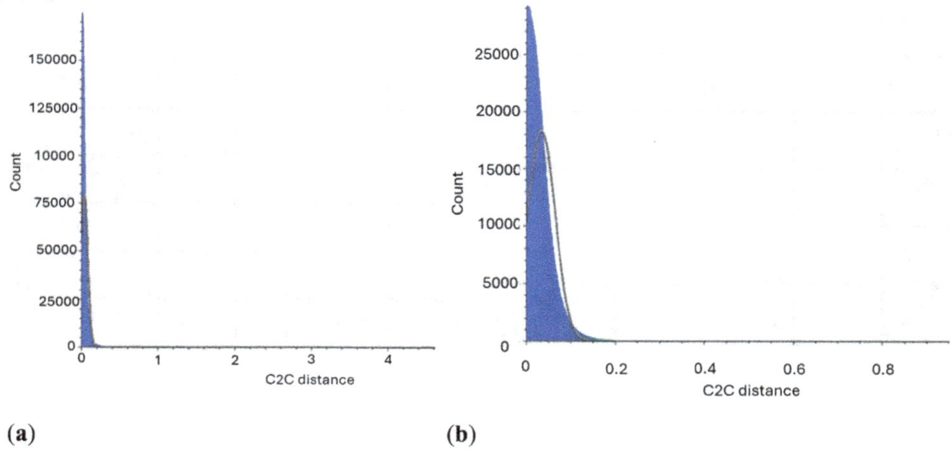

Towards a better point cloud assessment, the M3C2 plugin was used to compute the local distance of two-point clouds sideways the direction of normal surface, which tracks 3D variations in the orientation of surface, and to estimate for each measurement of distance a confidence interval relying on the roughness and registration error of the point cloud (Lague, Brodu, & Leroux, 2013). The point cloud obtained by Agisoft Metashape was as a reference cloud (Cloud#1) without any subsample. The orientation of the normal point cloud was set to "+X" and the core points' projection is achieved by retaining the original positions to generate a new point cloud that contains the differences between original point clouds. Figure 22 shows the heatmaps and the corresponding histograms. RMS is 0.5040 m and 0.5998 m before and after the noise removal, respectively, while the mean and SD are 0.0243 m and 0.6641 mm, respectively. According to the histograms, the values are concentrated close to zero indicating that point densities difference is negligible.

Figure 21. Point cloud distance after removal of the outliers. The histogram is a global indication of the error trend in the registration of the scans. Points represent the number of points on the Y-axis and Residuals represent the residual error on the X-axis (Distance between points).

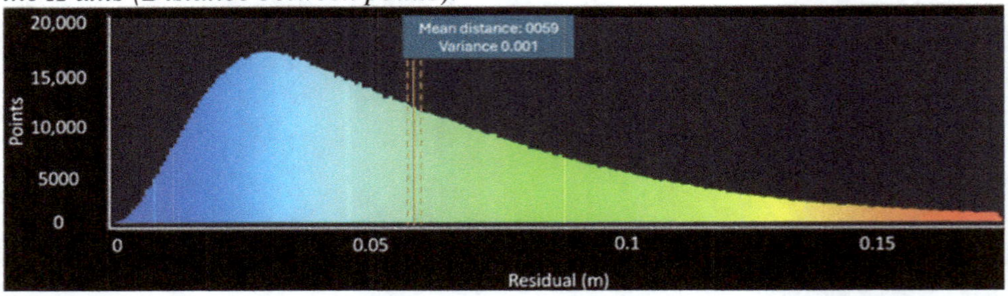

Figure 22. M3C2 distance calculation between two point clouds (a) before and (b) after the noise removal. The green color indicates the overlapping points while the grey color indicates the outliers, and the histograms (c) before and (d) after the noise.

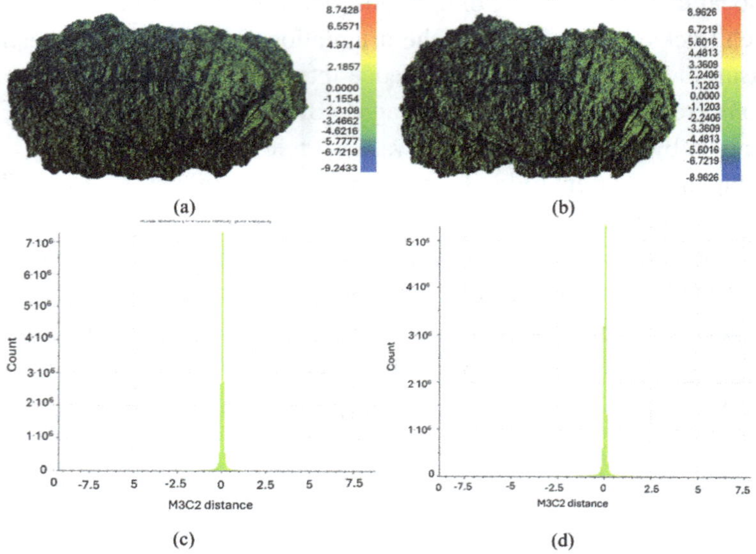

4.2.3. Roughness

Roughness is considered a negative measurement, and it indicates noise in the point cloud. It is used to conclude to the total of noise and features' over-smoothing in the reconstructions in comparison to the ground truth (Mastrorocco, Salvini, Esposito, & Seddaiu, 2016). Roughness calculation for all points is equal to the distance between each point and the optimal fitting plane calculated on its nearest neighbors. Figure 23 shows the roughness of point clouds in both programs. Color scale indicates the degree of roughness of the point clouds.

Figure 23. Roughness of point cloud in (a) Agisoft Metashape and (b) 3DF Zephyr.

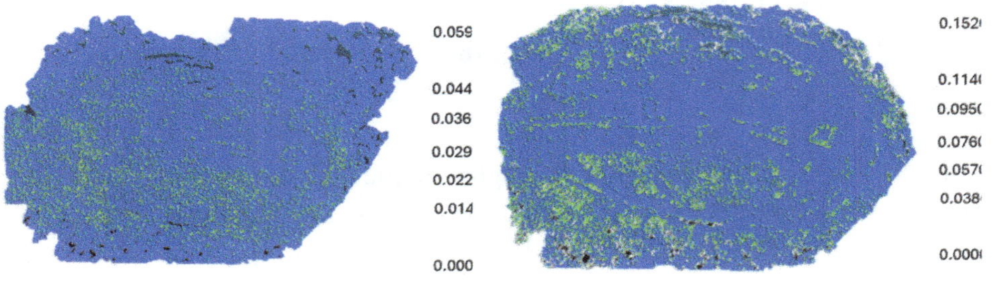

(a) **(b)**

In the scalar bar of a heatmap the values begin from zero, the blue points prove the lack of noise, while the red color indicates the opposite. Hence, it is observed that a small area from the whole is free from noise. Roughness in Agisoft Metashape is 0.0649 while in 3DF Zephyr 0.1626 and the RMS is 0.006 m and 0.0153 m, respectively. This means that the model in Agisoft Metashape presents less noise compared to 3DF Zephyr, thus, it is better reconstructed. However, this is relative, since holes or zero values also need to be considered to provide a final assessment (Žabota, Berger, & Kobal, 2023). In addition, according to the histogram in Figure 24, the mean and the SD in Agisoft Metashape are 0.0036 m and 0.0043 mm, respectively, while for 3DF Zephyr are 0.0104 m and 0.0112 mm, respectively.

Figure 24. Histogram of the roughens of the point clouds in (a) Agisoft Metashape and (b) 3DF Zephyr.

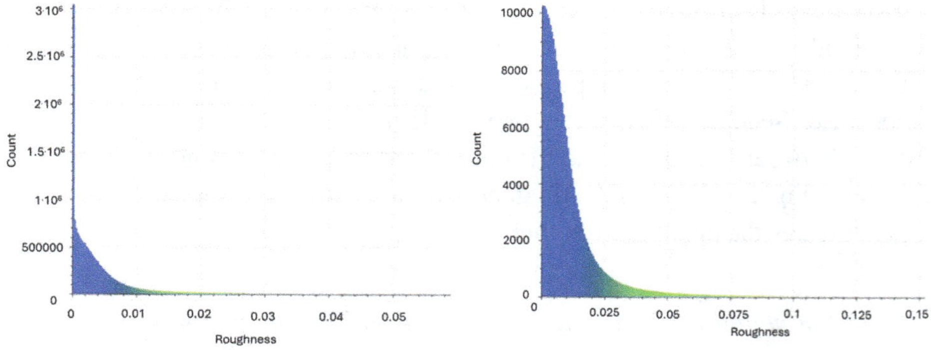

(a) **(b)**

With respect to the histogram of roughness, the distances having small values result in more reliable reconstructions, roughness-wise without inserting excessive noise, as showed by the smaller distances of histogram. 3DF Zephyr tends to over-smooth the surfaces as demonstrated by the figures. However, both Agisoft Metashape and 3DF Zephyr packages achieved equally high percentages of the "clear" 3D reconstructed models, up to 98.88% and 98.65%, respectively.

4.2.4. Mesh Performance Assessment

For mesh assessment, initially, meshes were registered using the Iterative Closest Point (ICP) algorithm and a 4×4 transformation matrix was computed, corresponding to the applied translation. Then Cloud to Mesh (C2M) distance was applied, and mesh vertices were compared. This step is an investigation between two meshes, i.e., from Agisoft Metashape and 3DF Zephyr, related to the similarity of shapes. Hence, in this step, ground coordinates were not used. Figure 25 shows the outcomes from the comparison between meshes obtained by both software packages. According to the diagram, the distribution is uniform. The mean distance is 0.0144 m and the SD is 0.0569 mm, while the RMS is 0.0587 m. The differences between models are negligible and, thus there is a non-notable difference related to the distance and displacement between them.

Figure 25. (a) Heatmap of the C2M distance between meshes and (b) the corresponding Gauss histogram.

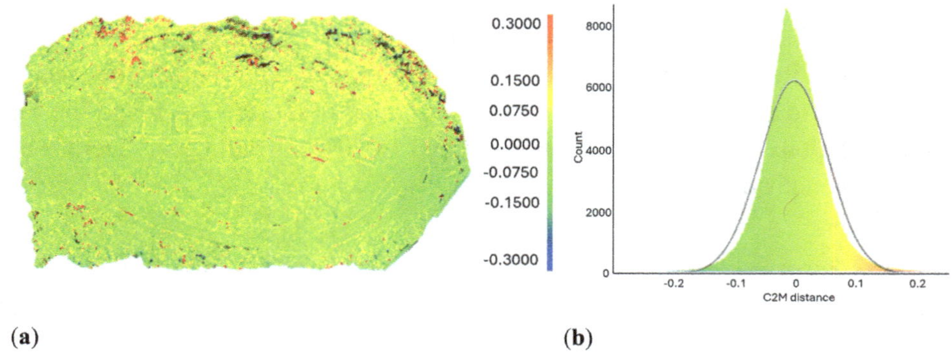

4.2.5. Digital Surface Model (DSM) Assessment

For the DSM assessment, the QGIS software package was applied. Initially, the two DSMs were subtracted to calculate the elevation differences on a cell-by-cell level. Figure 26 shows the elevation distance results. The models gave a range of elevation from 642 m to 674 m and from 644 m to 672 m, in Agisoft Metashape and 3DF Zephyr, respectively. According to the color bar, most elevations of the region coincide; however, the range of the difference from -2 m to 2 m is due to the existing vegetation and calibration issues referring to the camera's focal length.

The UAV used in this work is endowed with a non-metric, consumer-grade digital camera, and a camera self-calibration has not been feasible due to this limitation. Such cameras are known to have lens distortions that cannot be ignored, influencing the precision and robustness of aerial triangulation. Therefore, the collected UAV images can result to the "bowl effect" with camera self-calibration and the reconstructed model bends due to the inaccurate estimation of distortion parameters. Therefore, self-calibration within the two software due to the failure in the camera was employed by using more than three GCPs. The value of focal length was similar for both software (3482 for Agisoft and 3666 for Zephyr), however, Zephyr has a higher focal length than Agisoft.

Distortion refers to a third-order aberration that increases with the third power of the field's height for simple lenses; greater field of views (resulting from low magnification or short focal length) is more vulnerable to distortions than smaller field of views (resulting from high magnification or long focal length). However, without the existence of a test field, we cannot know which of the two software provides a less good calibration. It is probably Agisoft with the shorter focal length that has more error in focal and therefore its RMS and DSM have more errors. This is the reason why Figure 26, clearly indicates a clear pattern than one of the DSMs has bent like a bowl. This is a known phenomenon that is reported to happen when the camera's focal length has been solved inaccurately.

It is noted that the comparison of DSMs was limited to the core of the archaeological site to isolate most of the vegetation. In addition, the mean and the SD of the elevations' differences are 0.0135 m and 1.0704 mm and hence, there is no great difference observed.

Figure 26. Elevations' differences between two DSMs obtained by Agisoft Metashape and 3DF Zephyr; the DSM from Agisoft Metashape was subtracted from that of 3DF Zephyr, so as to check if the two software packages produce similar DSMs.

After that, the surface's area and volume were calculated. In general, the area of a pixel is multiplied by the value of the respective pixel that contains the height information. However, a DSM comprises many pixels and the total volume is the sum of each pixel area multiplied by the respective pixel value. Nevertheless, the pixel area relies on the pixel dimension i.e., the resolution. The volume is referenced from the corresponding datum (Bonczak & Kontokosta, 2019). In our case study, the volumes were calculated from the above base level, setting the minimum raster value (height value) of the DSM data (Báčová, Krása, Devátý, & Kavka, 2019). The area of the region was 11.655 m^2 and 10.845 m^2 in Agisoft Metashape and 3DF Zephyr, respectively, while the volume was 7.684.786 m^3 in Agisoft Metashape and 7.152.482 m^3 in 3DF Zephyr. Between the models, the number of both area and volumes do not differ particularly, and this confirms partially the minor distance be-tween them. Large volume numbers in both cases are attributed to the fact that the water level is used as the bottom, and our data are at a high altitude.

Finally, profile elevation was calculated (Kedron, Zhao, & Frazier, 2019). A small subarea was measured (Figure 27) from both models and the difference in the elevation was assessed, as presenterd in Figure 28. Figure 28 indicates a systematic elevation error which is attributed to the lack of ICP registration.

Figure 27. An example of subarea measurement from the archaeological site. The triangle in the middle of the line indicates the direction of its design.

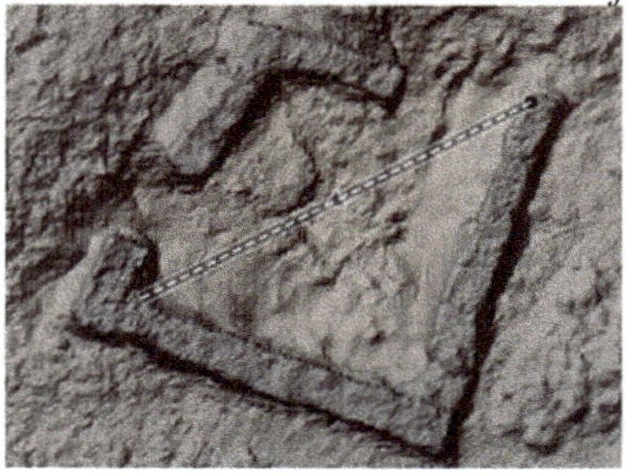

Figure 28. Diagram of the differences of elevation regarding the measurement of the object. The object is around 7 m.

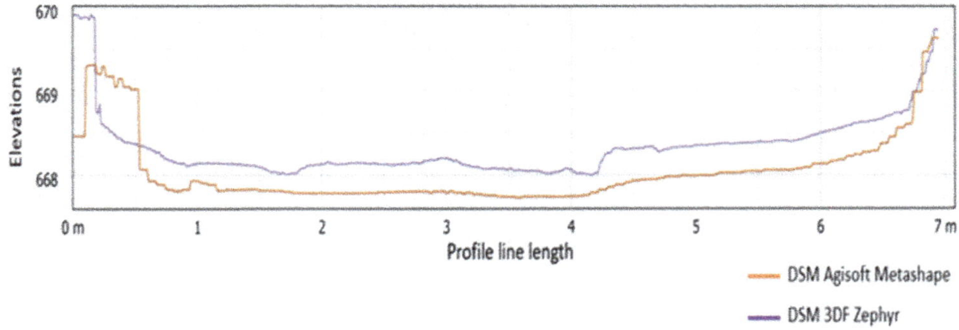

5. GEOVISUALIZATION FORMS

The term "Geographic Visualization" or "Geovisualization" concerns the visualization of geographical information including data georeferenced, visual methods and tools to spatial data and supporting applications. Geovisualization intends to visually investigate, the analysis, composition, and yields of geospatial information through Cartographic methods and Geographic Information System (GIS) techniques (Menno-Jan Kraak, 2003). The complex visual analysis of heterogeneous data focuses on dynamic maps and geographical design techniques (Edsall, 2005). The term "Geovisualization" is synonymous with the term Mapping (M.-J. Kraak, 2009). Nevertheless, a research group uses the term "Geovisual Analytics" (G. Andrienko et al., 2007; Gennady Andrienko et al., 2005) which agrees with the Geovisualization definition, i.e., "Geovisualization is the procedure where various visual methods and techniques are developing to geographical information supporting" (Nelson, 1996). In the next subsections, various geovisualization forms are presented, emphasizing the 3D model representation.

5.1. Static 3D Mapping Visualization and Animation

Maps are tools for spatial data visualization. Spatial data are stored in a database and exported for various purposes. Figure 29 shows the static 3D DSM map, while Figure 30 shows the Orthophotomap of the archaeological site under study, using GIS Arc Pro Esri software. Both Agisoft Metashape and 3DF Zephyr support animations, defining key frames either manually or automatically. However, in Agisoft Metashape the automatic procedure is achieved initially by horizontal or vertical camera position separately, and then it can be integrated. In 3DF Zephyr, camera positions are along the study area, and then the scene recording continues perimetrically. In addition, the recording can follow the GCPs. Figure 31 shows the position of cameras on the scene during the recording. In the 3DF Zephyr model, more details are captured, and they are more clearly depicted, compared to Agisoft Metashape.

Figure 29. 3D static map of Digital Terrain Model.

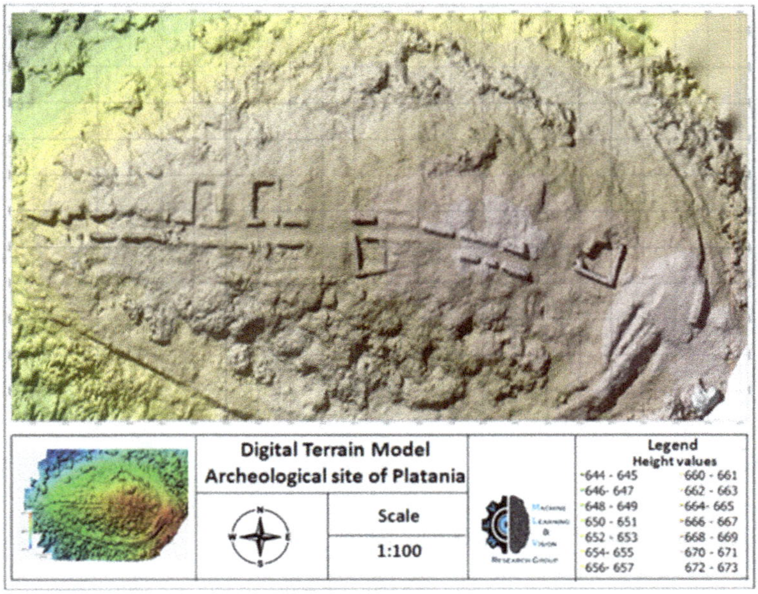

Figure 30. 3D static map of Orthophotomap.

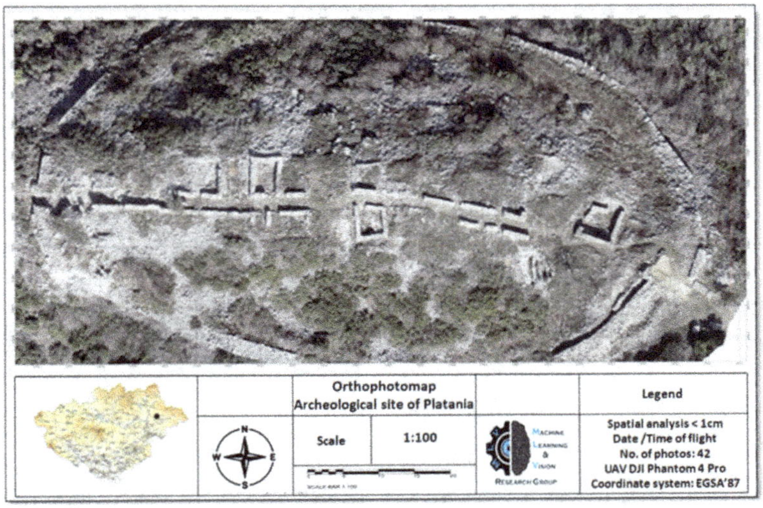

Figure 31. Camera positions for keyframes in (a) Agisoft Metashape and (b) 3DF Zephyr.

(a) (b)

5.2. Virtual and Augmented Reality (VR/AR)

In this subsection 3D point cloud obtained by Agisoft Metashape were used for virtual and augmented reality applications (Grewal, Sehra, Singh, & Rai, 2020). Initially, the model was viewed on the Sketchfab platform, which is based on WebGL and WebXR, allowing the users to set animated images or text information. Moreover, the users can activate the VR function and view the model on their cellphone or computer. Users insert the model into the physical world, and they can interact with it, by navigating inside and around the scene freely using wearable and smart devices (Ellenberger, 2017). Figure 32 shows the 3D point cloud representation in the Sketchfab platform and the environment of virtual reality.

Figure 32. 3D point cloud representation in (a) the Sketchfab platform and (b) the virtual reality environment via the Sketchfab platform.

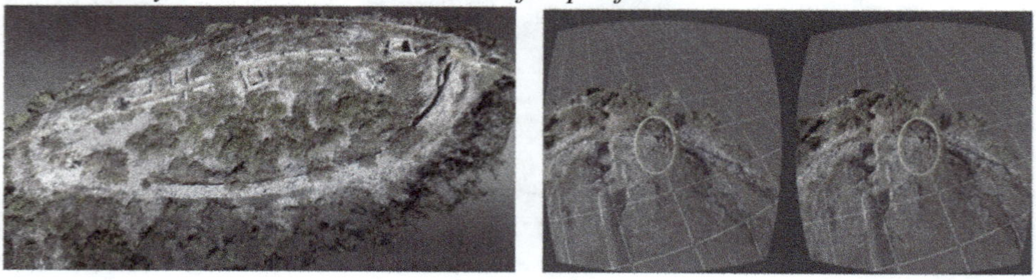

(a) (b)

Another geovisualization form is that of Augmented Reality (AR). Nowadays there are various platforms for this purpose. In this study, Vuforia Engine software combined with the Model Target Generator (MTG) and Unity3D were employed (Hameed, Hussein, Ahmed, & Omar, 2022). Unity3D supports the editor and Visual Studio for the identification of objects which appear in the application and the compilation of the corresponding code. Figure 33 shows the Model Target Generator environment and the result of the augmented model.

Figure 33. (a) Model Target Generator in Vuforia; (b) Augmented model using Unity3D.

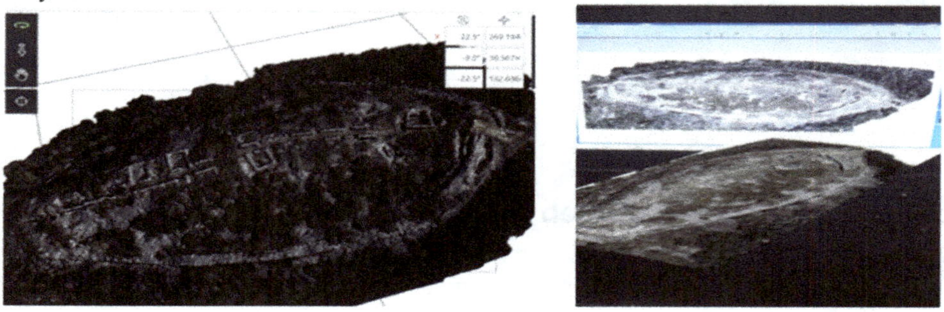

(a) **(b)**

5.3. Holography

Holography refers to the process of creating real 3D images. Holograms can be generated by computer, or the two wavefronts can be modeled and added together digitally. The digital image that results is subsequently printed on an appropriate film or mask and illuminated by a proper source towards reconstructing the wavefront under focus. In this study the hologram of the archaeological site under study was created in Blender software, which provides 3D graphics modeling, simulations, animations, rendering and video games, and then a transparent pyramid was constructed. It is noted that by using suitable projectors, the model can be illustrated in the physical or virtual world. Figure 34 illustrates the hologram design for the application of the archaeological site.

Figure 34. (a) Design of the archaeological site hologram in Blender environment; (b) Representation of model via pyramid.

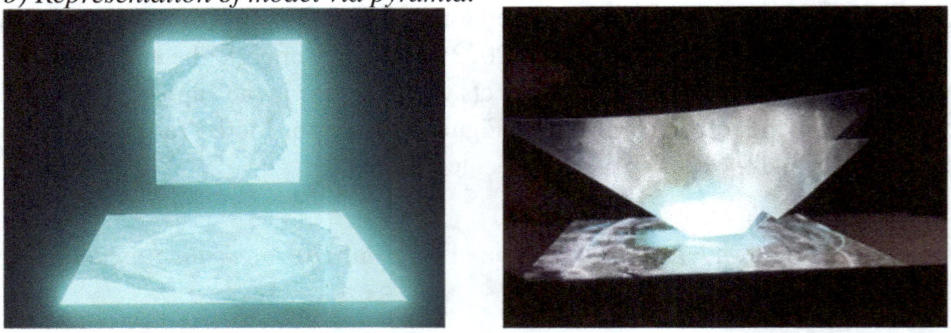

(a) **(b)**

5.4. Web Mapping and Web Build App

Online web mapping refers to a procedure for the posting and use of maps through Geographic Information Systems (GIS) on the Internet. The mapping includes a web program that allows interaction with the individual user. This is achieved by creating interactive maps (Veenendaal, Brovelli, & Li, 2017). Web GIS is a low-cost and easy way to propagate geospatial data providing processing tools. Web development and Internet expansion provide two significant benefits. Web development allows to visually interact with data and the creation of a web server that the users could use to generate maps. Nevertheless, Web GIS has some drawbacks such as internet connection speeds, converting the data to specific formats, and copyright issues (Agrawal & Gupta, 2017). In this study, data are analyzed either spatially or converted to maps or web services combined with different layers. The layers include features of the area such as streets, contours, and multimedia (video, images, storytelling, graphics). The basic map and layers are hosted and shared through ArcGIS Online. Then a barcode is created and scanned, and the application is tested on a mobile telephone. Figure 35 shows the web mapping and the web app on a mobile telephone.

Figure 35. (a) Web mapping; (b) Web app in mobile telephone.

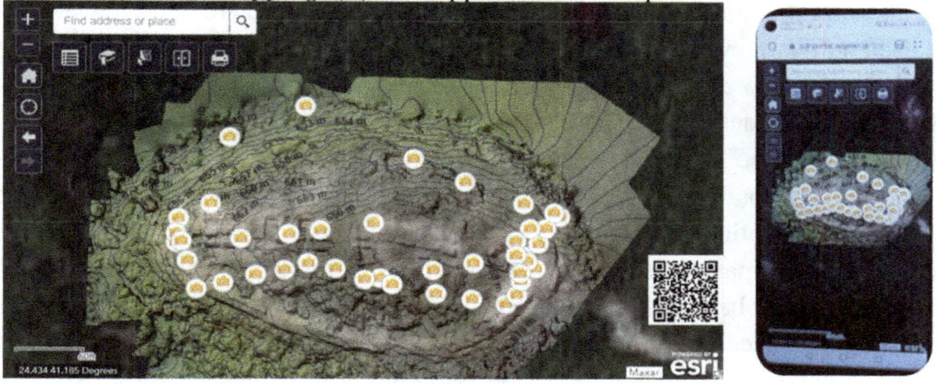

(a) (b)

6. OVERALL RESULTS, DISCUSSION, AND OPEN ISSUES

In this section, valuable conclusions are drawn, focusing on the enhancement of photogrammetry technology, 3D models evaluation and further data processing for modern geovisualization forms, highlighting simultaneously, merits, limitations, and open issues.

6.1. Results and Overall Evaluation

Both photogrammetric software packages provide similar automatic procedures related to the 3D reconstruction model with negligible processing time for its size. In addition, they do not present interruptions or technical errors in a medium quality. However, this also depends on the technical characteristics of the used computer. Moreover, they ensure the prospect of integrating various image datasets, captured from different viewpoints and ultimately with diverse focal distances, and of picking GCPs, more or less easily. Although these photogrammetric software generate high-resolution models of spatial accuracy between 1 to 2 cm, they provide some differences regarding the structure and the algorithms due to their competitive nature. Thus, the outcomes are quite or less different both qualitatively and quantitively.

In terms of the software structure, 3DF Zephyr is characterized as more structured and excels in the sense of the ability to back a previous stage processing directly creating clones so as not to lose the original model. In addition, it provides more filters and geometrical calculations among the models than Agisoft Metashape. Although the algorithms applied to the 3D reconstructions are relying on the SfM, developers of each software evolved the algorithms embedding more capabilities. Note that Agisoft offers more details to the model and appears denser without holes (Figure 10) since it provides a bigger number of generated points compared to 3DF Zephyr.

Comparing the 3D reconstruction processing and the model first of all, alignment of image, total spatial errors (mm), and reprojection errors (pix) were utilized towards investigating the capability of the software packages for aligning images and generating spatially accurate 3D models. Regarding the number of aligned images, it should be noted that all images that were inserted in both software have been aligned successfully. For model generation, 3DF Zephyr relies on the automatic finding of thousands of homologous points (keypoints) among images and then concluded with the 3D point creation, while in Agisoft Metashape alignment includes Aerial Triangulation (AT) and Bundle Block Adjustment (BBA), where images' feature points were detected and matched across images into tie points. In addition, 3DF Zephyr can find the position of camera for every image and is able to refine the calibration parameters of the camera by estimating both external and internal camera orientation parameters. Agisoft Metashape displayed better ability to align images at greater spatial scales than 3DF Zephyr. Moreover, in Agisoft Metashape the total processing of 3D model reconstruction was performed in about 39 min, in contrast to 3ZF Zephyr which was about 67 min. Regarding the number of model points, Agisoft Metashape was able to produce much better results than 3DF Zephyr; however, this does not guarantee better quality and visualization.

Despite the photogrammetric software being based on an efficient SfM technique, the performance of models obtained by each software package was investigated by applying digital measurements to assess the models' efficacy and spatial accuracy. The results of reconstruction were assessed compared to ground truth objects based on measurements of distance and comparison of roughness. The quantitative rating based on the heatmaps and histograms of distances revealed that both software packages exhibit extremely little deviations from the ground truth, while the roughness which represents the noise remains at a low level. To better understand the overall outcomes between the 3D models' comparison, mean error distance, SD of distance and RMSE were calculated. Figure 36 summarizes the mean distance error outcomes after the comparison of the models generated by both software packages. It is noted that the first four measurements refer to point clouds, the fifth to meshes and the last to DSMs.

Figure 36. Mean distance and roughness error from the comparison of models generated by Agisoft Metashape and 3DF Zephyr.

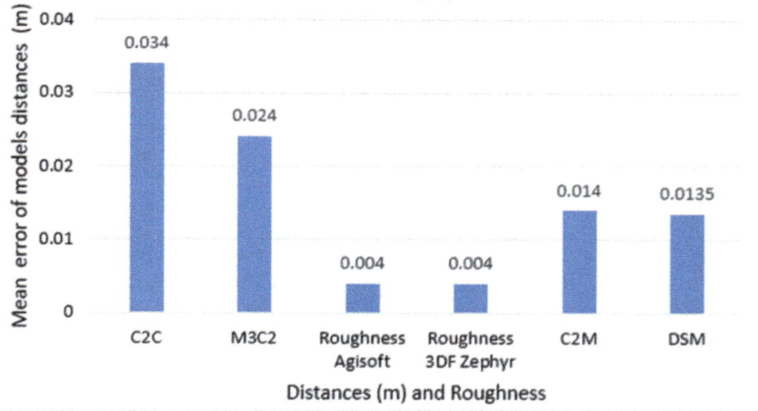

According to Figure 36, mean values displayed non statistically significant differences. Point clouds reach similar quality in terms of metrics such as accuracy, point density and roughness. Regarding the number of meshes, great differences do not exist, although visually the point cloud by Agisoft Metashape presented more outliers, while the median errors average between Agisoft Metashape and 3DF Zephyr DSM was at about 1 m. Thus, a systematic vertical shift between them exists, attributed to the dense vegetation that can obstruct the view of the ground surface and due to the complex terrain with steep slopes, and cliffs, or due to the rugged topography leading to the occlusions and shadows making it challenging to accurately estimate the ground elevation and thus, to accurately capture the elevation information.

In Figure 37 the SD of models' differences in distance and roughness are illustrated, while Figure 38 shows the corresponding RMS errors.

Figure 37. SD of models' difference distance and roughness.

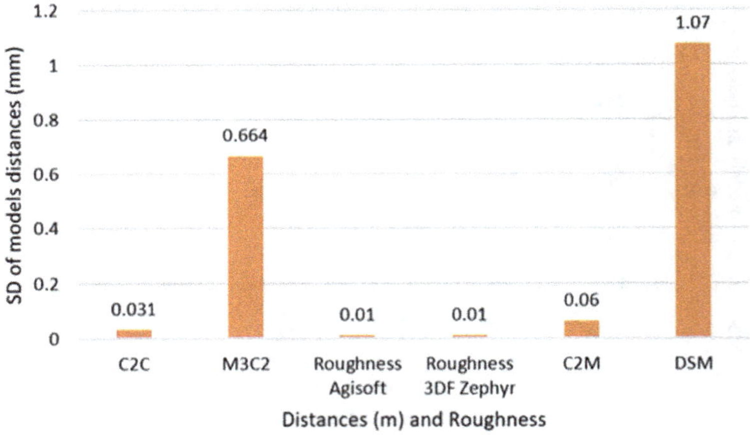

Figure 38. RMS error between extracted models by Agisoft Metashape and 3DF Zephyr concerning the distance and roughness.

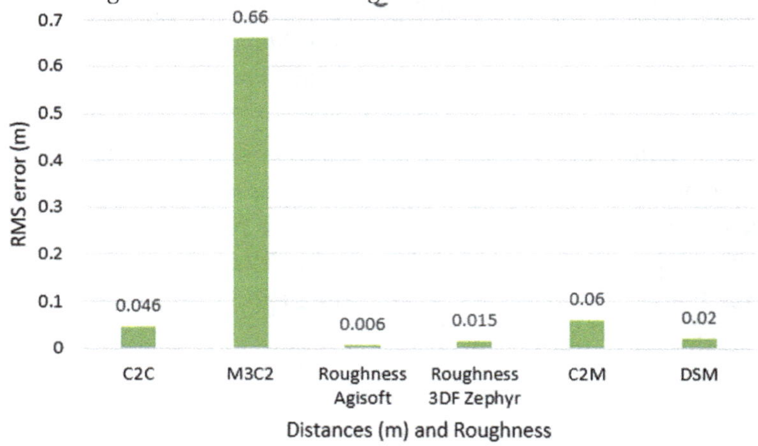

RMSE measure is widely used to assess the conformity between two models. The RMSE and SD values are similar to the C2C, M3C2 and C2M methods and indicate sufficient roughness performance. Hence, the differences among two-point clouds created by Agisoft Metashape and 3DF Zephyr is not significant i.e., they provide similar point densities and high-level smoothing of points. In addition, a height difference is observed in the case of DSM. This may occur due to the abnormal distribution of errors and outliers or systematic errors (bias) between two DSMs. However, the mean error of height difference is near zero.

Regarding geovisualization forms, different platforms have been developed providing many powerful tools to the users both to visualize and process various 3D models as well as to design their desired environments.

6.2. Advantages and Open Issues

Drone-based photogrammetry provides many advantages stemming from UAVs' technological advancements, and is related to acquisition of data, photogrammetric processing of data, and visualization aspects. Below the benefits of the UAVs' utilization and data processing through photogrammetric software are summarized:

1. UAV technology and data acquisition
    - UAVs can easily fly over rugged terrain, mountains, dense forests and rivers, allowing data acquisition even from environments where it is difficult to reach with other technologies, such as terrestrial methods, such as areas that are enrolled in the Natura 2000.
    - UAVs are equipped with progressive cameras and sensors that allow for capturing images and videos of high-resolution, efficiently and accurately with low costs.
    - UAVs can succed coverage of large areas in relatively short time ensuring that the data is consistent and reliable.
    - Data do not suffer from occlusions among high image overlaps.
    - 2D-image information can augment to 3D data points.
    - UAV images have GSD sufficient to provide good geometry and precise results with no external intervention.
2. Data processing leveraging photogrammetric software
    - Single workflow within aerial image-based applications.
    - Methodology is quick, flexible, low cost, and can produce accurate and high-quality models, even for complex objects.
    - Photogrammetric software packages are familiar to beginner users due to the automatic or semi-automatic workflow and user-friendly graphic interface, endorsed by predetermined default options.
    - Geometrical block images provide accuracy at the sub-pixel level and error checking using redundant observations as a system-inherent verification.
    - Reconstructed models contain a great number of points and density. In addition, resolution and accuracy correspond to the original images' ground sampling distance (GSD).

- Orthomosaic has a metric value leading to a better interpretation of the scene.
- The ability to automatically extract "real orthophotos" is an asset. Even for completely autonomous systems, an essential non-negligible manual aspect is that of data capturing.
- Data are permanent and can be easy to re-process and re-evaluate without missing information.
- Although the workflow is completely autonomous, one more benefit is the manual aspect because the users can specify their desired parameters.

3. Geovisualization
   - UAVs' Geospatial data enables them to model the real world. Accurate predictions can lead to better decision-making.
   - Through geovisualization, geospatial information and data exploration, the users gain a better understanding of the studied area.
   - Geovisualization facilitates visualization and communication between the spatial data and information in a clear, attractive, and interactive way, using maps, charts, graphs, animations, and web applications.
   - Generate context by utilizing colors and shapes to modify the visual focus for identifying problems, tracking changes, understanding trends, and performing forecastings associated to specific times and places.
   - Geospatial visualizations highlight the physical connection between data points.
   - Users can access the virtual environment and pick various views of phenomena to immerse them partially or entirely in artificial worlds.
   - It provides dynamic environments allowing the change of the display based on the input of the human operator, interaction, and navigation to virtual worlds.

However, corresponding limitations also exist:

1. UAV technology and data acquisition
   - Weather conditions and signal interference can affect drone flights and are vulnerable to various risks, such as cyberattacks, which can cause malfunctions or disruptions.
   - Drones have inadequate battery, capacity of payload, and range, consequently limited flight time and distance.
   - High-quality drones can be expensive.
   - Data acquisition is not possible during the night.
2. Data processing leveraging photogrammetric software

- Agisoft Metashape and 3DF Zephyr software packages are commercial; however, temporary authorizations may be granted.
- The measurements are not accurate when the GCPs are not visible due to the dense vegetation.
- Measurements' accuracy depends highly on the flight height.
- 3D model accuracy and quality can be affected by the quality of the input image leading also to motion blur.
- The image blocks need to obtain the correct images' number: a low images number could result in bad geometry and insufficient data, while too many images can make the procedure very slow.

3. Geovisualization
    - Geovisualization requires an enormous amount of data inputs for some tasks, leading easily to errors.
    - Geographic Information System (GIS) tools are expensive.
    - It provides spatial relationships but does not provide absolute solutions.
    - Data quality and availability may vary depending on the source, scale, resolution, accuracy, and reliability of the data.
    - Visualization and communication may be influenced by the choice of data, symbols, colors, scales, and formats, creating biases, misinterpretations, or confusion among the audience.

According to the current adaptation of 3D reconstruction models using photogrammetric software, the comparison models and the geovisualization forms, the subsequent conclusions can be made concerning the issues that arise from the Photogrammetry leverage from data acquisition to data post-processing.

UAVs Drone Technology

I. One of the primary encounters of the use of drones is that they need to obey to several standards and regulations that control the use of airspace, security, and privacy.

II. Larger and more long-lasting batteries can be used.

III. Data overload means larger amounts of data leading to the design of stronger software capabilities to overcome.

IV. As technology continues to advance, UAVs are assumed to increasingly participate in data collection and field research. The development of artificial intelligence and machine learning has the potential to enhance the capabilities of UAVs, enabling processing and analyzing data in real-time. This advancement could lead to more efficient and accurate data collection, as well as the advent of new and innovative applications in various fields.

Photogrammetric Software

I. Photogrammetric software initiatives for multi-view 3D reconstruction can be rated by considering the reconstruction quality and the runtime. However, since they are widely applied, it is not feasible to provide an overall ranking because it depends on the employed algorithms for generating the point cloud as well as on the quality of the images.
II. Photogrammetric software packages are capable of high-quality visualization; yet, improvements are needed, especially regarding calibration and accurate modelling, while the algorithms could be further refined.
III. For extracting high-resolution models, the processing time may turn very long since the image matching software demands a lot of time and potent hardware resources.
IV. The image blocks need to obtain the correct number of images: small number results in poor geometry; on the contrary, many images can make the procedure very slow, while it is not easy to locate defected/bad images in case the process fails.

Geovisualization

I. Development of extensible tools and methods that could facilitate understanding, and derive understandings, as gradually more large and complex geospatial data sets are available.
II. There is a lack of development of a novel generation of geovisualization techniques and tools for supporting group work (multiuser capabilities) instead of individuals, and issues of usability of visual display facilitated dialogues either between humans or between humans and computers.
III. Lack of a human-centered method to geovisualization aiming to integrate effort on technological developments leading to more usable and powerful tools working on human spatial cognition and the potentials of visual representations to empower thinking, learning, problem-solving, and decision-making.
IV. Protecting sensitive information, ensuring data integrity, and safe-guarding communication networks and infrastructure are critical challenges for the geospatial intelligence community.
V. Geospatial data often originate from diverse sources, formats, and standards.

VI. Computer graphics such as color and shading, geometric modeling and transformation, texture mapping and hardware acceleration are limited.

7. CONCLUSIONS AND FUTURE DIRECTIONS

In this work, the performances of two photogrammetric software packages were examined using data obtained by UAVs. The quality and accuracy of the reconstructed 3D models were evaluated by applying geometric metrics to compare distance and elevation differences as well as the levels of noise of point clouds, meshes, and DSMs obtained by the two software packages. In addition, for the first time geovisualization forms based on the geoinformatics and informatic were presented to demonstrate the potential of photogrammetry technology outcomes for further processing.

Photogrammetry integrates various data with high texture quality, and it is considered an excellent mean for the accurate recording of various scenes, specifically suitable for complex environments and environments with repeating patterns, such as an archaeological site. Even in cases of lower resolution of the raw data, this is compensated by the prospect of the software to acquire a final highly accurate model. In addition, it is appropriate for web-based applications. UAVs are powerful tools for data acquisition with high-resolution aerial imagery. In addition, uniform image scale and close capture geometry generally can assure accurate camera alignment. This technology continues to advance, opening up new possibilities and opportunities for researchers.

Regarding the 3D reconstruction model in Agisoft Metashape and 3DF Zephyr, it was proved that the two modeling approaches are sometimes equivalent in 3D model reconstruction regarding their quality, as they operate under the same prism of the SfM. Nevertheless, the final model depends on the employed algorithms used for generating the point cloud and on the images' quality. Although there are some differences in terms of quantity outcomes and runtime, the two software packages cannot be ranked clearly.

The results from further calculations between the two 3D models showed that the differences were relatively negligible. Especially, in Gaussian curvature key points that denote characteristics and recognizable geometric structures of the archaeological site were identified. In addition, although roughness also affects the estimation of meshes that may lead to low- or over-estimation of the distance between the two clouds, the meshes may smooth out details that could be significant for assessing the roughness local properties. The errors related to roughness of the surface were measured systematically, indicating minimal differences. Regarding DSM differences, missing data interpolation introduced minimal uncertainties; how-ever, it was difficult to quantify them.

Concerning the geovisualization forms, visual data offer flexibility, creativity and an in-depth understating to the user during the exploration. In addition, due to the dynamic environments, the users can interact by combining multi-independent visualizations and isolating desired regions for further processing, leveraging the suitable available tools.

In the future, it would be useful to develop approaches to combine TLS and photogrammetric methods for point cloud generation. In this way, gaps in the models will be covered without filtering. Another future work could be the development of methods to determine the bases of uncertainties in comparing point cloud data. In addition, a constant evaluation and re-evaluation of digital photogrammetry software as well as adding new tools, would be useful to keep up with the most current practices providing also more robust outcomes. Moreover, the integration of artificial intelligence into photogrammetric processing will open new prospects for more robust models identifying simultaneously key areas for future research. Finally, regarding geo-visualization, future research could be focused on the integration of Machine/Deep learning methods, the incorporation of tools to create and manipulate time series animation, the development of new tools that could support extensibility in the sense of users being able to create their own features, change their scale and rotate them, using interactive techniques with these tools. Another research direction could be texture mapping including planar 2D or volumetric 3D textures facilitating the disambiguation of interior and external structure of objects, as well as the development of tools capable of extracting and visualizing data derived from disparate databases. Finally, it would be useful to investigate the integration of Geovisualization and Geocomputation leading to the simulation interactively through dynamic methods aiming towards understanding quantitative relationships.

To this end, it cannot be clearly concluded if Agisoft Metashape or 3DF Zephyr generate the best 3D model; this depends on the desired work and the purpose. Nevertheless, both reconstructed 3D models should be further assessed for verification. Further geovisualization methods could provide sufficient visualization and interaction; however, further improvements are needed for better simulations and processing, leading to better decision-making, keeping pace with the evolution of Computer Vision, and responding to emerging challenges.

Statements and Declarations

Funding: No funding was received for conducting this study.

Data Availability Statement: The dataset involved in this paper (42 images and GCPs) will be available through a GitHub repository upon the acceptance of the paper.

Acknowledgments: This work was supported by the MPhil program "Advanced Technologies in Informatics and Computers", hosted by the Department of Informatics, Democritus University of Thrace, Kavala, Greece.

Competing Interests: The authors have no relevant financial or non-financial interests to disclose.

REFERENCES

Adami, A., Cerato, I., D'Annibale, E., Demetrescu, E., & Ferdani, D. (2014). Different photogrammetric approaches to 3D survey of the Mausoleum of Romulus in Rome. *2014Eurographics Workshop on Graphics and Cultural Heritage, GCH 2014*, 10. DOI: 10.2312/gch.20141300

Agarwal, S., Furukawa, Y., Sna, N., Curless, B., Seitz, S. M., & Szeliski, R. (2010). Reconstructing Rome. *Computer*, 43(6), 40–47. DOI: 10.1109/MC.2010.175

Agrawal, S., & Gupta, R. D. (2017). Web GIS and its architecture: A review. *Arabian Journal of Geosciences*, 10(23), 518. DOI: 10.1007/s12517-017-3296-2

Aguilera, D. G., & Lahoz, J. G. (2006). Laser scanning or image-based modeling? A comparative through the modelization of San Nicolas Church. *International Archives of Photogrammetry and Remote Sensing*, 36(B5).

Ahmad Fuad, N., Yusoff, A. R., Ismail, Z., & Majid, Z. (2018). Comparing the performance of point cloud registration methods for landslide monitoring using mobile laser scanning data. *The International Archives of the Photogrammetry Remote Sensing and Spatial Information Sciences, XLII*, 4(W9), 11–21. DOI: 10.5194/isprs-archives-XLII-4-W9-11-2018

Alidoost, F., & Arefi, H. (2017). Comparison of uas-based photogrammetry software for 3d point cloud generation: a survey over a historical site. *ISPRS Annals of the PhotogrammetryRemote Sensing and Spatial Information Sciences, IV*, 4(W4), 55–61. DOI: 10.5194/isprs-annals-IV-4-W4-55-2017

Andrienko, G. Andrienko, N., Dykes, J., Mountain, D., Noy, P., Gahegan, M., … Theus, M. (2005). Creating Instruments for Ideation. In *Exploring Geovisualization* (pp. 101–125). Elsevier. DOI: 10.1016/B978-008044531-1/50423-1

Andrienko, G., Andrienko, N., Jankowski, P., Keim, D., Kraak, M.-J., MacEachren, A., & Wrobel, S. (2007). Geovisual analytics for spatial decision support: Setting the research agenda. *International Journal of Geographical Information Science*, 21(8), 839–857. DOI: 10.1080/13658810701349011

Antova, G. (2019). Application of Areal Change Detection Methods Using Point Clouds Data. *IOP Conference Series. Earth and Environmental Science*, 221, 012082. DOI: 10.1088/1755-1315/221/1/012082

Báčová, M., Krása, J., Devátý, J., & Kavka, P. (2019). A GIS method for volumetric assessments of erosion rills from digital surface models. *European Journal of Remote Sensing, 52*(sup1), 96–107. DOI: 10.1080/22797254.2018.1543556

Baltsavias, E. P. (1999). A comparison between photogrammetry and laser scanning. *ISPRS Journal of Photogrammetry and Remote Sensing*, 54(2–3), 83–94. DOI: 10.1016/S0924-2716(99)00014-3

Bayram, B., Nemli, G., Özkan, T., Oflaz, O. E., Kankotan, B., & Çetin, İ. (2015). Comparison Of Laser Scanning And Photogrammetry And Their Use For Digital Recording Of Cultural Monument Case Study: Byzantine Land Walls-Istanbul. *ISPRS Annals of the Photogrammetry, Remote Sensing and Spatial Information Sciences*, 5(W3), 17–24. DOI: 10.5194/isprsannals-II-5-W3-17-2015

Bonczak, B., & Kontokosta, C. E. (2019). Large-scale parameterization of 3D building morphology in complex urban landscapes using aerial LiDAR and city administrative data. *Computers, Environment and Urban Systems*, 73, 126–142. DOI: 10.1016/j.compenvurbsys.2018.09.004

Brown, M., & Lowe, D. G. (2005). Unsupervised 3D Object Recognition and Reconstruction in Unordered Datasets. *Fifth International Conference on 3-D Digital Imaging and Modeling (3DIM'05)*, 56–63. IEEE. DOI: 10.1109/3DIM.2005.81

Bulgakov, A., Sayfeddine, D., Bock, T., & Fares, A. (2020). Generation of Orthomosaic Model for Construction Site using Unmanned Aerial Vehicle. *Proceedings of the 37th International Symposium on Automation and Robotics in Construction, ISARC 2020: From Demonstration to Practical Use - To New Stage of Construction Robot*. DOI: 10.22260/ISARC2020/0124

Burns, J. H. R., & Delparte, D. (2017). Comparison of commercial structure-from-motion photogrammety software used for underwater three-dimensional modeling of coral reef environments. *The International Archives of the Photogrammetry.Remote Sensing and Spatial Information Sciences, XLII*, 2(W3), 127–131. DOI: 10.5194/isprs-archives-XLII-2-W3-127-2017

CloudCompare. (2024). CloudCompare. Retrieved March 27, 2024, from https://www.danielgm.net/cc/

Cramer, M., Stallmann, D., & Haala, N. (2000). Direct georeferencing using gps/inertial exterior orientations for photogrammetric applications. *International Archives of Photogrammetry and Remote Sensing, 33*(B3/1; PART 3), 198–205.

Dayal, K. R., Raghavendra, S., Pande, H., Tiwari, P. S., & Chauhan, I. (2017). Comparative analysis of 3d point clouds generated from a freeware and terrestrial laser scanner. *The International Archives of the Photogrammetry. Remote Sensing and Spatial Information Sciences, XLII*, 4(W2), 67–71. DOI: 10.5194/isprs-archives-XLII-4-W2-67-2017

3DFLOW. (2024). 3DF ZEPHYR. Retrieved March 27, 2024, from https://www.3dflow.net/

Edsall, R. M. (2005). Challenging Conventions for Geovisualization Interface Design: Implications for Exploratory Data Analysis. *Ninth International Conference on Information Visualisation (IV'05)*, 642–647. IEEE. DOI: 10.1109/IV.2005.31

Ellenberger, K. (2017). Virtual and Augmented Reality in Public Archaeology Teaching. *Advances in Archaeological Practice*, 5(3), 305–309. DOI: 10.1017/aap.2017.20

Eltner, A., & Schneider, D. (2015). Analysis of Different Methods for 3D Reconstruction of Natural Surfaces from Parallel-Axes. *The Photogrammetric Record*, 30(151), 279–299. DOI: 10.1111/phor.12115

Eskicioglu, A. M., & Fisher, P. S. (1995). Image quality measures and their performance. *IEEE Transactions on Communications*, 43(12), 2959–2965. DOI: 10.1109/26.477498

Fassi, F., Fregonese, L., Ackermann, S., & De Troia, V. (2013). Comparison between laser scanning and automated 3d modelling techniques to reconstruct complex and extensive cultural heritage areas. *The International Archives of the Photogrammetry.Remote Sensing and Spatial Information Sciences, XL*, 5(W1), 73–80. DOI: 10.5194/isprsarchives-XL-5-W1-73-2013

Fonstad, M. A., Dietrich, J. T., Courville, B. C., Jensen, J. L., & Carbonneau, P. E. (2013). Topographic structure from motion: A new development in photogrammetric measurement. *Earth Surface Processes and Landforms*, 38(4), 421–430. DOI: 10.1002/esp.3366

Gabara, G., & Sawicki, P. (2023). CRBeDaSet: A Benchmark Dataset for High Accuracy Close Range 3D Object Reconstruction. *Remote Sensing (Basel)*, 15(4), 1116. DOI: 10.3390/rs15041116

Gagliolo, S., Ausonio, E., Federici, B., Ferrando, I., Passoni, D., & Sguerso, D. (2018). 3D cultural heritage documentation: a comparison between different photogrammetric software and their products. *The International Archives of the Photogrammetry.Remote Sensing and Spatial Information Sciences, XLII*, 2, 347–354. DOI: 10.5194/isprs-archives-XLII-2-347-2018

Grewal, K. K., Sehra, S. S., Singh, J., & Rai, H. S. (2020). An augmented reality based tool for contribution to OpenStreetMap. *International Journal of Information Technology : an Official Journal of Bharati Vidyapeeth's Institute of Computer Applications and Management*, 12(2), 345–352. DOI: 10.1007/s41870-019-00403-w

Grimm, A. (1980). Der Ursprung des Wortes PHOTOGRAMMETRIE. *International Archives of Photogrammetry, 23*(Part B10, Commission VI), 323–330.

Haala, N., & Cavegn, S. (2016). High density aerial image matching: state-of-the-art and future prospects. *ISPRS - International Archives of the Photogrammetry. Remote Sensing and Spatial Information Sciences, XLI*, B4, 625–630. DOI: 10.5194/isprsarchives-XLI-B4-625-2016

Hameed, Q. A., Hussein, H. A., Ahmed, M. A., & Omar, M. B. (2022). Development of Augmented Reality-based object recognition mobile application with Vuforia. *Journal of Algebraic Statistics*, 13(2), 2039–2046.

Hansman, R. J., & Ring, U. (2019). Workflow: From photo-based 3-D reconstruction of remotely piloted aircraft images to a 3-D geological model. *Geosphere*, 15(4), 1393–1408. DOI: 10.1130/GES02031.1

Hassani, F. (2015). Documentation of cultural heritage; techniques, potentials, and constraints. *The International Archives of the Photogrammetry. The International Archives of the Photogrammetry, Remote Sensing and Spatial Information Sciences*, 5(W7), 207–214. DOI: 10.5194/isprsarchives-XL-5-W7-207-2015

Kadobayashi, R., Kochi, N., Otani, H., & Furukawa, R. (2004). Comparison and evaluation of laser scanning and photogrammetry and their combined use for digital recording of cultural heritage. *The International Archives of the Photogrammetry, Remote Sensing and Spatial Information Sciences*, 35(5), 401–406.

Kaminsky, R. S., Snavely, N., Seitz, S. M., & Szeliski, R. (2009). Alignment of 3D point clouds to overhead images. *2009 IEEE Computer Society Conference on Computer Vision and Pattern Recognition Workshops*, 63–70. IEEE. DOI: 10.1109/CVPRW.2009.5204180

Kazhdan, M., & Hoppe, H. (2013). Screened poisson surface reconstruction. *ACM Transactions on Graphics*, 32(3), 1–13. DOI: 10.1145/2487228.2487237

Kedron, P., Zhao, Y., & Frazier, A. E. (2019). Three dimensional (3D) spatial metrics for objects. *Landscape Ecology*, 34(9), 2123–2132. DOI: 10.1007/s10980-019-00861-4

Kim, T., & Im, Y.-J. (2003). Automatic satellite image registration by combination of matching and random sample consensus. *IEEE Transactions on Geoscience and Remote Sensing*, 41(5), 1111–1117. DOI: 10.1109/TGRS.2003.811994

Kingsland, K. (2020). Comparative analysis of digital photogrammetry software for cultural heritage. *Digital Applications in Archaeology and Cultural Heritage*, 18, e00157. DOI: 10.1016/j.daach.2020.e00157

Kraak, M.-J. (2003). Geovisualization illustrated. *ISPRS Journal of Photogrammetry and Remote Sensing*, 57(5–6), 390–399. DOI: 10.1016/S0924-2716(02)00167-3

Kraak, M.-J. (2009). Geovisualization. In *International Encyclopedia of Human Geography* (pp. 468–480). Elsevier., DOI: 10.1016/B978-008044910-4.00033-X

Lague, D., Brodu, N., & Leroux, J. (2013). Accurate 3D comparison of complex topography with terrestrial laser scanner: Application to the Rangitikei canyon (N-Z). *ISPRS Journal of Photogrammetry and Remote Sensing*, 82, 10–26. DOI: 10.1016/j.isprsjprs.2013.04.009

Lee, J.-H., & Sull, S. (2019). Regression Tree CNN for Estimation of Ground Sampling Distance Based on Floating-Point Representation. *Remote Sensing (Basel)*, 11(19), 2276. DOI: 10.3390/rs11192276

Long, N. Q., Goyal, R., Bui, L. K., Cuong, C. X., Van Canh, L., Minh, N. Q., & Bui, X. N. (2023). Optimal Choice of the Number of Ground Control Points for Developing Precise DSM Using Light-Weight UAV in Small and Medium-Sized Open-Pit Mine. *Archives of Mining Sciences*. https://doi.org/DOI: 10.24425/ams.2021.138594

Luo, K., Pan, H., Zhang, Y., & Guan, T. (2019). Partial bundle adjustment for accurate three-dimensional reconstruction. *IET Computer Vision*, 13(7), 666–675. DOI: 10.1049/iet-cvi.2018.5564

Marelli, D., Morelli, L., Farella, E. M., Bianco, S., Ciocca, G., & Remondino, F. (2023). ENRICH: Multi-purposE dataset for beNchmaRking In Computer vision and pHotogrammetry. *ISPRS Journal of Photogrammetry and Remote Sensing*, 198, 84–98. DOI: 10.1016/j.isprsjprs.2023.03.002

Mastrorocco, G., Salvini, R., Esposito, G., & Seddaiu, M. (2016). 3D point cloud analysis for surface roughness measurement: Application of UAV photogrammetry. *Rendiconti Online Della Società Geologica Italiana*, 41, 313–316. DOI: 10.3301/ROL.2016.156

Maurya, R. K., & Kulkarni, S. T. (2021). Multidimensional spatial clustering and visualization of 3D topographic relief data. *International Journal of Information Technology : an Official Journal of Bharati Vidyapeeth's Institute of Computer Applications and Management*, 13(2), 581–592. DOI: 10.1007/s41870-020-00595-6

Metashape. (2024). Agisoft. Retrieved March 27, 2024, from https://www.agisoft.com/

Meydenbauer, A. (1867). Die photometrographie. *Wochenblatt Des Architektenvereins Zu Berlin*, 1(14), 125–126.

MicMac. (2024). MicMac. Retrieved March 27, 2024, from https://micmac.ensg.eu/

Milenkovic, M., & Ressl, C. (2013). Surface Roughness from Point Clouds-A Multi-Scale Analysis. *EGU General Assembly*, 15, 13570.

Murtiyoso, A., Grussenmeyer, P., Börlin, N., Vandermeerschen, J., & Freville, T. (2018). Open Source and Independent Methods for Bundle Adjustment Assessment in Close-Range UAV Photogrammetry. *Drones (Basel)*, 2(1), 3. DOI: 10.3390/drones2010003

Nebel, S., Beege, M., Schneider, S., & Rey, G. D. (2020). A Review of Photogrammetry and Photorealistic 3D Models in Education From a Psychological Perspective. *Frontiers in Education*, 5, 144. DOI: 10.3389/feduc.2020.00144

Nelson, E. S. (1996). How Maps Work: Representation, Visualization, and Design. *Cartographic Perspectives*, (24), 27–30. DOI: 10.14714/CP24.757

Nex, F., Gerke, M., Remondino, F., Przybilla, H.-J., Bäumker, M., & Zurhorst, A. (2015). ISPRS benchmark for multi-platform photogrammetry. *ISPRS Annals of the Photogrammetry.Remote Sensing and Spatial Information Sciences, II*, 3(W4), 135–142. DOI: 10.5194/isprsannals-II-3-W4-135-2015

Nex, F., & Remondino, F. (2014). UAV for 3D mapping applications: A review. *Applied Geomatics*, 6(1), 1–15. DOI: 10.1007/s12518-013-0120-x

Niederheiser, R., Mokroš, M., Lange, J., Petschko, H., Prasicek, G., & Elberink, S. O. (2016). Deriving 3D point clouds from terrestrial photographs - comparison of different sensors and software. *ISPRS - International Archives of the Photogrammetry.Remote Sensing and Spatial Information Sciences, XLI*, B5, 685–692. DOI: 10.5194/isprsarchives-XLI-B5-685-2016

Nikolov, I., & Madsen, C. (2016). Benchmarking Close-range Structure from Motion 3D Reconstruction Software Under Varying Capturing Conditions. In *Lecture Notes in Computer Science (including subseries Lecture Notes in Artificial Intelligence and Lecture Notes in Bioinformatics)* (pp. 15–26). DOI: 10.1007/978-3-319-48496-9_2

Oliensis, J. (2000). A Critique of Structure-from-Motion Algorithms. *Computer Vision and Image Understanding*, 80(2), 172–214. DOI: 10.1006/cviu.2000.0869

Oniga, V.-E., Breaban, A.-I., & Statescu, F. (2018). Determining the Optimum Number of Ground Control Points for Obtaining High Precision Results Based on UAS Images. *International Electronic Conference on Remote Sensing*, 352. MDPI. DOI: 10.3390/ecrs-2-05165

Padró, J.-C., Muñoz, F.-J., Planas, J., & Pons, X. (2019). Comparison of four UAV georeferencing methods for environmental monitoring purposes focusing on the combined use with airborne and satellite remote sensing platforms. *International Journal of Applied Earth Observation and Geoinformation*, 75, 130–140. DOI: 10.1016/j.jag.2018.10.018

Pádua, L., Adão, T., Hruška, J., Marques, P., Sousa, A., Morais, R., & Peres, E. (2018). UAS-based photogrammetry of cultural heritage sites. *Proceedings of the International Conference on Geoinformatics and Data Analysis*, 72–76. New York, NY, USA: ACM. DOI: 10.1145/3220228.3220243

Papadopoulou, E.-E., Vasilakos, C., Zouros, N., & Soulakellis, N. (2021). DEM-Based UAV Flight Planning for 3D Mapping of Geosites: The Case of Olympus Tectonic Window, Lesvos, Greece. *ISPRS International Journal of Geo-Information*, 10(8), 535. DOI: 10.3390/ijgi10080535

PIX4D. (2024). PIX4D. Retrieved March 27, 2024, from https://www.pix4d.com/

Ponto, K., & Tredinnick, R. (2022). Opportunities for utilizing consumer grade 3D capture tools for insurance documentation. *International Journal of Information Technology : an Official Journal of Bharati Vidyapeeth's Institute of Computer Applications and Management*, 14(6), 2757–2766. DOI: 10.1007/s41870-022-01040-6

Remondino, F., Spera, M. G., Nocerino, E., Menna, F., & Nex, F. (2014). State of the art in high density image matching. *The Photogrammetric Record*, 29(146), 144–166. DOI: 10.1111/phor.12063

Reulke, R., & Döring, T. (2006). Computer Vision im Kontext von Photographie und Photogrammetrie. In Reisig, W., & Freytag, J.-C. (Eds.), *Informatik* (pp. 315–328). Springer Berlin Heidelberg., Retrieved from http://link.springer.com/10.1007/3-540-32743-6_13 DOI: 10.1007/3-540-32743-6_13

Sankaranarayanan, J., Samet, H., & Varshney, A. (2007). A fast all nearest neighbor algorithm for applications involving large point-clouds. *Computers & Graphics*, 31(2), 157–174. DOI: 10.1016/j.cag.2006.11.011

Schöning, J., & Heidemann, G. (2015). Evaluation of Multi-view 3D Reconstruction Software. In *Lecture Notes in Computer Science (including subseries Lecture Notes in Artificial Intelligence and Lecture Notes in Bioinformatics)* (pp. 450–461). DOI: 10.1007/978-3-319-23117-4_39

Seitz, L., Haas, C., Noack, M., & Wieprecht, S. (2018). From picture to porosity of river bed material using Structure-from-Motion with Multi-View-Stereo. *Geomorphology*, 306, 80–89. DOI: 10.1016/j.geomorph.2018.01.014

Setiyadi, S., Mukhtar, H., & Cahyadi, W. A. (2021). A Comparative Study of Affordable Photogrammetry Software for Reconstructing 3D Model of a Human Foot. *2021 IEEE 7th International Conference on Smart Instrumentation, Measurement and Applications (ICSIMA)*, 286–291. IEEE. DOI: 10.1109/ICSIMA50015.2021.9526314

Shen, Y., Lindenbergh, R., & Wang, J. (2016). Change Analysis in Structural Laser Scanning Point Clouds: The Baseline Method. *Sensors (Basel)*, 17(12), 26. DOI: 10.3390/s17010026 PMID: 28029121

Singh, S. P., Jain, K., & Mandla, V. R. (2014). 3D Scene Reconstruction from Video Camera for Virtual 3D City Modeling. *American Journal of Engineering Research*, •••, 140–148.

Singla, K., & Nand, P. (2023). Optimizing deep learning architectures for novel view synthesis: Investigating the impact of NeRF MLP parameters on complex scenes. *International Journal of Information Technology : an Official Journal of Bharati Vidyapeeth's Institute of Computer Applications and Management*. Advance online publication. DOI: 10.1007/s41870-023-01470-w

Snavely, N., Seitz, S. M., & Szeliski, R. (2008). Modeling the World from Internet Photo Collections. *International Journal of Computer Vision*, 80(2), 189–210. DOI: 10.1007/s11263-007-0107-3

Tenze, L., & Canessa, E. (2024). altiro3d: Scene representation from single image and novel view synthesis. *International Journal of Information Technology : an Official Journal of Bharati Vidyapeeth's Institute of Computer Applications and Management*, 16(1), 33–42. DOI: 10.1007/s41870-023-01590-3

Ullman, S. (1979). The Interpretation of Structure From Motion. In *The Interpretation of Visual Motion*. The MIT Press., DOI: 10.7551/mitpress/3877.003.0009

Vasileiadis, N., Tsoukala, E., & Kostopoulos, D. S. (2019). The late Miocene bovids from Platania (Drama Basin, Greece), with description of a new species of Palaeoryx. *Geobios*, 55, 57–76. DOI: 10.1016/j.geobios.2019.06.005

Veenendaal, B., Brovelli, M. A., & Li, S. (2017). Review of Web Mapping: Eras, Trends and Directions. *ISPRS International Journal of Geo-Information*, 6(10), 317. DOI: 10.3390/ijgi6100317

Wang, Y., Deng, N., Xin, B., Wang, W., Xing, W., & Lu, S. (2020). A novel three-dimensional surface reconstruction method for the complex fabrics based on the MVS. *Optics & Laser Technology*, 131, 106415. DOI: 10.1016/j.optlastec.2020.106415

Wang, Z., & Bovik, A. C. (2002). A universal image quality index. *IEEE Signal Processing Letters*, 9(3), 81–84. DOI: 10.1109/97.995823

Westoby, M. J., Brasington, J., Glasser, N. F., Hambrey, M. J., & Reynolds, J. M. (2012). 'Structure-from-Motion' photogrammetry: A low-cost, effective tool for geoscience applications. *Geomorphology*, 179, 300–314. DOI: 10.1016/j.geomorph.2012.08.021

Wu, C. (2024). VisualSFM: A Visual Structure from Motion System. Retrieved March 27, 2024, from http://ccwu.me/vsfm/

Xu, Z., Zhang, Y., & Xu, X. (2016). 3D visualization for building information models based upon IFC and WebGL integration. *Multimedia Tools and Applications*, 75(24), 17421–17441. DOI: 10.1007/s11042-016-4104-9

Žabota, B., Berger, F., & Kobal, M. (2023). The Potential of UAV-Acquired Photogrammetric and LiDAR-Point Clouds for Obtaining Rock Dimensions as Input Parameters for Modeling Rockfall Runout Zones. *Drones (Basel)*, 7(2), 104. DOI: 10.3390/drones7020104

Zhang, Y., Luo, X., Yang, W., & Yu, J. (2019). Fragmentation Guided Human Shape Reconstruction. *IEEE Access : Practical Innovations, Open Solutions*, 7, 45651–45661. DOI: 10.1109/ACCESS.2019.2905879

Chapter 5
Analyzing Traffic Congestion Patterns in M'diq:
A GIS Study

Ayoub Lahlouh
https://orcid.org/0000-0001-9119-3810
Abdelmalek Essaâdi University, Morocco

Nisserine Ben Driss
https://orcid.org/0009-0007-8915-2236
Abdelmalek Essaâdi University, Morocco

Sanaa Cheikh
Abdelmalek Essaâdi University, Morocco

ABSTRACT

This study integrates topographical, geographic, and infrastructure data into a GIS framework to investigate the complex dynamics of urban traffic congestion in M'diq. The research emphasizes common causes of congestion, including urban morphology, road network connectivity, public service distribution, and the impact of seasonal tourism. A composite heat map is utilized to identify heavily congested areas, such as Lalla Nezha and Nasr Streets. The study elaborates on the historical growth patterns, geographical difficulties, and randomness of the road networks, all of which exacerbate traffic issues. Oriented mainly toward the analysis of the causes of congestion, this research puts a spotlight on data-driven approaches to urban planning as a means to handle them. Findings feed into larger discussions of urban traffic management by underscoring the need for strategies in innovative planning, taking into account M'diq's unique topographical and infrastructural characteristics.

DOI: 10.4018/979-8-3693-6452-9.ch005

INTRODUCTION

The topic of urban transportation is complex and multifaceted, involving a blend of economic, social, spatial, and political dimensions. While numerous studies have addressed urban transportation, they often focus solely on economic, social, or technical analyses, overlooking the crucial spatial dimension that often hinders the sector's development (Banos & Thévenin, 2011). Transportation is a vital sector, significantly influencing a country's spatial organization and economic prosperity (L. Wang et al., 2018). The attractiveness of certain locations fundamentally requires accessibility, which naturally depends on the quality and type of transportation infrastructure.

In recent years, urban transportation has faced several challenges at the spatial, economic, and social levels. One of the most prominent issues is traffic congestion, which occurs during specific daily periods in major cities or seasonally in cities with high seasonal tourist activities. Zhao et al. (2023) emphasize the importance of understanding both local and global spatiotemporal dependencies in traffic data to address urban traffic congestion effectively (Zhao et al., 2023). Traffic congestion isn't just about infrastructure; it also involves the residents and visitors, as is the case in our study area of M'diq, a model for a small city. Despite its small size and geographically limited area due to natural constraints, M'diq has seen significant growth in recent years in terms of urban expansion and demographic growth.

M'diq is characterized by the concentration of its public services within a geographically limited area, attracting a large number of vacationers during the summer period. This leads to a significant increase in vehicles within the urban area, posing several quantitative, qualitative, and organizational problems for urban transport managers. Addressing these issues requires practical solutions that focus on optimizing the spatial organization of the road network, closely aligning with the distribution of public services within the city

Urban planning or geography and spatial planning specialists study traffic congestion problems leading to time wastage, stress, pollution, accidents, and lack of parking lots in city centers. They also explore how to connect the city's outskirts with the center and interconnect them (Zhou et al., 2020). To address the complex spatial challenges of urban transportation in M'diq, this research utilizes a Geographic Information Systems (GIS) framework. The methodology integrates various geospatial, topographical, and infrastructural datasets to conduct a comprehensive analysis of traffic congestion. By employing advanced geospatial analytical techniques, including the use of software like ESRI's ArcGIS and ERDAS (Rahman, 2019), the study examines the spatial distribution of public services, road network connectivity, and the impact of seasonal tourism. The methodology is designed to provide a detailed understanding of the spatial dynamics at play and to identify areas of improvement for urban transport planning in M'diq. This approach not only addresses the quantitative aspects of traffic data but also offers a spatially informed perspective, crucial for developing effective and sustainable solutions to the city's transportation challenges.

1 RESEARCH AREA

M'diq, nestled on the scenic Tetouan coast of Morocco (Figure 1), has undergone significant demographic shifts. Its population grew from a modest 2,412 in 1960 to 56,130 by 2014 (Figure 2), as reported by the Haut Commissariat au Plan (2014). This growth has been a cornerstone in supporting the urban development of M'diq, providing essential human resources at various developmental stages. Nevertheless, the decelerating growth rate and diverse demographic patterns necessitate a reevaluation of population movement data and an assessment of the area's future demographic outlook.

Figure 1. Position of the M'diq

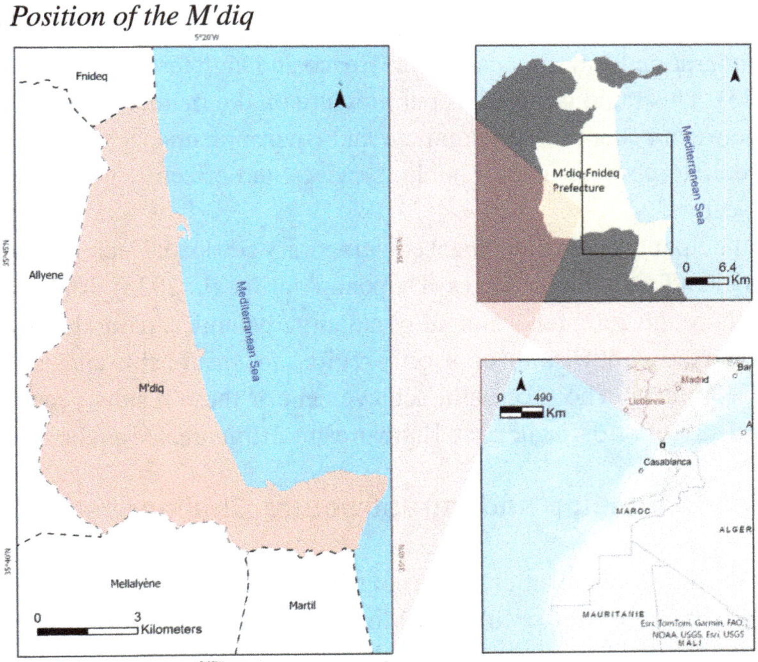

Figure 2. Population Trends in the M'diq (1960-2014) according to the High Commission for Planning

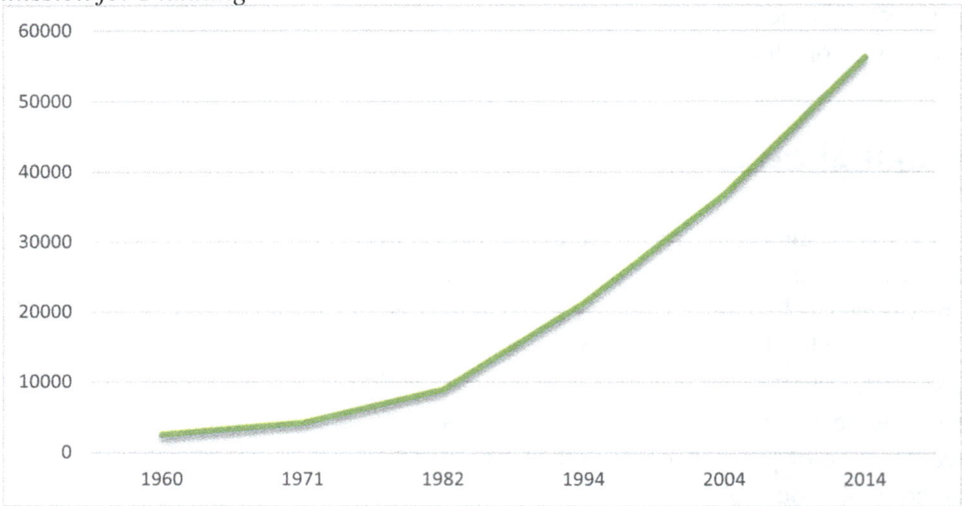

The population in M'diq has significantly increased within the past year mainly due to immigration from rural parts of Morocco and high fertility rates (Agharroud et al., 2023). This rapid growth has put pressure on the municipality to maintain a balance between economic development and environmental sustainability, particularly in delivering high-quality public services and effective governance across essential sectors.

The municipality has to implement contemporary geospatial technologies in order to properly handle these challenges (Damoah-Afari et al., 2023). It is necessary to possess a thorough comprehension of population dynamics, natural resource management, and the identification of priority regions for sustainable and economically viable development. The successful achievement of these results necessitates the deliberate distribution of financial and human capital to guarantee quality and efficacy.

1.1 Urban Expansion and Infrastructure Challenges

The fast pace of urbanization as the contributing factor to spatial growth in cites like M'diq is by no means peculiar to Morocco. In another major Moroccan city, Casablanca, a study identifies that one of the critical factors in achieving sustainable development within cities lies in understanding interdependencies between growth patterns and infrastructure versus socioeconomic indicators (2022). In this respect, one study recently used remote sensing data to estimate the past urban land use changes and predict future expansion in attempts to emphasize the role of geospatial techniques in improving the current state of urban planning. (Benchelha et al., 2022) Hence, this insight is very critical for M'diq moving into the future with its urban and demographic complexities.

These insights are of special significance to M'diq, which is struggling to personalize its own urban and demographic complexities. In this respect, remote sensing and GIS will introduce a new dimension in discovering how spatial growth contributes to infrastructure, public services, and environmental sustainability within M'diq.

1.2 Environmental and Economic Implications

With time and growth in the town of M'diq, the environmental and economic implications cannot be set aside. Being a city located in close proximity to both coastal and forested natural resources, the environmental impacts of urban sprawl are bound to hit it soon. Thus, sustainable development strategies have to cater for the preservation of those resources alongside satisfying the needs of a growing population and an ambitious economy for the city.

This is directly linked to its ability to keep attracting businesses and tourism. Without a doubt, this would be much dependent on the quality of infrastructure and public services. Efficient urban planning, founded on geospatial analysis and strategic resource allocation, will therefore become crucial in striking a balance between growth and sustainability for M'diq. Uncontrolled and unregulated urban sprawl may result in great environmental changes, as manifested in other fast-growing areas where urban sprawl was often coupled with farmland loss and green space loss that contributed to environmental degradation (Fan et al., 2022).

Moreover, sustainable urban planning that integrates economic development is necessary for fostering a competitive urban economy. This would mean infrastructures for business activities and public services that will enhance such activities, while ensuring that such development does not come at the expense of environmental sustainability.

2 PROBLEM STATEMENT

The coastal town of M'diq experiences a dramatic transformation each summer as it becomes a magnet for vacationers. This combination, along with the demographic explosion that has turned the town into a critically overpopulated area, has brought traffic-related challenges. The narrow, winding streets of M'diq, made for a much smaller, more sedate pace of life, now host extended travel times, more pollution, and basically a lowering of the quality of life for residents and visitors in the summer months. Due to geographical constraints and an overemphasis on concentrated public service sites, the existing road network and urban infrastructure are increasingly unable to handle the seasonal spikes in both vehicular and pedestrian traffic.

This congestion doesn't just cause inconvenience but also poses severe disadvantages. From an economic perspective, it puts local businesses and the tourism industry major lifelines for the town's economy at risk. Environment-wise, it moves toward contributing to more significant emission rates in the view of air quality and public health (Downs, 2004). Socially, it impacts the daily lives of residents, leading to frustration and a decrease in the overall livability of the area. The decelerating growth rate and various demographic patterns further complicate the understanding and prediction of population movement, resulting in layers of complexity to an already pressing issue.

The challenge lies in addressing the multifaceted nature of traffic congestion in M'diq, particularly during the critical summer period. Understanding the causes and contributing factors to this status form the bases for developing a coherent set of strategies, aiming at decongestion, sustainability, and livability of the town in general. Studies have shown that smart growth and strategic urban planning can significantly impact traffic patterns and congestion (Karimi et al., 2022). This paper tries, through the discussion of these broader principles applied to the unique context of M'diq, to define and discuss the extent of these challenges to set the stage for a detailed investigation of potential solutions.

3 MATERIALS AND METHODS

In the scenario of facing this multicriteria problem of traffic congestion in M'diq, especially during the summer season, the following is the multipronged approach scheme with multiple tools and methods to be used. The work has been divided into a number of key components, each with a different focus, contributing to the holistic understanding of the problem. Here are the materials and methods that have been used:

3.1 Remote Sensing, Aerial Photography, and Data Integration for Urban Analysis:

Objective: The main purpose of the component is to outline the historical and current urban development of M'diq and further assess its impact on traffic conditions. Such work can be done by integrating different data sources, such as remote sensing, aerial photography, and urban plans, into a complete dynamic geographic database that supports advanced spatial analysis (Table 1).

Tools & Data:

To achieve this objective, a diverse array of tools and data sources were employed:

- Satellite Imagery: Historical satellite images were retrieved from Landsat 5, 7, and 8 and Sentinel-2, spanning several decades from 1974 to 2020. These images gave a temporal dimension to urban expansion and land use changes.
- Aerial Photography: High-resolution aerial photographs from 2006 were acquired from the local authorities. The images gave detailed views of the urban layout, which was very critical in validating the satellite data.
- Digital Maps: Maps at different scales from various sources, like the OpenStreetMap global site (2020), have enabled the retrieving of a very detailed and updated cartographic representation of M'diq.
- Urban Development Plans and Topographic Maps: Historical and recent urban development plans, such as SDAULT from 1998 and urban plans from 2001, 2010, and 2017, and topographic maps, like Tetouan 1:50,000 scale from 1973, were vital in understanding the planned versus real urban growth and infrastructure development.

Method:

- **Data Integration:**

This was then followed by a systematic approach to packaging all the collected data satellite imagery, aerial photography, maps, and urban development plans into a GIS system. The process of integration can be summarized as follows:

- Georeferencing: each map, aerial photograph was georeferenced so that the spatial location details and accuracy were secured. The operation was carried out through the ground control points within the maps in high-resolution imagery.
- Construction of Database: The integrated data were constructed in a unified geo-database format within the GIS system. It comprised land-use patterns, road networks, public service, and demographic layers that allow in-depth, multilayered spatial analysis.
- Data Harmonisation: Data harmonization techniques were applied in order to ensure consistency across different time periods and sources of data. This may include modifications to the resolution of satellite images and ensuring all the provided maps and plans use the same spatial reference.
  - **Remote Sensing and Classification:**

Time-series analysis for the classification of satellite imagery and its analysis in ERDAS Imagine and ArcGIS Pro was conducted in order to detect built-up areas, vegetation cover, and other land uses across time. The following steps were involved in the classification process:

- Pre-processing: radiometric and atmospheric correction of all satellite images to enable data comparison between the different time periods.
- Classification Algorithms: Supervised classification algorithms were employed where training data was selected based on known land cover types such as urban, vegetation, and water areas.
- Change detection: will be performed using the post-classification comparison method to compare the land-use and built-up areas at different times. This provided an opportunity for the exact quantification of urban expansion and tries to shed light on how M'diq's spatial footprint has evolved.

- **Aerial Photography Analysis:**

Ground data was extracted from the high-resolution aerial photographs to validate the classification results obtained from the satellite images. This was done by:

- Visual Interpretation: The results of these classifications were subjected to the process of visual interpretation to identify key features such as roads, buildings, and public spaces, which will serve in cross-verification of the classified satellite data.
- Feature Extraction: Specific features, such as new developments or changes in the road networks, were digitized with the addition of more layers onto the GIS database. This provided detailed ground level information to enrich the spatial database that was not reflected in the satellite images.
- Sample Features Cross Referencing with Urban Plans The sample features were cross-checked with the urban development plans to discover the extent of similarities or differences between the planned and actual developments.

Table 1. Primary Data Used in Constructing the Geographic Database for M'diq

Category	Name	Date
Paper and Digital Maps	Topographic map of Tetouan scale 1:50000	1973
	Map titled "YEBALA 2" for M'diq scale 1:50000	1927
	Digital map of M'diq from the OpenStreetMap global site.	2020

continued on following page

Table 1. Continued

Category	Name	Date
Aerial Photos and Satellite Imagery	Aerial photo of the city of M'diq	2006
	Satellite imagery from Landsat 5,7,8	1974, 1984, 1994, 2003, 2016
	Satellite imagery from Digital Globe	2014
	Satellite imagery from Sentinel 2	2020
Plans and Designs	Master plan for the urban development of the Tetouan coastal tourist area (SDAULT)	1998
	Urban development plan for M'diq	2001
	Urban development plan for M'diq	2010
	Urban development plan for M'diq	2017
	Urban designs for M'diq	2001, 2006, 2014

This integrated methodological approach provided a robust framework for analyzing the historical and current urban development of M'diq and its direct influence on traffic congestion.

3.2 Integrated GIS and Traffic Data Analysis:

Objective: The primary objective of this integrated approach is to analyze the physical, infrastructural, and temporal elements contributing to traffic congestion in M'diq. The current research is focused on the integration of Geographic Information Systems with traffic data analysis in order to give insight into the spatial and temporal dynamics of congestion during peak periods like the summer tourist season.

Tools & Data:

A variety of tools and datasets were employed to conduct this analysis:

- GIS Software: ArcGIS Pro will be the prime tool applied to spatial analysis, network analysis, and data integration. It has strong analytics that permit examining many of the spatial factors contributing to congestion.
- Sources of Traffic Data: Real-time data and historical traffic data were sourced from Google Maps Traffic, ESRI Traffic services, and local authorities dealing with transportation. Such datasets provided the key information on the state of traffic flow, congestion hotspots, and travel times.
- Topographic and infrastructural data: Detailed data regarding the topography of M'diq, pre-existing road networks, public services, and land use patterns

were extracted from local authorities, open-source databases, and through field surveys. This information was very useful to understand how the physical and infrastructural elements affect the pattern of traffic.

- Field Surveys: Field data were collected using the **Mapit** Mobile Application, which enabled the precise recording of geographic locations, attributes, and conditions of road networks and public services. Additionally, **Mapillary** was utilized to collect and detect traffic signs across the city, providing a comprehensive view of the traffic infrastructure and potential regulatory issues.

Method:

- **Data Integration and Preprocessing:**

All data collected from topographical maps to road networks, public services locations, traffic data, and field survey data were integrated into a GIS system. Integration made it possible to carry out multi-layer analyses of several characteristics contributing to phenomena such as traffic congestion.

Temporal and spatial resolutions of the various datasets were homogenized to have consistency in analysis. For example, traffic data from various sources will be aligned with the GIS database using spatial interpolation techniques.

- **Spatial Analysis:**

Topographic Analysis: ArcGIS Pro was used to realize slope, aspect, and elevations through spatial analysis operations in the study area for establishing the physical constraints on the road network. Those analyses helped describe areas where topography might add to traffic bottlenecks, such as steep slopes or sharp turns.

Density of Public Services: Mapping of the density and spatial distribution of public services, such as school areas, hospital areas, and administrative offices, has been done in trying to locate probable areas with congestion arising from high demand for such services. The estimation used for density was by means of kernel density, whereby results showed areas of high concentration of public facilities that might add to the localized traffic problems.

- **Network Analysis:**

Evaluation of the Road Network: The M'diq road network was fully analyzed using the network analysis model provided within Esri in ArcGIS Pro. Running the model simulated the flow of traffic, estimated travel time, to fully understanding the dynamics of city traffic.

Bottleneck Identification: For this network analysis model, several traffic scenarios were run covering peak and off-peak hours. These runs have provided possibilities for finding the intersections and road sections that are prone to congestion. The shortest path algorithms were also applied in analyzing possible detours and their impacts on the general flow of traffic to identify critical bottlenecks within the network.

Traffic Sign Detection and Analysis: The data collected through Mapillary was integrated into the analysis to detect and map the distribution of traffic signs throughout M'diq. On this component of the study, it permitted gaining insight into the current state of regulatory compliance, visibility of traffic signage, and requirements that need to be fulfilled by additional signing to manage and improve traffic flow in the city.

- **Traffic Data Analysis:**

Real-Time and Historical Data Analysis: In interpreting the temporal patterns of congestion, an analysis was conducted of ESRI traffic data for peak travel times, recurrent traffic hotspots, and traffic conditions during different seasons in order to check for variability.

Comparative Analysis: Comparison was made between the normal and peak periods, focusing on summer months when M'diq sees its maximum visit by tourists. This brought out time discrepancies of traffic congestion and highlighted those periods that require intervention measures most critically.

Mapping of Congestion Hotspots: Using the traffic data, the congestion hotspots were mapped and cross-referenced with the outcomes of the spatial analysis. This was done to correlate the physical and infrastructural elements, as components, with real traffic patterns and to gain an overview of congestion in M'diq.

- **Field Survey Analysis:**

Mapit Mobile Application: The application was heavily used during the field surveys for capturing real-time data on the road conditions, placement of traffic signs, location of public services. Geolocation through GPS in the app provided accurate points for the survey data, which were then integrated into the GIS system for further analysis.

Ground Truthing: Data collected using Mapit were then used to validate the remote sensing and GIS analyses, ensuring that the accuracy of the spatial models developed in this study was adhered to.

3.3 Algorithmic Calculation of Indices:

The degree of connectivity expresses the level of accessibility without considering distance and is crucial for urban planning and development due to its direct impact on serving the residents of cities and villages. The analysis of connectivity in transport networks relies on simplifying the network into a topological diagram, which consists of several nodes representing stations and intersections(Xie et al., 2020), along with a set of links connecting these nodes or intersections. The theory dealing with nodes and links topologically is known as graph theory (Figure 3).

Figure 3. Graph theory

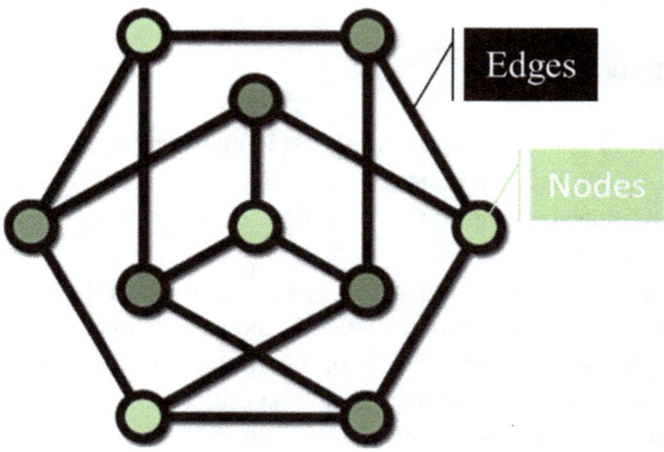

The following indices were calculated:

1. **Beta Index (β):** Measures the ratio of the number of links (L) in the network to the number of nodes (N). It indicates the degree of connectivity of a network.

$$\beta = \frac{L}{N}$$

2. **Gamma Index (γ):** Measures the ratio of the actual number of links in the network to the maximum possible number of links.

$$\gamma = \frac{L}{3(N-2)}$$

3. **Alpha Index (α):** Measures the degree of connectivity considering the number of circuits or loops in the network.

$$\alpha = \frac{L - N + 1}{2N - 5}$$

These indices were calculated using the spatial and traffic data collected.

4 RESULTS AND DISCUSSIONS

This section presents the findings from the comprehensive analysis conducted to understand the patterns and underlying causes of traffic congestion in M'diq, especially during the critical summer months. The primary objective of this research was to utilize a combination of remote sensing (Droj et al., 2023), GIS spatial analysis (Vikram & Bhardwaj, 2023), traffic data analysis, and algorithmic calculations to provide a detailed understanding of the historical, current, and predictive dynamics of traffic flow and urban development in M'diq. The results are crucial for informing future urban planning and traffic management strategies aimed at alleviating congestion and enhancing the overall livability of the town.

The findings are organized systematically, beginning with the historical evolution of M'diq's urban landscape, followed by the current state of its physical and infrastructural elements as revealed through GIS analysis. This is complemented by insights from real-time and historical traffic data, which shed light on patterns and peak congestion times. Additionally, the calculated indices provide a quantitative measure of the road network's characteristics. Together, these results offer a multifaceted view of the traffic congestion issue, grounded in empirical data and spatial analysis (X. Chen et al., 2023).

It is important to note that the results are presented objectively, without interpretation or bias. The subsequent Discussion section will delve into the implications, interpretations, and potential applications of these findings. The aim here is to provide a clear and factual account of the data collected and the outcomes of the various analyses conducted.

Visual aids such as maps, graphs, and tables are included throughout to illustrate the results and provide a more intuitive understanding of the data. Each visual is accompanied by a caption explaining its significance and how it relates to the study's objectives.

In presenting these results, we hope to provide a solid foundation for understanding the complex issue of traffic congestion in M'diq and to set the stage for meaningful discussions on potential solutions and strategies (Tanvir & Haque, 2023).

4.1 Historical Urban Growth and Land Use Changes

M'diq, located approximately 14 km from Tetouan, did not emerge as an urban center until the period of the Spanish Protectorate. It was initially established as a military base for the Protectorate authorities, particularly given its strategic location on the land route and railway connecting the cities of Tetouan and Ceuta. At that time, it was known as Rincón del Medick, or "Corner of the Strait." By the late 1920s, the civilian population of M'diq was about a thousand, most of whom were Spanish and worked as fishermen.

The town of M'diq developed during the colonial period, following the road paving operations carried out by the Spanish after their capture of Fnideq. They paved the road extending from the center of Kandisa to Oued Negro, which they took over on May 23, 1911, and then the road from Oued Negro to the mouth of Oued Smir. The Spanish advanced towards M'diq and occupied it in April 1912, naming it Rincón.

4.1.1 Pre-Independence Period

The colonizers began organizing the occupied territory, starting with the construction of a series of military barracks, weapon caches, and facilities within the village. They also built houses, cafes, and shops for trading purposes, primarily located on Mohammed V Street (currently Lalla Nezha). In addition to using the area as a military base, the nearby plains (known as Asmir Plain or Bouzghlal) were exploited for agricultural activities, particularly tomato production. During this period, a series of road projects initiated by the colonizers exploited the country's resources in the northern region. M'diq was considered a transit area for the road axis constructed between Tetouan and Ceuta, extending approximately 42 km and passing through urban centers like Martil, Fnideq (Castillejos), and Ceuta (Ali, 2015).

The colonial authorities also constructed a railway line connecting the city of Tetouan towards Ceuta, covering 40.25 km and passing through Martil, Mellalyin, M'diq, and Fnideq. This line was a primary means of transporting goods from Tetouan to Ceuta and transporting raw materials for the local industry, with the goods transported via this line reaching 73,000 tons per year (Ali, 2015).

4.1.2 Post-Independence Period

This period began when Morocco gained independence, and several built-up areas emerged within the urban space of M'diq, especially after the construction of the port began in 1961. This event marked a turning point for the town, attracting a significant number of immigrants to work on the port construction, which commenced operations in 1966. The demand for labor increased, attracting workers from various rural areas to work as sailors, particularly from the Martil port, which had ceased operations during this period. On the other hand, a series of tourism projects were positioned as part of the five-year, three-year, and four-year tourism plans that covered various Moroccan coastal cities. From the mid-1960s, several institutions and investors (Maroc Touriste, SOMADET, Club Méditerranée) focused on M'diq's coastal area, undertaking several tourism projects, including the V.V.T complex near M'diq port, Holiday Club in M'diq, and the Kabila and Ristinga Smir tourist complexes(Ali, 2015).

Other factors also contributed to the development of M'diq's built-up area, including the road axes, which played a significant role in attracting people, economic activities, and social facilities. M'diq benefited greatly from being located on the main road axis number 13, connecting Tetouan and Ceuta, which had significant dynamism in the region. Additionally, the growth of economic activities in the town (fishing, smuggling trade) led to a strong urbanization movement, resulting in the consolidation of scattered buildings in the areas adjacent to the center. According to the M'diq Municipality (1991), this was the actual start of the center's expansion and the emergence of what would later be known as informal housing. This period also witnessed discord and disintegration in the urban landscape, exacerbating the housing crisis and real estate speculation, leading to the spread of unauthorized construction in many neighborhoods, including the western (Deradeb, and the lower and upper tracks) and southern (Zawia and Al Istiklal) areas. This type of housing represented 57% of the total housing and 67.2% of the total plots of land in 1991 (El Cheikhi & El Abdlaoui, 1997).

The dawn of the 21st century marked a pivotal era in M'diq's urban development, witnessing rapid expansion, particularly in previously undeveloped areas like the Bouzghlal plain. This growth, driven by lower real estate costs, encroached on agricultural lands and the "Marja Asmir" nature reserve. Informal construction, especially in neighborhoods like Salaam, Qalaa, and Cherifa, became the main urban expansion driver, despite challenging terrains and lack of basic facilities.

Recent urban planning and development initiatives have improved infrastructure and living conditions, attracting migrants and contributing to the city's growth. Additionally, the M'diq coast has seen significant tourist development since the 1960s, particularly under the "Tamuda Bay" initiative, with most tourism projects evolving into real estate developments, consuming vast coastal areas. These changes reflect the city's dynamic evolution and the state's focus on tourism as a developmental catalyst. (Figure 4)

Figure 4. Urban Change Distribution by Historical Period

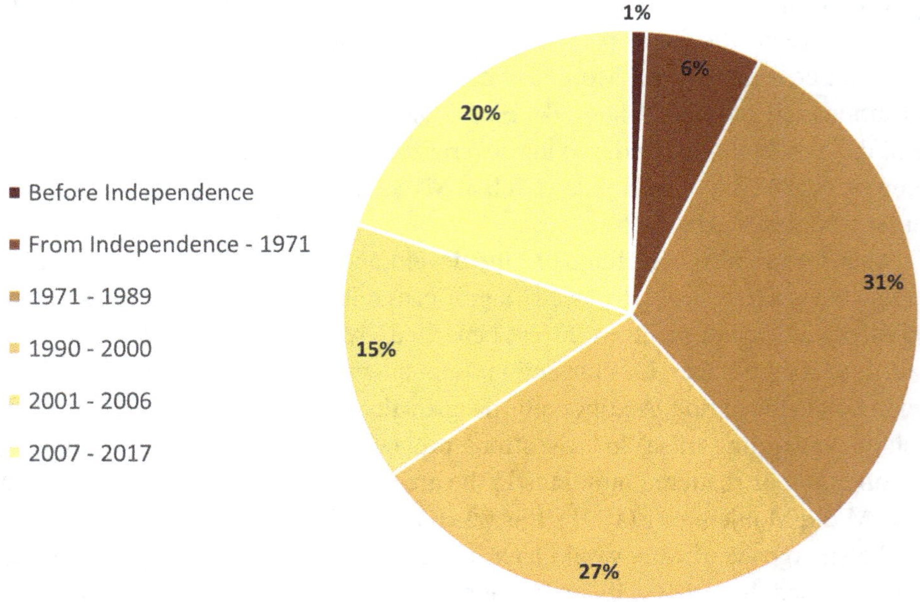

In the accompanying map 'The Historical Evolution of the Urban Area of M'diq from Independence to the Year 2020' (Figure 5) the historical and contemporary data sources detailed in the Methodology section have been utilized to visually represent the urban changes in M'diq over time. This map synthesizes the information from various paper and digital maps, aerial photos, satellite imagery, and urban plans and designs to illustrate the city's growth and transformation, complementing the narrative of M'diq's historical urban development discussed above.

Figure 5. The Historical Evolution of the Urban Area of M'diq from Independence to the Year 2020

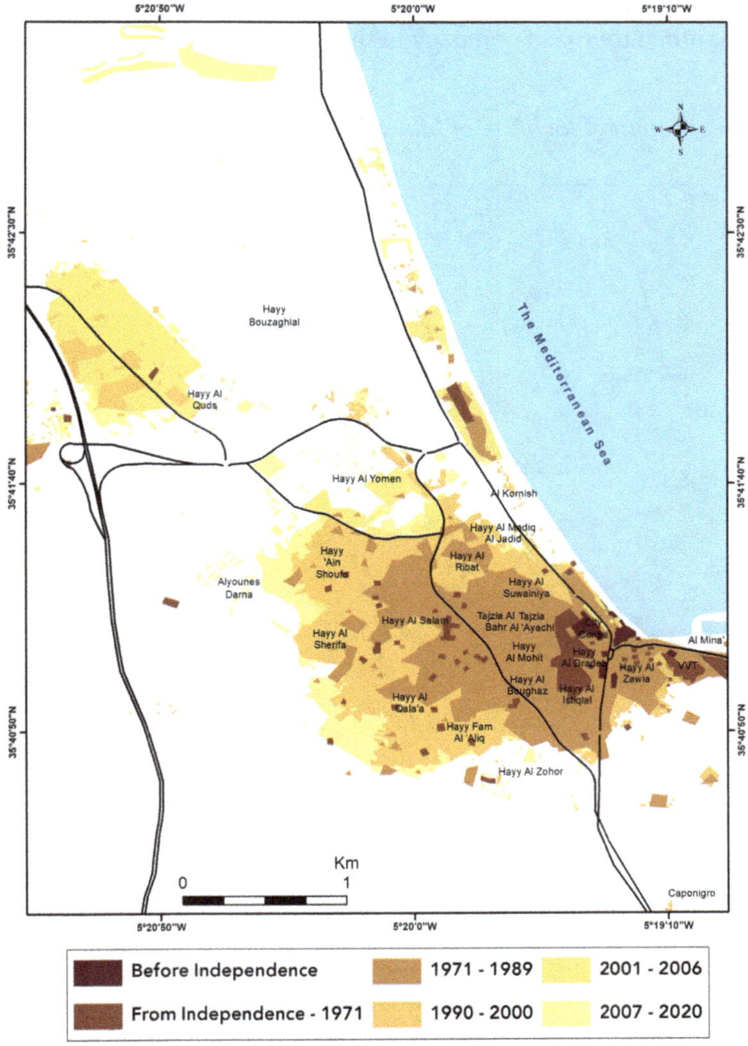

4.2 The Role of Geographical Location and Topography in the Urban Transport Issues of the M'diq Municipality

The M'diq municipality occupies an important geographical location that serves as a point of convergence and passage from Morocco to occupied Ceuta (Figure 6). It is also an important seasonal tourist destination during the summer period, which results in significant traffic movement on the main road axes, especially in the city center. This has led to difficulties in movement within and outside the city, negatively impacting various commercial sectors, particularly the commercial speed of the public urban transport company in the municipality(Shafiei et al., 2023).

Figure 6. Geographical location of M'diq Municipality

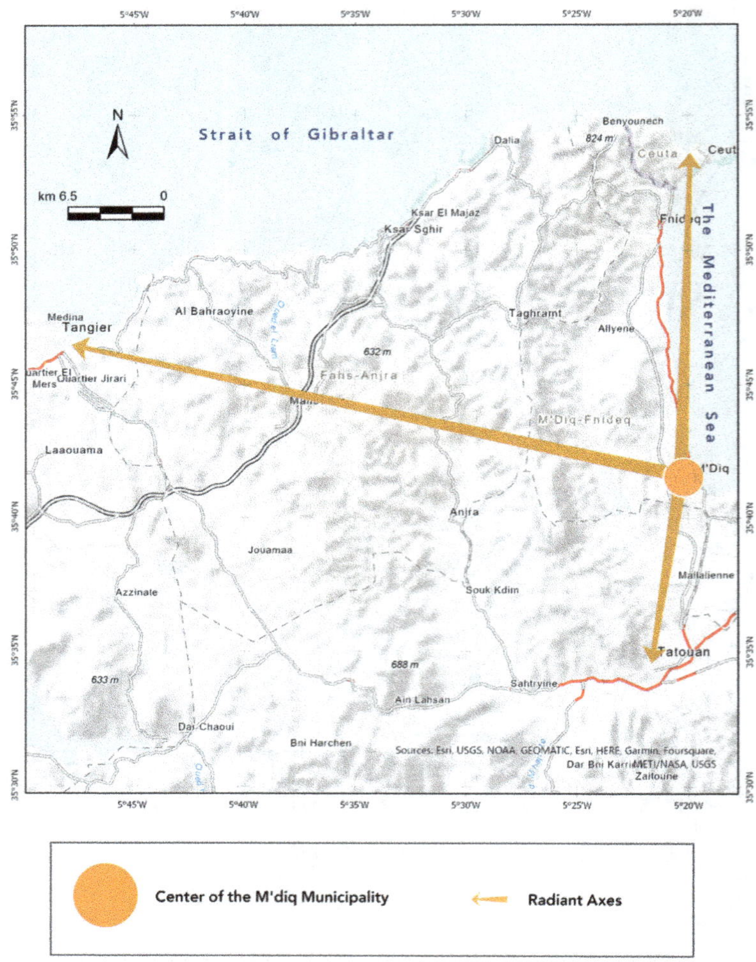

Topographically, the M'diq municipality is surrounded by a range of elevations on all sides except for the western side (Figure 7), which is bordered by the Mediterranean Sea. These elevations gradually decrease, tending to drop to sea level as one moves eastward. Despite the relatively low elevations in the studied area, the terrain is characterized by very steep slopes (Figure 8). The proximity to the base level, the impact of tectonics, and the density of the hydrographic network all indicate the ruggedness of the terrain (Bennani et al., 2020).

Figure 7. The spatial distribution of elevation levels in M'diq

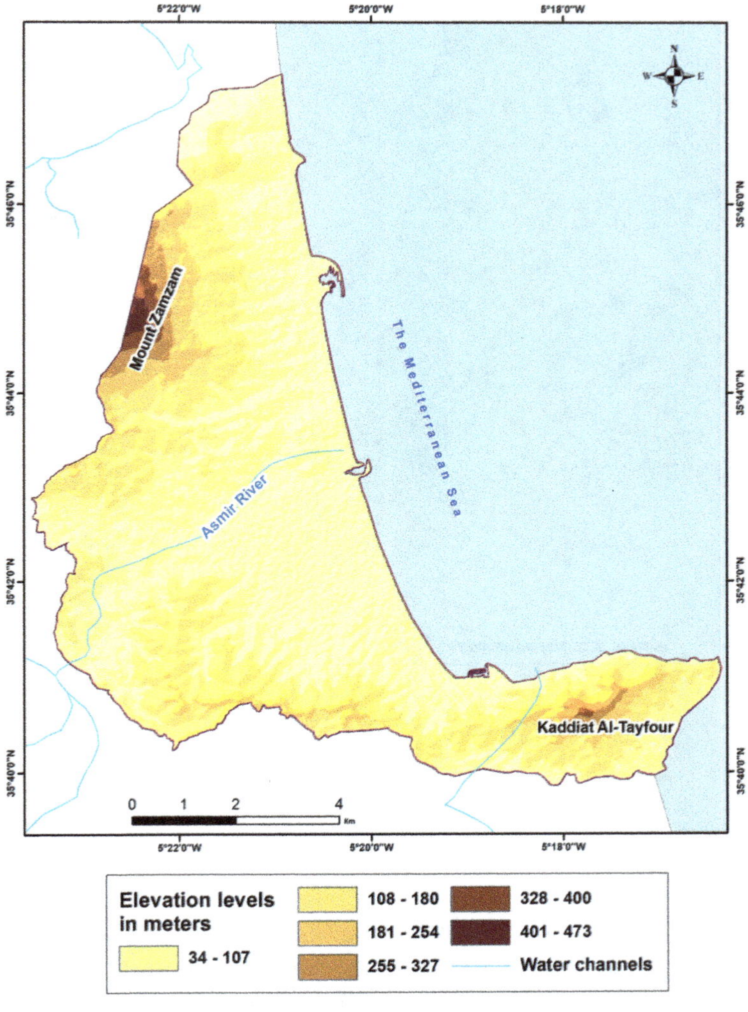

Figure 8. The spatial distribution of slopes in M'diq

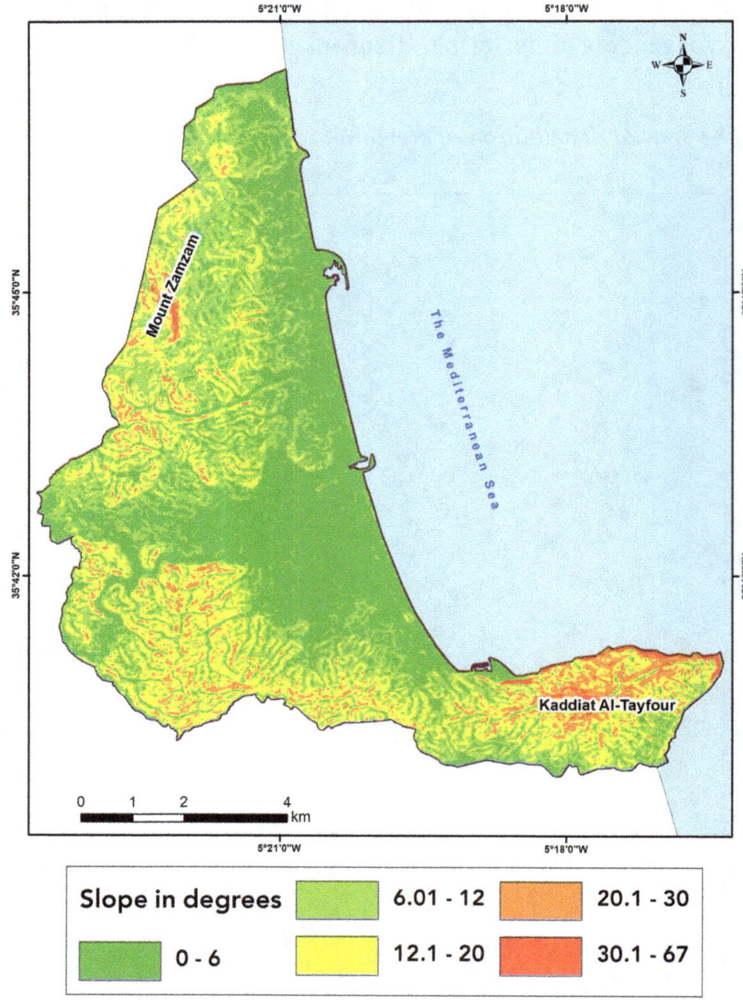

These topographic barriers stand as an obstacle to the development of the road network connecting different parts of the city, especially the development of suitable transverse roads, which are often replaced by stairs designated for pedestrians or steep, narrow roads (Figure 9).

Figure 9. Use of stairs for pedestrians in the Fom Al-Ouliq neighborhood and Houmat Al-Sharifa in M'diq due to the steepness of the slope

The recent urban planning initiatives and the development of tourism infrastructure, such as the construction of hotels and tourist complexes, have contributed to the city's growth and expansion (Jiachen et al., 2021). However, the rapid urbanization and the consumption of previously vacant areas for tourism and real estate development projects have led to increased traffic congestion, particularly during peak tourist seasons. The challenge of managing traffic flow and providing adequate transportation infrastructure is further complicated by the city's topography and the intense urban activity along the coast (Pujol, 2019).

In summary, the geographical location and topography of M'diq play a significant role in shaping its urban transport issues. The municipality's position as a key transit point and tourist destination, combined with its challenging terrain, creates a complex environment for traffic management and urban planning. Understanding these factors is crucial for developing effective strategies to address the traffic congestion and improve the overall urban mobility in M'diq.

4.3 City Morphology and the Road Network Dominated by Randomness

Roads are considered the main engine of the city, being the fundamental link between various services within the city, and they play a crucial role in defining the quality of urban planning according to the model or form of the prevailing roads in the city. Therefore, we find some designations such as cities with circular, arterial, or radial planning (T. Wang et al., 2022).

The urban morphology of the M'diq municipality is the result of a historical process the region has experienced. It was characterized by the absence of monitoring and a lack of forward-looking vision from both the residents and the parties concerned with planning and monitoring urban development in the area. This has resulted in a city with fragile random urbanization, consisting of neighborhoods with geometric shapes lacking a sense of harmony. In addition to the concentration of various services and administrative facilities in the city center, there is also a dominance of a random road network that does not meet the needs of the residents and hinders the work of administrative services, especially security, civil protection, and postal services (Zang et al., 2023).

The inability of street networks in many cities to meet their traffic requirements is due to the limited capacity, the abundance of undesirable intersections, the lack of parking spaces, bridges, tunnels, and other factors that lead to traffic congestion and accidents. This has demonstrated the need for a hierarchical gradation of city streets and their classification according to their functional importance. This is to organize the relationship and achieve a balance between different land uses and the traffic movement on the streets serving these uses (Mohammed Almatar, 2023).

There are several classifications for the city's road network. The more efficient the activities and the various uses of urban land, the better and more efficient the transport networks will be, achieving ease of movement and speed of transport. This has prompted planners to establish classifications for the city's street network to organize the relationship between transport and different land uses. Therefore, the need arose to achieve a hierarchical gradation of the streets and classify them according to the criteria of rank and capacity (Z. Chen et al., 2020).

The length of the road network covering the urban area of M'diq is about 188.38 km, distributed unevenly and in different geometric shapes. Functionally, it is classified according to the services it provides, its strategic location, in addition to its capacity, size, and degree of meeting services and connecting different aspects of the city into streets, alleys, passages, and paths. (Table 2)

Table 2. Distribution of Road Types in M'diq's Urban Area

Road Classification	Count
Street	184
Alley	482
Passage	1021
Dead-end Passage	202

continued on following page

Table 2. Continued

Road Classification	Count
Path	39
Unclassified	405

Based on the data presented in the table (Table 2) and (Figure 10), the road network in M'diq is predominantly composed of passages, followed by alleys, dead-end passages, and streets. The high proportion of passages and alleys highlights the fragility of the road network in this area, where there is a notable deficiency in main roads that can adequately meet the needs of the population. This situation contributes to several spatial issues, including the crisis of spatial congestion. The fragility of the road network in M'diq is primarily due to the challenging topography of the area and the dominance of random construction across various neighborhoods of the city.

Figure 10. Distribution of Road Types in M'diq's Urban Area

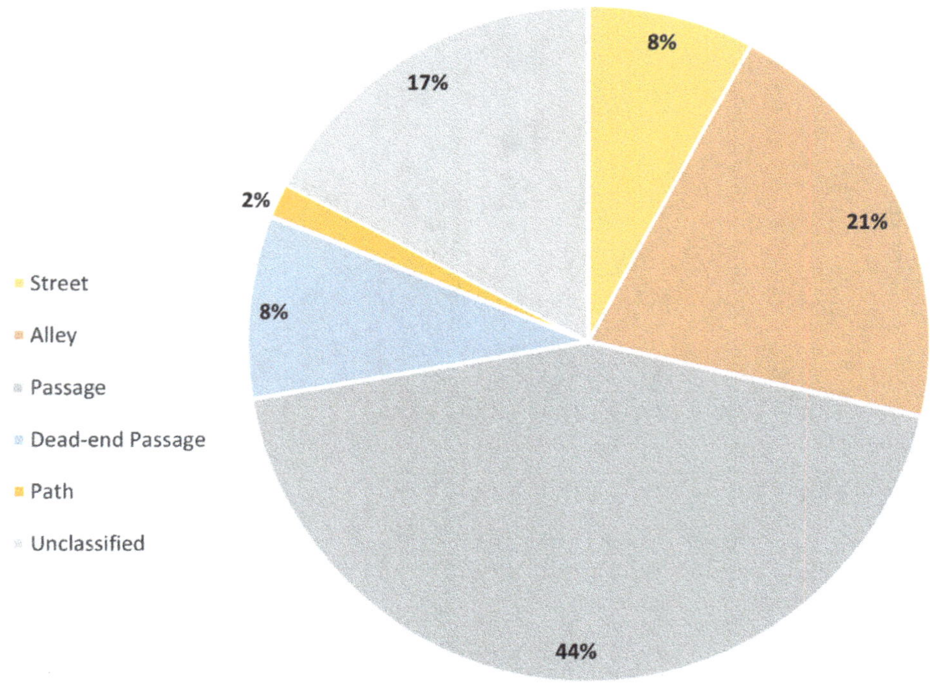

This analysis underscores the urgent need for a comprehensive urban planning strategy that addresses these infrastructural deficiencies, prioritizes the development of main roads, and mitigates the spatial congestion challenges currently faced by M'diq.

4.4 Degree of Road Network Connectivity: Abundance of Roads and Weak Interconnections

As illustrated in (Figure 11), the spatial distribution of nodes and links in M'diq's road network reveals the structural complexity that influences the connectivity indices discussed above. The map visually demonstrates the concentration of nodes in certain areas and the relative sparsity in others, which directly impacts the Beta, Gamma, and Alpha indices.

Figure 11. Spatial Distribution of Nodes and Links in M'diq's Road Network.

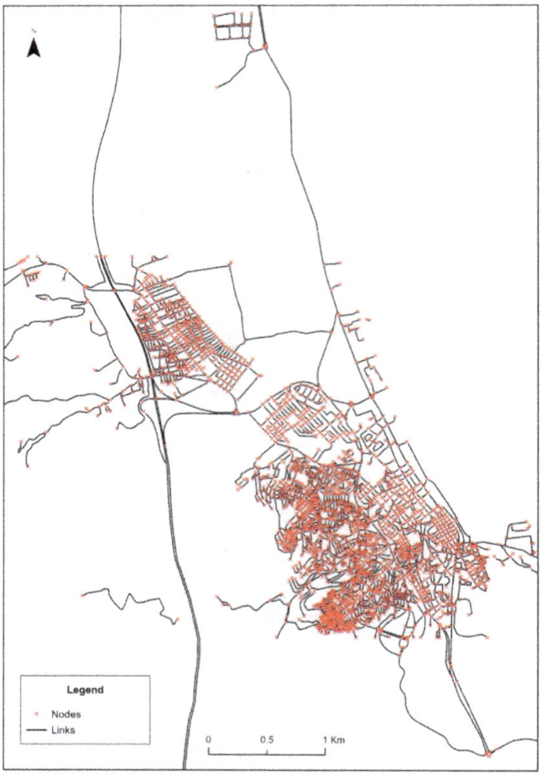

The M'diq road network comprises 3906 links and 2650 nodes, which provide the basis for calculating various connectivity indices. These indices offer insights into the overall structure and connectivity of the network, assessing its ability to support urban mobility and development.

While the current signage helps regulate traffic and ensure safety, the data suggests that there may be a need for additional signs or modifications to existing ones in response to evolving traffic patterns. For instance, expanding parking facilities and enhancing traffic control at key intersections could further alleviate congestion and improve the overall efficiency of the city's transportation network.

4.4.1 Beta Index (β):

The Beta Index, calculated using the 3906 links and 2650 nodes, is 1.47. This value indicates a more than developed road network. However, given the complexity of M'diq's road network in a narrow urban area, this index might not fully capture the network's connectivity.

4.4.2 Gamma Index (γ):

The Gamma Index was found to be 0.49, indicating that the road network has weak connectivity due to the lack of line intersections, which reduces the number of links in the network.

4.4.3 Alpha Index (α):

The Alpha Index was calculated to be 0.24, indicating that the degree of connectivity of the road network in M'diq is weak. This is attributed to the weak intersection of lines and the abundance of dead-end paths that cover the city's road network.

These findings underscore the need for a strategic approach to urban transport mobility planning, focusing on improving mobility, accessibility, and road network functionality while reducing congestion, accidents, and environmental pollution.

4.5 Traffic Sign Distribution and Its Impact on Traffic Flow

The map (Figure 12) illustrates the distribution of various traffic signs across the city of M'diq as of 2020. Such traffic signs equal a big portion of what regulates and guides in the flow of traffic for the city, which often finds itself highly congested during tourist peak seasons. From an analysis of these signs, several important insights can be drawn.

Figure 12. Distribution of Traffic Signs and Regulatory Measures in M'diq (2020)

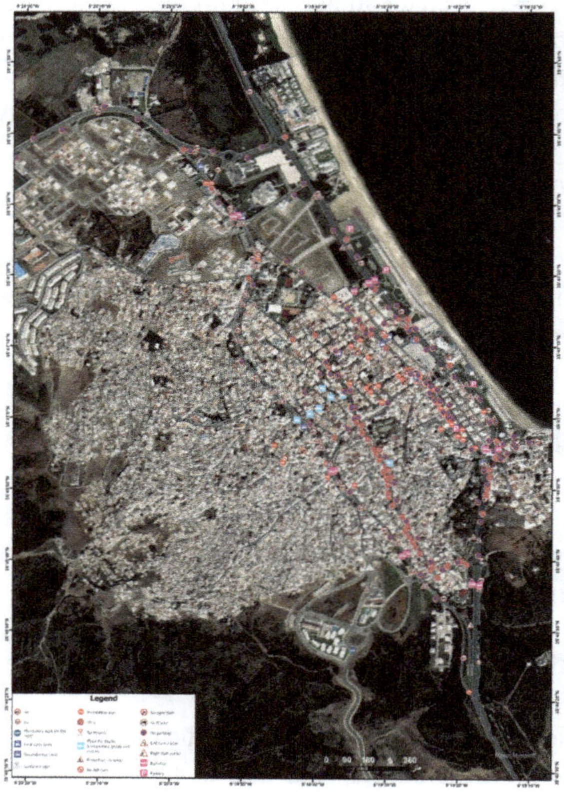

Overabundance of Traffic Signs in the City Center:

The city center especially along the main commercial axes like Lalla Nezha Street and Nasr Street reveals a greater number of informatory traffic signs. There are regulatory, parking restriction, and speed limit information, as well as directional directions. As a matter of fact, the density of setup corresponds to areas with maximum congestion of traffic in the city, justifying many more rigid controls over the movement in those areas.

Parking and No-Parking Areas:

The map shows various indicated parking areas and no-parking zones, in particular, near public service buildings and commercial areas. In this case, such strategic placement is targeted at controlling the high demand for parking within the city center, especially during peak hours. Nevertheless, there are limited parking spaces indicated on the map, which implies that this can be an area for improvement with regard to urban planning in a bid to reduce congestion.

Speed Limits and Traffic Flow:

The city is well-marked with speed limits, with lesser extents up to 60-40 km/h in the major towns. These are meant to keep more pedestrian and motorist safety in mind, and the fulfillment of these therefore becomes very critical in more high pedestrian areas.

Critical Intersections and Choke Points:

The map visualizes principal junctions and choke areas, specifically along the city's principal arterial. These are choke areas of traffic where it moves really slow due to many fusions of various roads or crosswalks. It would be in these areas that directional and regulatory signs are really important to minimize traffic conflict and manage the movement of vehicles with as much normal flow as possible.

Traffic Control and Enforcement:

Strategic distribution of the traffic control signs at junctions, such as stop and yield signs, are visible in sufficient numbers at all sensitive points. In essence, these placement signs determine whether or not a traffic backlog will be accrued, and particularly in the city's road arteries, they help in allowing movement in a civilized manner. Nevertheless, these effectiveness levels will be influenced by consistent enforcement, particularly in peak tourist seasons.

Pedestrian Safety and Signage:

Both mainly occur in areas that are considered to be very pedestrian-dense, majorly around schools, markets, and public transport terminals. The human safety signs of pedestrian crossings help reduce the very many pedestrian accidents in very crowded urban centers.

4.6 The Increase in the Number of Summer Visitors as a Primary Factor in Traffic Congestion

The significant impact of vacationers on the performance of the road network is manifested through the pressure they exert on the main commercial axes, especially Lalla Nezha axis (Figure 13). This latter separates the city from the main tourist area represented by Corniche No. 1, and Corniches No. 2 and 3, overlooking the sea. It also serves as a public space for celebrating various national occasions and holidays. This leads to a doubling of the movement of vacationers throughout the day until late at night between the city center and the tourist area, contributing to the pressure on pedestrian pathways that are strongly distributed along this axis and alongside its intersection points with the rest of the road network. Consequently, the result obtained is the obstruction of traffic movement and its stoppage for several minutes. This is also the case for other commercial axes, especially the Nasr Street axis, which separates the most densely populated neighborhoods from the city center.

Figure 13. Traffic Congestion on Lalla Nezha Axis

To better understand the extent of this congestion, (Figure 14) illustrates the traffic situation in M'diq for July 2023 based on data from ESRI. The map shows the varying levels of traffic flow on different streets, with Lalla Nezha Street and surrounding areas experiencing the most severe congestion, marked in red.

In addition to the general traffic conditions, the distribution of traffic volume by vehicle type across major streets is critical to understanding the dynamics of congestion. (Figure 15) shows the daily average number of vehicles passing through the main road. This map highlights the heavy burden on Lalla Nezha Street, where private cars dominate the traffic, exacerbating congestion during peak periods.

Figure 14. Traffic Situation in M'diq for July 2023 Based on data from ESRI

Figure 15. The daily average number of vehicles passing through the main road

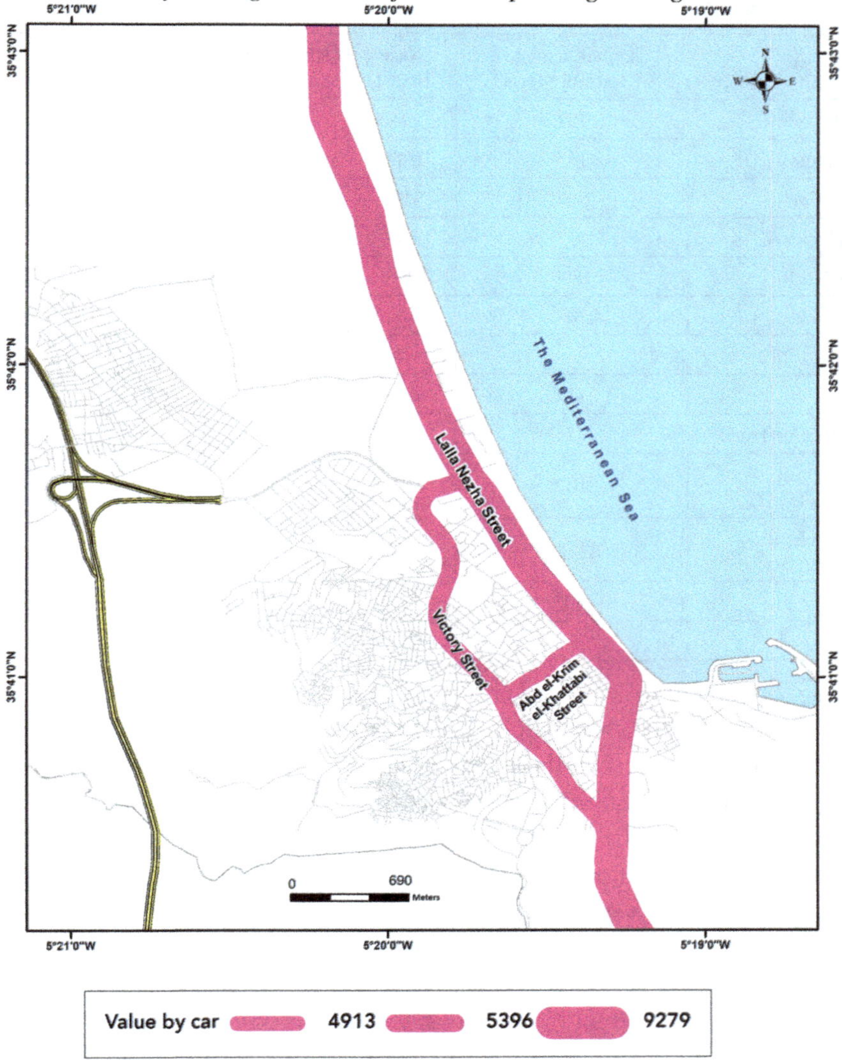

To better understand the traffic congestion, (Table 3) below presents the traffic volume for different types of vehicles on major streets from 9:00 AM to 6:00 PM. This data highlights the distribution of various vehicle types across the key roadways in M'diq, emphasizing the significant pressure on Lalla Nezha Street, Nasr Street, and Abdelkarim Al Khattabi Street.

Table 3. Traffic Volume by Vehicle Type on Major Streets in M'diq (9:00 AM - 6:00 PM)

Vehicle Type	Abdelkarim Al Khattabi Street	Nasr Street	Lalla Nezha Street	Total	Average Traffic Volume
Bicycle	101	94	87	282	94
Motorcycle	145	282	250	677	255
Private Car	1480	3176	4366	9022	3007
Small Truck	937	945	544	2426	809
Large Truck	1899	216	3062	5177	1726
Bus	0	4	91	95	32
Minibus	3	5	77	85	28
Truck	112	210	263	585	195
Large Trailer Truck	1	59	115	175	58
Three-Wheel Motorcycle	107	97	114	318	106
Other (School Transport, etc.)	128	308	310	746	249
Total	4913	5396	9279	19588	6559
Percent	25%	28%	47%	100%	

As illustrated in the chart below (Figure 16) the traffic pressure in M'diq is predominantly concentrated on three main streets. Lalla Nezha Street alone captures 47% of the traffic, while Nasr Street accounts for 28%, and Abdelkarim Al Khattabi Street captures 25%. These streets serve as critical arteries within the city's road network, bearing the majority of the traffic load, while other roads contribute comparatively less to the overall traffic volume.

Figure 16. Distribution of Traffic Pressure on Major Streets in M'diq

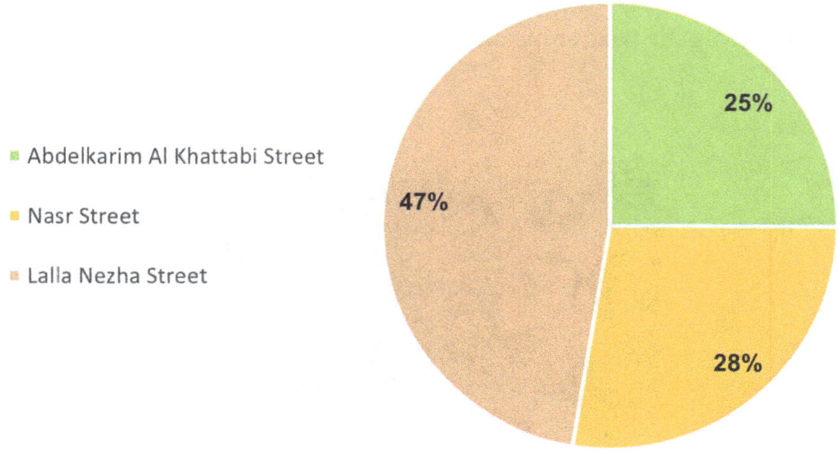

The critical role of these arteries, particularly during the summer influx of tourists, is further underscored by the comparative analysis of travel times conducted using ArcGIS Network Analysis. This analysis compares travel times between two locations on January 19, 2024, and August 19, 2024, at 2:58 PM.

- January 19, 2024: The travel time between Location 1 and Location 2 was 5 minutes, reflecting the relatively free-flowing traffic conditions typical of the off-peak season (Figure 17).

Figure 17. Route and Travel Time Analysis for January 19, 2024

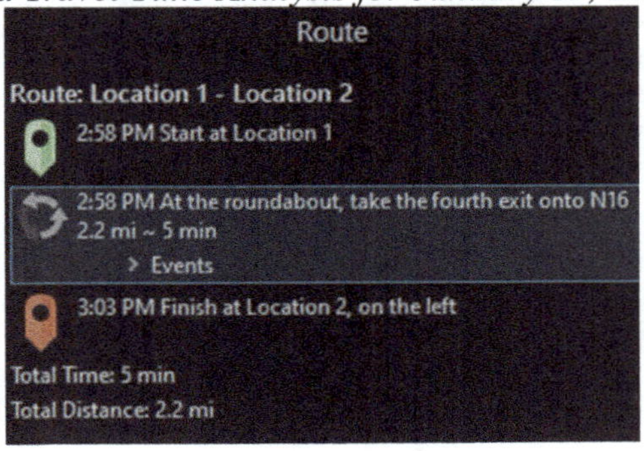

- August 19, 2024: In stark contrast, the same route on August 19, 2024, took 21 minutes to traverse, indicating a substantial increase in congestion and delay during the peak summer period (Figure 18).

Figure 18. Route and Travel Time Analysis for August 19, 2024

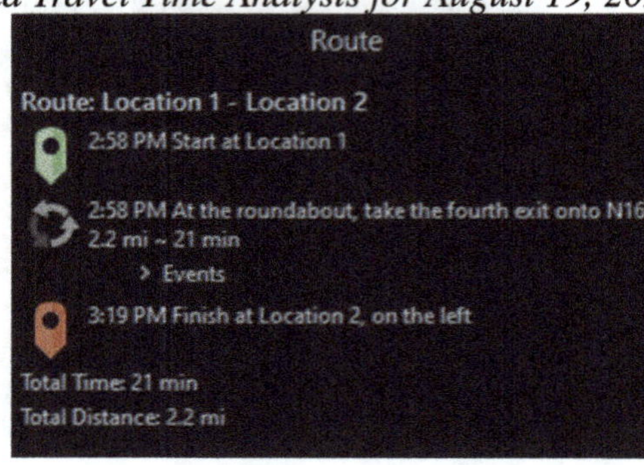

Figure 19. Lalla Nezha Route Analyzed for Travel Time Comparison Between January 19, 2024, and August 19, 2024

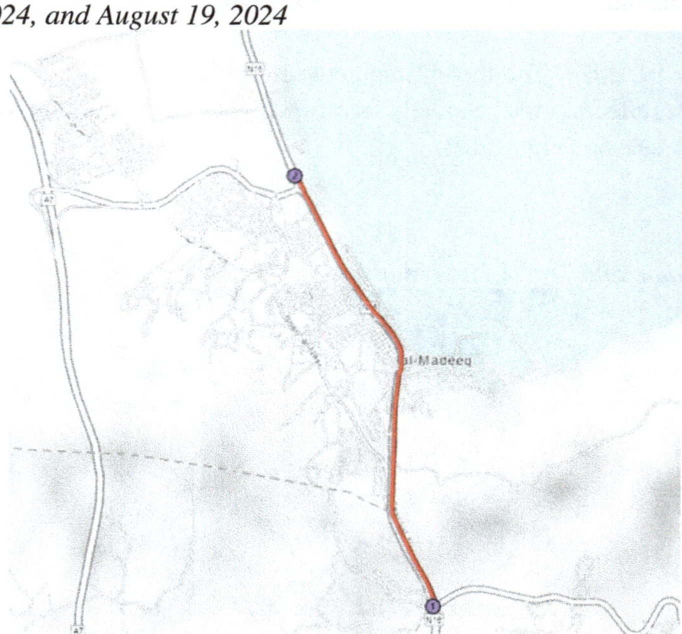

The difference in travel times between these two dates is further illustrated in the bar chart below (Figure 20). The chart vividly captures the dramatic increase in travel time from 5 minutes in January to 21 minutes in August, underscoring the seasonal variation in traffic conditions.

Figure 20. Comparative Travel Times on Lalla Nezha Route (January 19, 2024, vs. August 19, 2024)

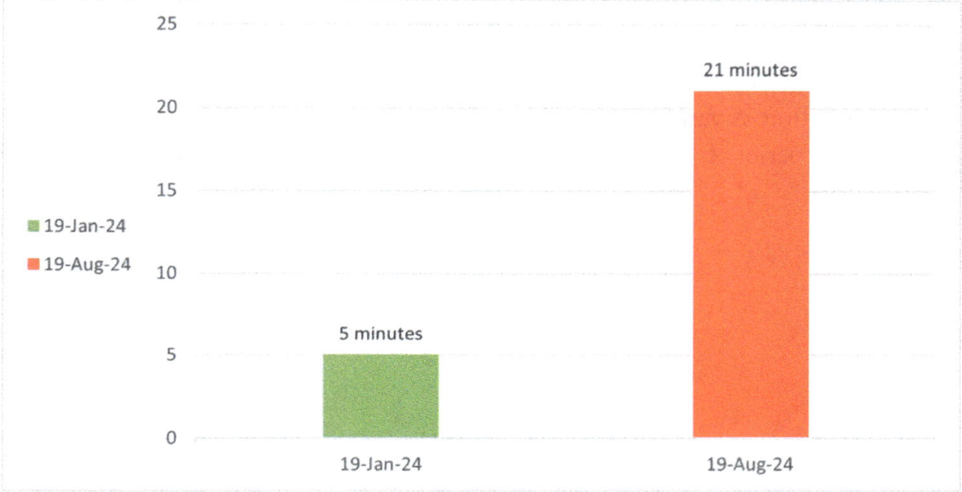

This visual comparison highlights the profound impact of tourist influx on the efficiency of the road network during peak seasons. The increase in travel time suggests that the Lalla Nezha route, while sufficient during off-peak periods, becomes highly congested and inefficient during peak periods. The chart reinforces the need for targeted infrastructure improvements or alternative traffic management strategies to accommodate the seasonal increase in traffic volume.

This analysis of traffic distribution underscores the critical need for targeted interventions to manage congestion on M'diq's busiest streets. However, traffic pressure is only one aspect of the broader issue of spatial congestion, which is also significantly influenced by the scarcity and random distribution of parking spaces within the city.

4.7 Scarcity and Random Distribution of Parking Spaces as a Primary Factor in Spatial Congestion

Estimates indicate that about 30% of traffic congestion within cities is caused by people searching for parking spaces, translating into countless amounts of wasted fuel and carbon emissions (Ben-Dor et al., 2022). In M'diq, the number of official parking spaces does not exceed 11, covering an area of approximately 3.81 hectares. This is a very small area that barely meets the needs of the population during normal times of the year, let alone the summer period when the number of cars in M'diq quadruples. The scarcity and random distribution of parking spaces, coupled with the concentration of various transport stations collectively contribute to the spatial congestion experienced in M'diq, especially during the peak summer months.

Figure 21. The Impact of the Concentration of Parking Spaces and Transport Lines on Traffic Congestion

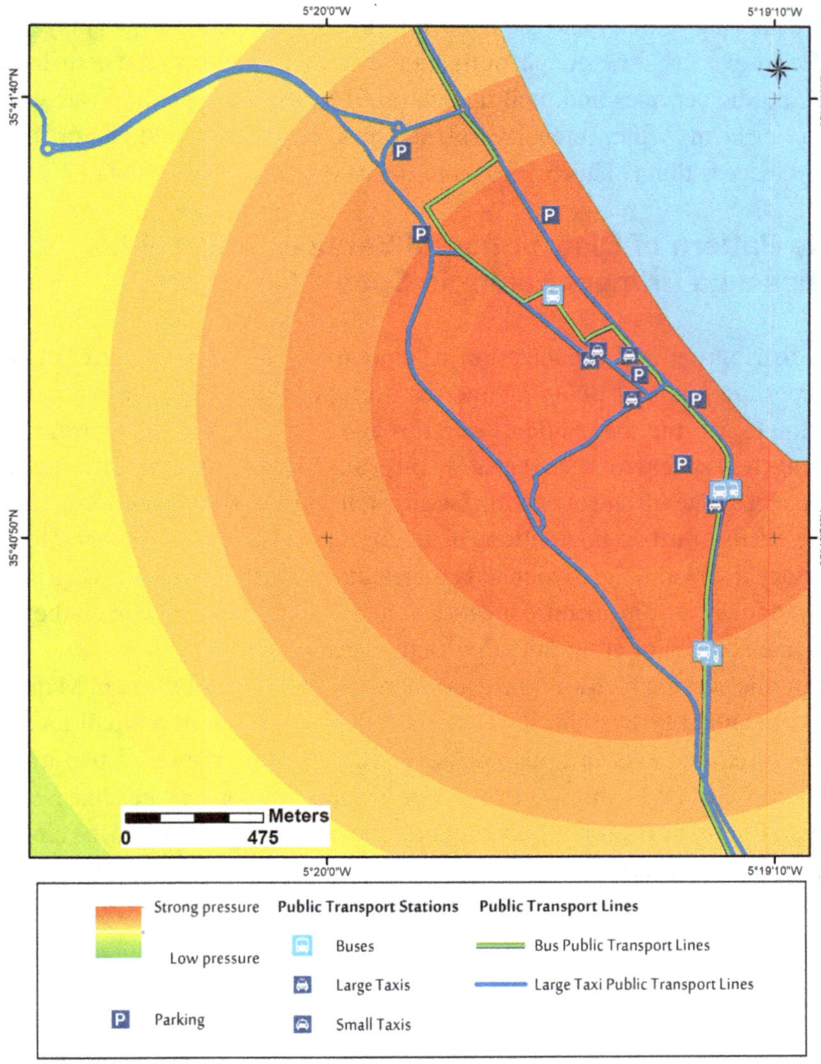

As shown in (Figure 21), the areas of high pressure (indicated in red) where parking scarcity is most acute are particularly concentrated around the main transport stations. The map highlights how the few available parking spaces (marked with 'P') are unevenly distributed, leading to significant pressure in certain zones. This uneven distribution is compounded by the locations of public transport stations for buses, large taxis, and small taxis, which are concentrated in areas that already experience strong congestion. The overlay of transport lines on the map further emphasizes how these factors converge to exacerbate congestion in the city center and along key routes.

While parking scarcity and the concentration of transport stations play a significant role in exacerbating congestion, another critical factor is the distribution of essential public services and facilities within the city. The spatial arrangement of these services can further intensify traffic issues, as residents and visitors navigate the city to access them. This brings us to the next key consideration.

4.8 The Pattern of Distribution of Services and Public Facilities as a Primary Factor in Spatial Congestion

The distribution and organization of essential public services are among the most important goals of planning and territorial development, due to their direct connection to serving the residents of cities and villages, meeting their needs, and facilitating their access to these services(Liu et al., 2020). Properly distributed public services, such as health and education facilities and administrative buildings, not only serve the needs of urban populations in an efficient manner but also tend to reduce unnecessary travel and the associated congestion. However, if poorly distributed or too over-concentrated in specific locations, such services may turn in on themselves and become sources of congestion, particularly at peak hours.

The public service facilities have a remarkable poor distribution in M'diq. Even though their quantity is small, they are quite concentrated in a small area of the city. This generates a lot of bottlenecks, especially along the city's two important axes, where the traffic is already dense. For instance, major services like healthcare facilities, administrative centers, and schools are concentrated around the city center, making the place choked with heavy human and vehicular traffic all day (Figure 22). This is further aggravated by the absence of an efficient public transport system and the relatively inadequate public parking spaces, leaving inhabitants and visitors with no option but to rely seriously on private vehicles.

Figure 22. Geographic Distribution of Public Services in M'diq.

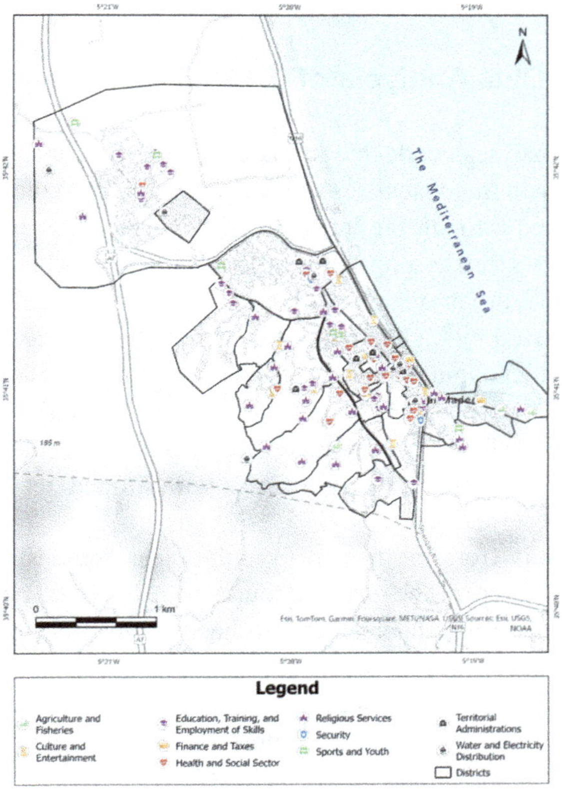

This means that the geographical clustering of such services enhances the geographical already-existing congestion between the main through-fares. Inefficiency in the dispersion of services leads to multiple trips and, subsequently, extended travel time, which enhances spatial congestion.

To face these challenges, therefore, urban planners in M'diq should see decentralization of not only the different dependent public provision but planning for public transport and more parking placed around. Public services would be redistributed within the city, and accessibility through various transport modes would be improved to relieve the global pressures on the road network a much less unbalanced and inefficient city.

From the above, it then can be deduced that the congestion of movements that are exhibited from the central axes of M'diq Road is a problem that is faced on various fronts: these are majorly the reliance on private vehicles, poor presence of public means, and the lack of parking spaces, which are also poorly distributed.

Addressing such interlinked factors will be vital for improvements in mobility and quality of life in the city.

4.9 Integrated Data Analysis of Urban Traffic Patterns in M'diq

Based on the analyses conducted in the preceding sections, we have produced a composite heat map that visualizes traffic congestion in M'diq (Figure 23). This heat map, developed through the integration of diverse data sets, reveals critical insights into the city's traffic congestion patterns. The results highlight pronounced congestion in specific urban areas, notably along Lalla Nezha and Nasr Streets. These areas are characterized by high commercial activity and a dense concentration of public services, which significantly contribute to traffic volume (Y. Wang et al., 2022). Furthermore, the map underscores the impact of M'diq's unique topography on traffic flow. Steep slopes and irregular road layouts are identified as key factors exacerbating congestion, particularly in areas with limited accessibility (Zang et al., 2023).

A notable finding from the heat map is the clear demonstration of seasonal variation in traffic patterns. The influx of tourists during the summer months markedly intensifies congestion, underscoring the need for adaptive traffic management strategies that can accommodate these seasonal fluctuations (Chan & Chye, 2023). This seasonal aspect of traffic congestion presents a unique challenge for urban planners in M'diq, requiring a dynamic approach to traffic management that goes beyond conventional static models (Jin et al., 2022).

The discussion of these results highlights the interplay between urban infrastructure, service distribution, and human behavior in shaping traffic dynamics. The concentration of public services in specific areas, coupled with inadequate public transport and parking infrastructure, leads to increased traffic pressure on main roads. This situation is further complicated by the city's challenging topography, which limits the feasibility of certain infrastructural expansions or modifications.

These insights have significant implications for urban planning and traffic management in M'diq. They suggest a need for a strategic review of urban planning practices, particularly in terms of service distribution and road network design. The findings advocate for the decentralization of public services to distribute traffic more evenly across the city and the enhancement of public transportation options to reduce the reliance on private vehicles. Additionally, the unique seasonal traffic patterns observed necessitate the development of flexible traffic management systems capable of adapting to the city's changing needs (Jin et al., 2022).

In conclusion, the composite heat map serves as a valuable tool in understanding the complexities of traffic congestion in M'diq. The insights gained from this analysis provide a foundation for informed decision-making in urban planning and traffic management, highlighting the need for innovative, adaptive strategies to improve urban mobility in the city.

Figure 23. Composite Heat Map of Traffic Congestion in M'diq

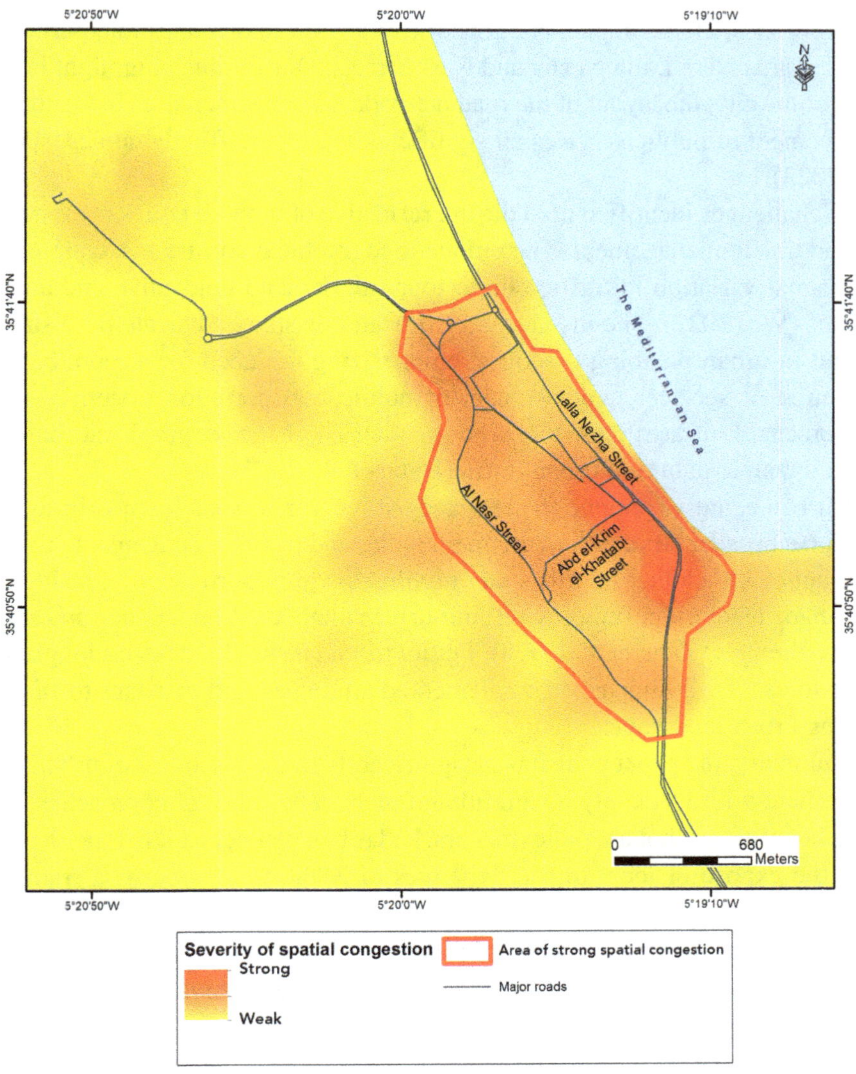

CONCLUSION

This research provides a detailed GIS analysis of urban traffic congestion in M'diq, using a comprehensive approach that blends geographical, topographical, infrastructural, and data-driven insights. We created a GIS-enhanced heat map that effectively shows the complex patterns of traffic flow and congestion, particularly in critical areas like Lalla Nezha and Nasr Streets. Our findings highlight how the design of the city, the layout of the road network, seasonal increases in visitors, and the placement of public services all significantly affect traffic dynamics. (Prajwal et al., 2023).

The challenges identified in M'diq are reflective of broader issues in urban planning and traffic management. The unique topographical constraints, coupled with the seasonal variation in traffic due to tourism, demand innovative and adaptive solutions (Wu, 2022). The insights gained from this study advocate for a strategic overhaul in urban planning practices, emphasizing the need for a more balanced distribution of services, enhancement of public transportation systems, and the development of infrastructure that accommodates both the geographical challenges and the dynamic nature of urban traffic(Dalal et al., 2023).

As urban centers continue to grow and evolve, the lessons learned from M'diq can inform broader strategies in urban planning and traffic management. The integrated approach adopted in this study highlights the importance of considering a wide range of factors and employing data-driven methods to understand and address urban challenges (Zang et al., 2023). Future research should focus on longitudinal studies to assess the impact of implemented strategies and to adapt to the ever-changing urban landscapes.

In summary, the paper contributes significantly to the discourse of urban traffic management with its deeply insightful and practically relevant approaches in this regard, benefiting scholars, policymakers, and urban planners alike. It neither stops at just the exposé of local traffic dynamics of M'diq nor proposes a framework adaptable to other cities to redress related issues. This research furnishes the ground for more data-driven and effective solutions to urban congestion with a GIS-based analysis that integrates geographic, infrastructural, and seasonal factors in the city, hence becoming an important reference in dealing with complexities relevant to traffic management in growing cities.

REFERENCES

Ali, B. (2015). *Territorial Policies of the Spanish Colonial Authorities in Northern Morocco 1912-1956*. Al-Hidaya Press.

Banos, A., & Thévenin, T. (Eds.). (2011). *Geographical information and urban transport systems*. ISTE.

Ben-Dor, G., Ogulenko, A., Klein, I., & Benenson, I. (2022). Modeling the Effect of Congestion Charge and Parking Pricing on Urban Traffic: Example of Jerusalem. *AGILE: GIScience Series*, 3, 1–4. DOI: 10.5194/agile-giss-3-26-2022

Benchelha, N., Bezza, M., Belbounaguia, N., Benchelha, S., & Benchelha, M. (2022). Modeling Dynamic Urban Growth Using Cellular Automata and Geospatial Technique: Case of Casablanca in Morocco. *International Journal of Geoinformatics*. Advance online publication. DOI: 10.52939/ijg.v18i5.2369

Bennani, M., Jawab, F., Amegouz, D., Hani, Y., & Elmhamedi, A. (2020). The impact of congestion on the cost of transport in urban areas: Application for the location of an urban distribution center. *2020 IEEE 13th International Colloquium of Logistics and Supply Chain Management (LOGISTIQUA)*, 1–6. DOI: 10.1109/LOGISTIQUA49782.2020.9353949

Chan, J. J., & Chye, S. W. (2023). Impact of Smart City Initiatives on Urban Planning Strategies in Singapore: An In-Depth Analysis of Technology-Driven Solutions and Their Influence on Sustainable Development and Quality of Life. *Journal of Strategy and Management*, 7(7), 11–21. DOI: 10.53819/81018102t4200

Chen, X., Zhang, K., Chuai, G., Gao, W., Si, Z., Hou, Y., & Liu, X. (2023). Urban Area Characterization and Structure Analysis: A Combined Data-Driven Approach by Remote Sensing Information and Spatial–Temporal Wireless Data. *Remote Sensing (Basel)*, 15(4), 1041. DOI: 10.3390/rs15041041

Chen, Z., Jiang, Y., & Sun, D. (2020). Discrimination and Prediction of Traffic Congestion States of Urban Road Network Based on Spatio-Temporal Correlation. *IEEE Access : Practical Innovations, Open Solutions*, 8, 3330–3342. DOI: 10.1109/ACCESS.2019.2959125

Dalal, Y. M., Naveen, S., & Um, A. K. (2023). Unveiling Traffic Patterns for Effective Traffic Management System. *2023 International Conference on Network, Multimedia and Information Technology (NMITCON)*, 1–6. DOI: 10.1109/NMITCON58196.2023.10276327

Damoah-Afari, P., Osei, J. D., Yevugah, L. L., Boakye, L., & Amaning, K. (2023). Integrating GIS and Satellite Remote Sensing to Assess the Geospatial Dynamics of Urban Footprints in the Dormaa Central Municipality in Ghana. *Journal of Energy Engineering and Thermodynamics(JEET) ISSN 2815-0945, 3*(02), Article 02. DOI: 10.55529/jeet.32.22.39

Downs, A. (2004). *Still stuck in traffic: Coping with peak-hour traffic congestion.* Brookings Institution Press.

Droj, G., Droj, L., Badea, A.-C., & Dragomir, P. I. (2023). GIS-Based Urban Traffic Assessment in a Historical European City under the Influence of Infrastructure Works and COVID-19. *Applied Sciences (Basel, Switzerland)*, 13(3), 1355. DOI: 10.3390/app13031355

El Cheikhi, N. E., & El Abdlaoui, M. (1997). Social, Economic, and Spatial Transformations in Urban Centers (Martil, M'diq, and Fnideq) on the Northern Tetouan Coast and Adjacent Areas. *Faculty of Arts and Human Sciences Journal, 8.*

Fan, P., Chen, J., Fung, C., Naing, Z., Ouyang, Z., Nyunt, K. M., Myint, Z. N., Qi, J., Messina, J. P., Myint, S. W., & Peter, B. G. (2022). Urbanization, economic development, and environmental changes in transitional economies in the global south: A case of Yangon. *Ecological Processes*, 11(1), 65. DOI: 10.1186/s13717-022-00409-6 PMID: 36397837

Jiachen, L., Mengqing, M., Xin, X., & Wenhui, R. (2021). The Spatial Effect of Shared Mobility on Urban Traffic Congestion: Evidence from Chinese Cities. *Sustainability (Basel)*, 13(24), 14065. DOI: 10.3390/su132414065

Jin, J., Rong, D., Pang, Y., Ye, P., Ji, Q., Wang, X., Wang, G., & Wang, F.-Y. (2022). An Agent-Based Traffic Recommendation System: Revisiting and Revising Urban Traffic Management Strategies. *IEEE Transactions on Systems, Man, and Cybernetics. Systems*, 52(11), 7289–7301. DOI: 10.1109/TSMC.2022.3177027

Karimi, H., Ghadirifaraz, B., Shetab Boushehri, S. N., Hosseininasab, S.-M., & Rafiei, N. (2022). Reducing traffic congestion and increasing sustainability in special urban areas through one-way traffic reconfiguration. *Transportation*, 49(1), 37–60. DOI: 10.1007/s11116-020-10162-4

Liu, Q., Liu, Z., Xu, Y., Xiong, W., Yang, J., & Wang, Q. (2020). Toward Identifying the Urban Community Structure from Population Flow and Public Services Distribution. *2020 IEEE International Conference on Big Data (Big Data)*, 3133–3142. DOI: 10.1109/BigData50022.2020.9378225

Mohammed Almatar, K. (2023). Traffic congestion patterns in the urban road network: (Dammam metropolitan area). *Ain Shams Engineering Journal*, 14(3), 101886. DOI: 10.1016/j.asej.2022.101886

Prajwal, V., Abhay, R., Rohith Gowda, D., & Savitha, G. (2023). AI-Driven Urban Traffic Optimization to Assess Complex Traffic Patterns for Public Traffic Control and Mobility. *International Journal for Research in Applied Science and Engineering Technology*, 11(11), 794–798. DOI: 10.22214/ijraset.2023.56630

Pujol, B. (2019). *Machine learning for early detection of traffic congestion using public transport traffic data*. https://api.semanticscholar.org/CorpusID:199011854

Rahman, M. N. (2019). Urban Expansion Analysis and Land Use Changes in Rangpur City Corporation Area, Bangladesh, using Remote Sensing (RS) and Geographic Information System (GIS) Techniques. *Geosfera Indonesia*, 4(3), 217. DOI: 10.19184/geosi.v4i3.13921

Shafiei, S., Gu, Z., Grzybowska, H., & Cai, C. (2023). Impact of self-parking autonomous vehicles on urban traffic congestion. *Transportation*, 50(1), 183–203. DOI: 10.1007/s11116-021-10241-0

Tanvir, M. R., & Haque, A. (2023). Assessment of Development Plans for Small Urban Areas of Bangladesh: Remote Sensing and GIS-based Approach. *Applied Spatial Analysis and Policy*, 16(4), 1591–1625. DOI: 10.1007/s12061-023-09520-3

Vikram, L., & Bhardwaj, M. (2023). Assessment of Urban Wastelands using GIS and IoT as Tools for Spatial Data Analysis. *Current World Environment*, 18(2), 893–903. DOI: 10.12944/CWE.18.2.35

Wang, L., Xue, X., Zhao, Z., & Wang, Z. (2018). The Impacts of Transportation Infrastructure on Sustainable Development: Emerging Trends and Challenges. *International Journal of Environmental Research and Public Health*, 15(6), 1172. DOI: 10.3390/ijerph15061172 PMID: 29874785

Wang, T., Hussain, A., Sun, Q., Li, S. E., & Jiahua, C. (2022). The Prediction of Urban Road Traffic Congestion by Using a Deep Stacked Long Short-Term Memory Network. *IEEE Intelligent Transportation Systems Magazine*, 14(4), 102–120. DOI: 10.1109/MITS.2021.3049383

Wang, Y., Wu, B., & Li, L. (2022). *Urban Redevelopment and Traffic Congestion Management Strategies*. Springer Nature Singapore. DOI: 10.1007/978-981-19-1727-1

Wu, W. (2022). Research on Urban Congestion Control Strategy Based on Smart Traffic Management System. *Journal of Humanities and Social Sciences Studies*, 4(1), 89–96. DOI: 10.32996/jhsss.2022.4.1.9

Xie, Z., Lv, W., Huang, S., Lu, Z., Du, B., & Huang, R. (2020). Sequential Graph Neural Network for Urban Road Traffic Speed Prediction. *IEEE Access : Practical Innovations, Open Solutions*, 8, 63349–63358. DOI: 10.1109/ACCESS.2019.2915364

Zang, J., Jiao, P., Liu, S., Zhang, X., Song, G., & Yu, L. (2023). Identifying Traffic Congestion Patterns of Urban Road Network Based on Traffic Performance Index. *Sustainability (Basel)*, 15(2), 948. DOI: 10.3390/su15020948

Zhao, W., Zhang, S., Wang, B., & Zhou, B. (2023). Spatio-temporal causal graph attention network for traffic flow prediction in intelligent transportation systems. *PeerJ. Computer Science*, 9, e1484. DOI: 10.7717/peerj-cs.1484 PMID: 37547406

Zhou, L., Zhang, S., Yu, J., & Chen, X. (2020). Spatial–Temporal Deep Tensor Neural Networks for Large-Scale Urban Network Speed Prediction. *IEEE Transactions on Intelligent Transportation Systems*, 21(9), 3718–3729. DOI: 10.1109/TITS.2019.2932038

Chapter 6

The Role of Remote Sensing in Monitoring the Shortcomings of Urban Policy:
The Case of Kenitra City Morocco

Fahd Sabrou
https://orcid.org/0009-0005-4798-2595
Chouaib Doukkali University, Morocco

Mostafa Chbada
https://orcid.org/0009-0001-5342-294X
Faculté des Sciences Humaines et Sociales - Kénitra, Morocco

ABSTRACT

Throughout the history of Kenitra City, the urban policy concept has taken on two main dimensions. First, it is the mechanisms through which the former French colonial authorities deliberately perpetuated an ethnic and spatial divide between Europeans and Moroccan citizens, which was imposed by economic, social, and security motives. Second, it was a political choice that came after independence and remained confined between the sectoral approach to management and the urgent treatment of problems. These two dimensions have led to the fragmentation of the city and the absence of integration and cohesion between its various components. More than a century after the adoption of the urban policy in Morocco, the city of Kenitra is now at a unique turning point in its history. It is at a stage where it will soon decide, in the coming generations, whether the new development experiences will create a barrier and disintegration with the classical urban policy, or continue

DOI: 10.4018/979-8-3693-6452-9.ch006

down a path that will not guarantee decent living prospects. For this reason, based on remote sensing techniques, this work seeks to determine the logic governing urban decision-making and reveal the flaws in urban planning and the city policy followed in Kenitra. This will be achieved by monitoring and analyzing the exacerbations of many urban, social, economic, and environmental problems over recent decades.

INTRODUCTION

Urban policy is closely linked to the issues of development and urbanization. It encompasses "the processes aimed at reducing the divide of spatial disparities within cities, ensuring social justice in fragile and sensitive neighborhoods, and improving the living conditions for their residents." (Merlin, 2010) This concept emerged in 1913 to reflect intervention methods for "addressing" cities with imbalances or those facing specific developmental challenges, ultimately making cities more inhabitable. However, these interventions, regardless of their extent, can vary significantly depending on the surrounding political, economic, and social conditions. This means that the objectives on which the concept is based may not necessarily match what exists in real life.

Perhaps the experience of Kenitra city is the clearest example of this after the urban policy followed by the colonial authorities at the beginning of the last century adopted a logic that was governed by security and ethnic considerations, a logic that made Moroccan cities dual urban clusters that included two types of neighborhoods which remained divided geographically and socially, although they shared some functions. This resulted in the emergence of two cities within one city. The impact of this discrimination increased by adopting an urban policy after the independence, which did not constitute a rupture with colonialism for most cities, given that it did not prevent the exacerbation of the socio-spatial imbalances inherited from the colonial era.

Kenitra, a modern city in northwestern Morocco, developed from nothing to become one of Morocco's major cities within a century. Today, "it plays a regional role" (Nasraoui and Benzidiya 2024), attracting and radiating influence. The city's growth has been fueled by significant "migration from its regional surroundings and various parts of Morocco" (Aliaâ et al. 2017; Loulad et al. 2023), expanding rapidly but often without a comprehensive vision for its future. This has led to a pronounced fragmentation, both urban and social.

This study aims to contribute to a comprehensive approach by linking space with its social content, considering the influence of time and urban policies. It seeks to answer several questions related to the transformation process of Kenitra, the issue of city unity, and the centralization of urban areas. The study measures and analyzes the city's urban expansion paths and the factors driving this expansion in two phases: the colonial period, which defined the city's initial development, and the last twenty years, marked by the adoption of the only urban planning design for the city after the independence. This analysis utilizes Landsat satellite imagery and Geographic Information Systems (GIS), wich is "essentially a collection of computers and programs that combine to collect, record, store, retrieve, analyse, manipulate, and display spatial information on screen". (de Blij, 2012) The study addresses two main questions:

- What urban policy led to the current reality in Kenitra?
- How do stakeholders in city planning contribute to the shortcomings of urban policy?

To avoid neglecting the dynamics of space and society and their interactions, two hypotheses are proposed:

- Security and ethnic concerns during the colonial period imposed an urban policy that resulted in urban duality in Kenitra, preventing the application of the principle of city unity.
- The post-independence expansion of Kenitra's area occurred rapidly and often without a comprehensive vision for the city, leading to a pronounced fragmentation, both urban and social.

2. METHODOLOGY

The study employed a historical and analytical approach to assess the impact of urban policy on the spatial development of Kenitra City, Morocco. The methodology combined the analysis of historical urban policy decisions with the use of remote sensing techniques to monitor and evaluate urban expansion and its socio-spatial consequences. Remote sensing "provides data globally by satellites, and locally by airplanes or, more recently, smaller unmanned aerial vehicles (UAVs, commonly known as drones) and handheld devices" (Van der Linden, 2018), and it "underlies authoritative geospatial datasets, which are important for urban studies and planning authorities" (Wellmann, 2020). Furthermore, The notion of urban duality marked by "the juxtaposition of two strongly differentiated, is used for poor countries: colonial city/indigenous city, formal city/informal city" (Rieucau, 2004), and that's why urban fragmentation can be defined as "a coup [party or absolute] at the parties of the city, on social, economic and political plans." (Gervais-Lambony, 2001).

2.1 Data Collection

Satellite imagery from the Landsat program was utilized for the years 2000, 2010, and 2022. To obtain images of acceptable quality, satellite images from all months of these years were downloaded. Despite some variability in results due to empty fields or vegetation cover, the images were processed to ensure they were suitable for the study. The images were obtained from the United States Geological Survey (USGS) Earth Explorer platform and included data captured by the Thematic Mapper (TM) and the Operational Land Imager and Thermal Infrared Sensor (OLI_TIRS). The specific characteristics of the images used are summarized in Table 1.

Table 1. Details of Satellite Images Used in the Study

Year	Satellite/Sensor	Bands	Spatial Resolution	Acquisition Date	Processing Date	Geodetic Reference System
2000	Landsat 5 TM	1-7	30 m (Bands 1-5, 7), 60 m (Band 6)	2000/07/28	2022/09/25	UTM Zone 30N, WGS 1984
2010	Landsat 5 TM	1-7	30 m (Bands 1-7)	2010/07/16	2022/02/25	UTM Zone 29N, WGS 1984
2022	Landsat 8 OLI_ TIRS	1-7	30 m (Bands 1-7)	2022/08/11	2022/09/22	UTM Zone 30N, WGS 1984

2.2 Data Processing

The satellite images were processed using ArcGIS, one of the most important GIS programs. The processing steps included:

- Image Preprocessing: Electromagnetic waves from various bands were combined into a single image for each year. This included correcting the spatial resolution of the 2000 image using the Panchromatic band to achieve a 15-meter resolution image of Kenitra City.
- Supervised Classification: The images were processed using supervised classification "to extract and categorize land uses" (Karmaoui et al. 2023; Moumane et al. 2022), including built-up areas, empty spaces, vegetation cover, and water networks. This classification allowed for the detailed analysis of land use changes over time.
- NDVI Calculation: The Normalized Difference Vegetation Index (NDVI) was "calculated to evaluate changes in vegetation cover throughout the study period. The NDVI values ranged from -1 (indicating no vegetation) to 1 (indicating dense vegetation)" (Moumane et al. 2021). Positive values were as-

signed green shades, while negative values were assigned gray, representing built-up areas. The classified land use data and NDVI values were analyzed to identify patterns of urban expansion and land use change. The study focused on the impact of urban policy decisions made during the colonial period and post-independence on the spatial distribution and development of Kenitra. The analysis included assessing the fragmentation of urban areas, the decline in agricultural and forest lands, and the implications of these changes on the city's socio-spatial fabric.

$$NDVI = (NIR - Red) / (NIR + Red)$$

NIR (Near-Infrared): Reflected infrared and near-visible electromagnetic waves.
Red: Reflective visual red electromagnetic waves.

3. RESULTS AND DISCUSSION

3.1. The Urban Policy During the Colonial Period: The Concern for Separating Europeans and Moroccans

The urban policy that was implemented by the French Resident-General, Marshal Lyautey (Hubert Lyautey: 1854-1934), since his arrival in Morocco with his team, is a combined product of the compulsions and goals of the capitalist mode of production (road and rail infrastructure, ports, industrial zones, business centers...) with the formulation of a societal project that Lyautey partially realized in some Moroccan cities like Casablanca, Rabat, Fez, Meknes, Agadir, and Kenitra. This community project aimed to separate the European and Moroccan populations to “preserve the traditions and customs of the locals from the shock of European civilization” (Sabrou, 2020) as Lyautey put it. The first urban design for Kenitra in 1915, supervised by Lyautey, clearly reflects this policy.

Figure 1. Kenitra City Land Use Planning Map of the year 1923

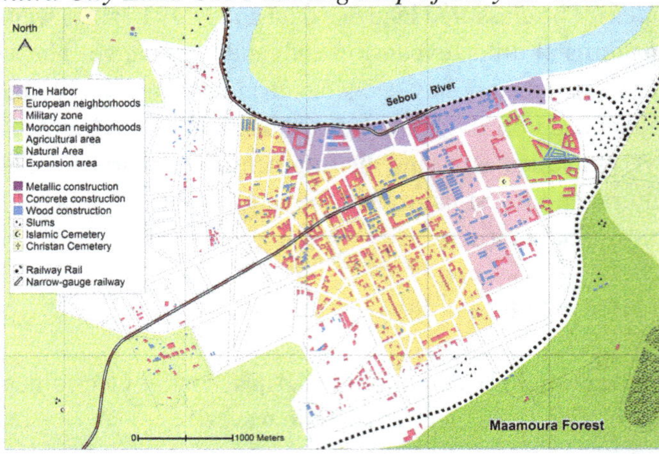

Source: Plan of Kenitra, Geographic Service of Morocco, Rabat October 1923 (Adapted by the author)

Given that the city is a space for development and political interactions, this followed policy, which has long been criticized, and as a result of which Lyautey was accused of apartheid, was intended to create a "besieged" space for the people so that it is not possible for Moroccans to learn, assimilate and own political matters. During that period, with near-total neglect of these neighborhoods, the French focused investment on European neighborhoods, providing them with urban functions and rare activities, which quickly transformed them into attractive centers. This resulted in a distinctive morphological structure in these areas, characterized by wide streets for cars and carriages, in contrast to Moroccan neighborhoods' narrow and closed alleyways, leading to an irregular urban fabric.

While "colonial urbanization" dominated the European city, "symbolized by prosperity and modernity" (Sabrou, 2020), the colonial authorities controlled Moroccan neighborhoods by designating specific areas for Moroccan housing that were enclosed and differed significantly from traditional Moroccan cities. These areas lacked essential urban characteristics that form the basis of social identity, such as open squares and spaces, with poor infrastructure and basic amenities. However, the colonial authorities justified this by claiming that the designs prioritized neighborhood and community revitalization, as seen in neighborhoods like the Habbous in Casablanca and the Moroccan neighborhood in Kenitra, known today as Khabbazat.

Figure 2. A design of the Moroccan neighborhood in Kenitra City

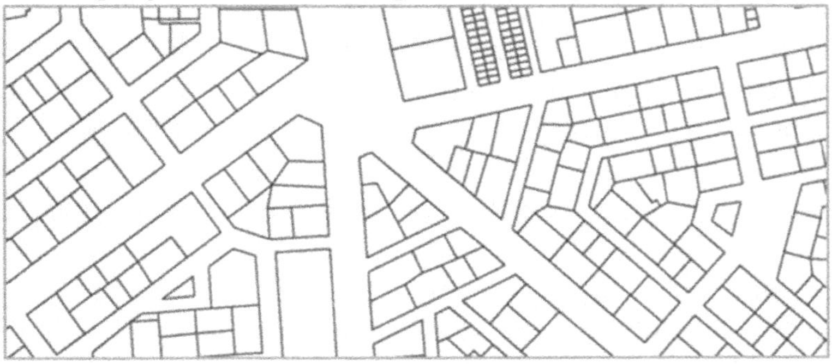

Source: The National Agency for Land Conservation, Cadastre and Cartography, Department of Cartography, Rabat 2006.(Adapted by the author)

Colonial urbanization became the main driver of general development, not only for major cities but also for the entire national territory. The policy focused on connecting Morocco with the outside world through urban concentration on the Atlantic coast, using this concentration as a means of economic centralization. The aim was to shift Morocco's center of gravity from the interior to the coast and absorb the dominance of Spanish-controlled Moroccan ports in the northover foreign trade. This policy was implemented while the colonial authorities delegated urban space production to the private sector, explaining the limited achievements in housing, which had more economic than social objectives. Perhaps the most important thing that distinguished urban policy in that era was its reliance on a set of principles, which were directed towards linking urbanization to the economic market and ignoring the social character. This was done through:

- Making urbanization based on distinguishing between Moroccans and Europeans, reinforcing dual urbanism by restricting Moroccan urbanization and allowing European expansion. With this policy, the French authorities aimed to "destroy civil society, then accept its members individually into the new city, which was built by and for the benefit of foreigners" (Laroui, 1975);
- Monitoring shantytowns without fighting them, and constantly exposing them to tension and instability due to their illegal status, in order to establish social differentiation among Moroccans through housing, and to ensure the concentration of the labor force.
- Opening the door to the private sector and limiting state intervention in urbanization to make the housing crisis a market necessity (Chouiki, 2010).

Segregation indicates “the isolation and marginalization of a social group or individuals for racial, religious, cultural, social, sexual, or other reasons, resulting in discrimination and unequal treatment.” (Merlin, 2010) Although the French authorities, during their approach to this policy, had claimed that the desire to respect local customs and traditions, especially family privacy among “indigenous people,” was what produced this approach, the reality showed that it primarily aimed to gather Moroccans and isolate them spatially from the Europeans, Through urban planning driven by an obsession with ethnic and social discrimination between Europeans and Moroccans.

The best evidence of this is the adoption of a “reconstruction policy based on economic returns without social returns.” (Hensens, 1971) Such reconstruction often leads to the elimination of neighborhoods, as a social and spatial unit with an identity, within which a minimum level of solidarity prevails. Perhaps it is permissible to go as far as to say that it was in the interest of colonialism to destroy the social and cultural content of Moroccan neighborhoods, to avoid the presence of a force opposing it. Despite repeated assurances from the French in that period to preserve the quality of life in the neighborhoods within the framework of housing operations for Moroccans, this was not achieved, and perhaps the undeclared goal of this reconstruction was “to strip those neighborhoods of their form and content.” (Dethier, 1970).

3.2. Urban Policy After the Independence: Lack of Integration and Cohesion

The independence did not constitute a rupture with colonization for Kenitra city, as the urban policy followed did not prevent the exacerbation of socio-spatial imbalances inherited from that era, nor did it adopt a proactive urban planning policy due to the absence of visions for the future of cities and their development.

Thus, during the early independence period, national policy was marked by the obsession with controlling the situation inherited from colonization and solidifying the political system. The urban policy was no exception, as the state quickly intensified its interventions to tighten its control over urban life, a situation that persisted for several decades thereafter. Consequently, the principles upon which the state's strategy in urbanization was built can be summarized into three main directions (Sabrou, 2020):

- Reinforcing the political system by surpassing the economic dimension of urban life, considering urbanization and housing as means to achieve political, economic, and social balance;

- Intensifying levels and forms of urban space control, based on considering urbanization as a fundamental channel for its policies and for expanding its presence and influence;
- Harnessing the state's density of presence to serve private sector initiatives in urbanization.

Thus, since the independence, the Moroccan state has pursued a housing policy that has only managed to eliminate some shantytowns. The process benefited only a limited number of inhabitants, as housing programs were used as "a winning card for political and ideological purposes, rather than aiming to provide decent housing for the workforce, which settled in increasingly deteriorating housing" (Benzakour, 1978). Alongside the state's withdrawal from its financial commitments to provide housing, redirecting its attention to more profitable sectors opened the way for "the national bourgeoisie" to accumulate capital in construction and public works, which flourished since the 1960s. The state liberated its real estate reserves without infringing on private property, and encouraged loans, allowing a broad segment of the middle class to obtain housing. However, the real estate speculation quickly led to a sharp rise in land prices.

With its failure in housing and urban development policies, the state realized the necessity of integrating urban planning into a new urban policy, focusing on creating plans to address the crisis of urbanization and housing. These plans formed new tools for the urban policy to prepare cities, reinforced by a significant arsenal of statutory laws to ensure control of the property and implement the directions that will be included in the construction documents. The state's intervention, during this period, through creating economic housing neighborhoods and residential subdivisions constituted the main driver for the urban expansion. Nevertheless, this expansion did not alleviate the housing crisis, which worsened with the accelerating pace of demographic growth. The residential subdivision policy has often exacerbated construction problems due to inadequate or absent basic infrastructure.

It took Moroccan authorities about 23 years after the independence to begin planning for the development of Kenitra. This involved the guiding plan for urban development, which projected the city's growth and development horizon by the year 2000. The authorities aimed to adopt a new pattern of urban development to regulate the expansion and dispersion of residential areas on the city's outskirts, by creating residential subdivisions towards the west and east, and implementing some restructuring projects in the southeast of the city, filling in vacant areas in the industrial quarter.

Thus, this phase, which extended until the late nineties, witnessed a strong state framing of urban expansion in Kenitra, without limiting it or the role of the private sector in it. This led to a new distribution of roles between the state and the private sector, the two main controlling parties in urban expansion. This expansion occurred under state authority, either through direct intervention or by opening doors to private initiative. Meanwhile, the state's intervention increased the presence and benefits of the private sector, "making private initiative one of the controlling factors in urban production" (Chouiki, 2010).

Kenitra City had to wait for about 22 years after the guiding plan for urban development was established to obtain a development design approved by Decree No. 2.04.128 on March 31, 2004 (Sabrou, 2020). This design was developed according to the general guidelines set by the management plan for the year 1982. It determined: the rules for using the soil in the city, including agreements in the public interest; equipment for collective public facilities; consultative reconstruction sectors; sectors subject to special legal regulations; sectors from which real estate reserves can be created; agricultural lands and forest areas that must be preserved; and fragile areas... The development design also designated sectors distinguished by their density and functions, such as individual housing areas, group housing areas, entertainment areas, and activity areas.

The objectives of the Kenitra urban development design can be summarized in five points:

- Ensuring the unity of the city, by ensuring a balance between the Maamoura and Saknia communities with regard to infrastructure equipment and facilities;
- Strengthening the links between communities by breaking barriers imposed by Sebou River, Maamoura Forest, and the military district in the city center, through road development;
- Rearranging the railway network to connect the northern industrial zone and southwestern areas with the rest of the city sectors;
- Preparing spaces between neighborhoods to maintain urban fabric homogeneity;
- Addressing substandard housing areas.

However, twenty years after the approval of this design, reality shows that it has failed to achieve any of its goals. On the contrary, it expanded the urban area, exacerbated urban fragmentation, entrenched urban duality, failed to achieve urban fabric homogeneity, and did not combat substandard housing. The architectural style introduced marred the city's previous aesthetic and destroyed the environmental scenery, forest cover, and other green spaces.

The results derived from satellite imagery classification for the years 2000, 2010, and 2022 indicate significant urban dynamics in Kenitra during the study period. Figures (3-4-5) shows that the city's built-up area was 2019 hectares in the year

2000, equivalent to 18.30% of the total urban area of the city. By 2010, the built-up area increased to 3356 hectares, representing 30.41% of the total area, rising to 4592 hectares or 41.61% by 2022, with an annual expansion rate of 3.73% during the study period from 2000 to 2022. This urban expansion consumed surrounding empty, cultivated, and forested lands, which decreased from 27.94%, 27.33%, and 22.63% respectively in 2000 to 21.80%, 18.90%, and 14.64% in 2022. The water areas remained relatively constant due to water bodies like Sebou River and Fouwarat Pond, at a rate of 3.80% in 2000, increasing to 5.87% in 2010 due to floods in 2009, then decreasing to 3.05% in 2022.

Figure 3. Kenitra land use map of 2000: A city with a natural background

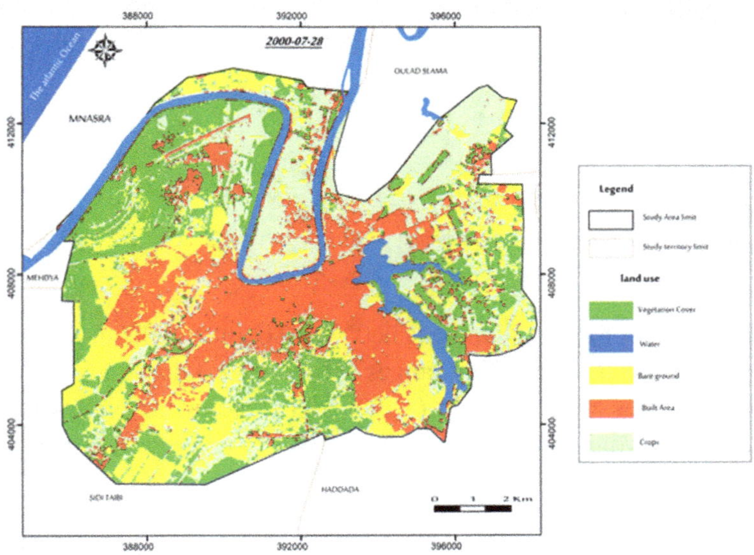

Source: Landsat satellite imagery classification of 2000.

As the built-up area of Kenitra more than doubled during the last two decades, this expansion came at the expense of agricultural lands, green spaces, and forests. This was confirmed by working with satellite imagery for the years 2000, 2010, and 2022 to extract the Normalized Difference Vegetation Index (NDVI).

The NDVI allows determining the distribution of vegetation cover and soil based on the characteristics of the spectrum reflected from green plants. NDVI is a simple index for analyzing remote sensing results, indicating whether an area contains vegetation or not.

Figure 4. Kenitra land use map of 2010: Expansion at the expense of bare ground and green spaces

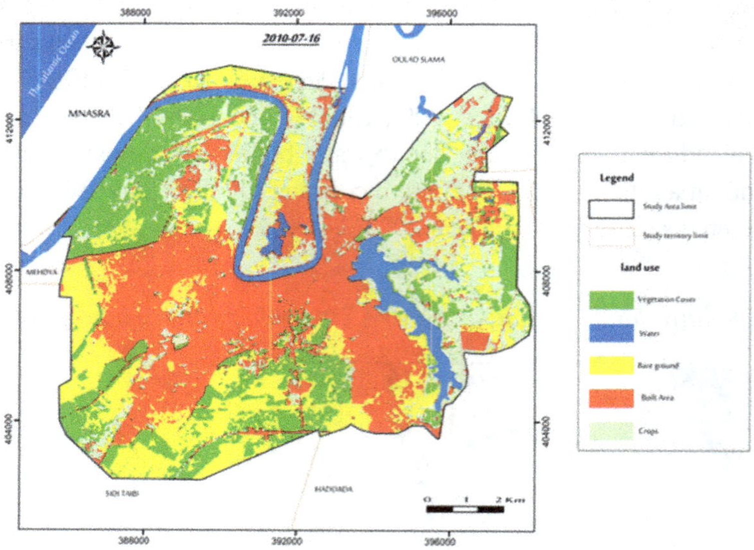

Source: Landsat satellite imagery classification of 2010.

The wavelength of Red light ranges between 0.6 μm and 0.7 μm, and the wavelength of NIR (Near Infrared) light ranges between 0.7 μm and 1.3 μm.

The NDVI values range between (1), where the area is covered with greenery, and (-1) in areas devoid of plants, as the latter absorbs colors and reflects green and near-infrared rays near the visible spectrum.

Accordingly, positive values were assigned green and its derivatives, while negative values were assigned gray, which mostly represents the spread of buildings in general.

Based on map readings and comparisons, it is evident that the areas colored in green and its shades, representing vegetation cover, significantly declined during the study period. In contrast, gray expanded significantly, indicating the spread of the built-up area.

Figure 5. Kenitra land use map of 2022: Thriumph of urban constructions over the environment

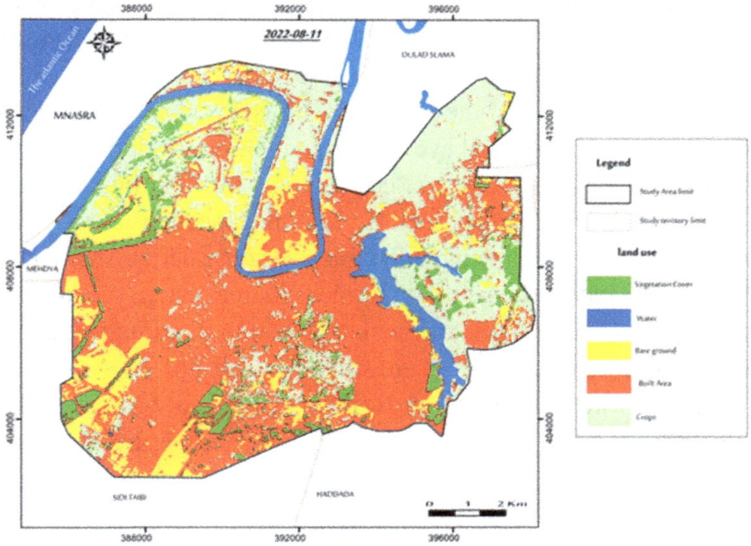

Source: Landsat satellite imagery classification of 2022.

3.3 Deficiencies of The urban policy in Kenitra city during the last two decades

Perhaps the most significant thing that allows us to talk about "shortcomings of the urban policy" in Kenitrais the absence of integration and cohesion in city management, as the prevailing approach focuses on addressing immediate problems that many attribute to the housing crisis, facilities, and equipment. Meanwhile, the city represents a point of convergence and overlap of several variable and unstable factors and challenges, requiring a comprehensive and integrated approach.

The deficiencies affecting the urban policy in Kenitra constitute a major obstacle to any leap or economic growth. The main indicators of this situation can be summarized in a set of negative aspects:

- **Urban areas' fragmentation:** The first decision of General Resident Lyautey, accompanied by his architect Ligiardi (Angelo), was to separate Moroccan neighborhoods from new European cities and move the city center from the old city to the European one. Since then, Moroccan neighborhoods have witnessed socio-spatial imbalances mainly related to investment weakness and the ineffectiveness of urban planning, while the European city (city center) has continued to uphold a social, economic, and architectural model that somewhat reflects welfare and prosperity.

- **Corruption and inefficiency of local actors:** "Bad policy puts place-making above helping people" (Glaeser, 2011) Due to issues related to the nature of local actors, and given that the real estate sector in the city is a field prone to conflicts of interest and the pursuit of undisclosed and illegitimate profits, local authorities turn a blind eye to many violations within the private sector involving real estate developers, contractors, and notaries in the construction and urban development sectors. This is especially concerning regarding tracking rules, monitoring, adherence to construction standards, liability registers, and land settlement. Not to mention real estate speculation, money laundering, and tax evasion. Further exacerbating the situation is the transformation of Kenitra into a city where stalled projects proliferate, mostly large-scale projects whether public or private, which are not completed due to technical or financial justifications but rather indicate the involvement of many actors associated with certain public interests such as municipal councils and private sectors like real estate developers in a real estate chaos beyond legal frameworks.

Figure 6. Examples of stalled projects in Kenitra City: The Cultural Complex

Left: from the 1990s, canceled; Right: since 2008
Source: Airbus, Maxar Technologies 2024.

- **Discontent in the urban practice:** "The urban practice encompasses various realistic and tangible methods, patterns, customs, and behaviors that reflect how urbanites utilize the city and alter the views of functional areas."

(BERTRAND, 1977) The difficulties hindering the proper Kenitra practice can be translated through:

- **Lack of an effective urban planning policy:** The overall condition of the city is unsatisfactory, evident in deteriorating sidewalks, public spaces, and green areas, in addition to growing problems related to urban transportation and security;
- **Absence of an attractive framework:** Kenitra suffers from a deficit in cultural facilities such as theaters, exhibition halls, and museums... leading to a stagnant urban image that lacks dynamism and fails to provide areas for relaxation, amusement, or entertainment, at a time when cultural activation and entertainment should be priorities in urban management and planning;
- **Deterioration of the architectural and historical heritage:** Kenitra possesses a rich and diverse architectural heritage, yet the absence of urban policies aimed at preserving and valuing this heritage, as an urban and cultural asset, leaves many historical buildings vulnerable to neglect and decay;
- **Lack of harmony among urban components:** It is evident in the spread of random housing and the deterioration of old buildings, along with poor distribution and deployment of communal facilities among different city neighborhoods.

These factors significantly and negatively impact the urban residents' sense of identity. While Kenitra experiences substantial growth, largely due to migration flows, this growth does not provide sufficient time for social and spatial settlement, andthus the emergence of an identity at the level of spatial social stratification. Through this, urban residents can clearly define the city's boundaries and absorb its practices.

- **Fragility and deterioration of the environmental framework:** Kenitra's environmental scene is marked by several deterioration factors attributed to the rise of urban pollution amid the decline and inadequacy of green spaces and prepared squares, particularly in densely populated areas;
- **Lack of a strategic and comprehensive development vision:** Despite having an arsenal of urban planning documents outlining urban development perspectives, Kenitra lacks any strategic design to promote social and economic sectors.

CONCLUSION

The National Charter for Territorial Development places urban issues among the most important for sustainable development, entrusting Moroccan cities with a driving role in rehabilitating and preserving national stability and competitiveness on the global stage. Today, major urban agglomerations exclusively possess the success factors for development and economic growth in the majority of countries, whether advanced or still in the growth phase. To ensure the fulfillment of this function, these urban agglomerations, which constitute the main centers for producing and strengthening wealth, are also required to ensure the elements of social and cultural integration and cohesion.

Regrettably, Kenitra, plagued by several issues like inflation and suburbanization density, remains unable to assume this role, as the imbalance affecting urban facilities obstructs this aim, especially concerning public facilities and the decline of the economic and urban fabric and the living framework. Thus, it accumulates a hub of exclusion, marginalization, economic and social vulnerability, and aggravating environmental and security challenges.

The urban policy, determining integrated and merged intervention by the state or local authorities and civil society, holds significant importance in this regard. The urban policy "combines urban planning and various interventions affecting social, educational, economic, and cultural areas to enhance the urbanites' quality of life" (Delarue, 2008). Therefore, it is necessarily multidisciplinary with a horizontal and specific approach, necessitating partnerships, sharing, contracting, and requiring institutional coordination and integration that can take several practical and executive forms. It requires national solidarity through robust state intervention and a confirmed commitment from local activities within a cohesive contractual framework. Hence, the urban policy primarily represents a political choice, especially regarding major urban planning issues posed by social integration and security issues nested, of course, in state-exclusive interventions.

Thus far, the urban policy pursued in Kenitra adheres to a "tackle emergencies" approach, devoid of a cohesive and comprehensive vision, considering the risks and responsibilities of urban residents, civil society, and economic stakeholders, both for the public and private sectors. It is a policy "characterized by the state's intervention at times with arbitrary measures that may lead to resentment and tension, which increases the dissonance between citizens and authorities" (Sabrou, 2020).

For instance, in the housing field, the state intervenes through shelter recovery or relocation operations and the establishment of residential fragmentation, which, even if it theoretically responds to basic needs, limits the problems that the city suffers from to the housing issue alone. Noting that the city did not provide a single job, such as housing, until it experienced problems related to housing alone. The city is a complex and compound body that is manufactured with the succession of time, away from cases of emergency and urgent intervention and through integrated operations. It cannot be restricted to the housing function alone, although it is the most important component. For example, the migrant does not seek the city to look for housing, but this is the same reason that drives him/her to accept temporarily living in a residence in very unhealthy and fragile conditions in several cases. By coming to the city, this migrant seeks, before everything, the services offered by the city, work, entertainment, and urban life quality, or in other words, he/she comes to "practice civilization".

Moreover, anyone scrutinizing the urban planning documents, prepared by local authorities, cannot overlook the multitude of flaws within them, how they are formulated, approved, and how their implementation is monitored. The prolonged preparation and approval period of urban planning documents constitutes a form of evasion from addressing urban development issues, leading to an exacerbation of the problems. Currently, the approval of most urban planning documents only occurs after their diagnosis of reality has become outdated, "after the solutions it contains become transcendent as a result of the emergence of a new reality." (Chouiki, 2010) Despite Morocco not yet having widely adopted sectoral urban planning through collective service plans, which allow for the fragmentation of issues within a single city, through integrated urban projects enabling the spatial division of urban issues and their comprehensive and participatory handling, continuing to draw inspiration from the colonial legacy based on socio-spatial discrimination contributes to "reproducing fragmented approaches to urban development issues." (Sabrou, 2020).

The success of Kenitra City increasingly depends on abstruse decisions made by zoning boards and preservation committees. It certainly makes sense to control construction in dense urban spaces. And for that, limiting building can be done with "three rules" (Glaeser, 2011):

First, the city should replace the current lengthy and uncertain permitting process with a simple system of frees;

Second, historic preservation should be limited and well defined;

Finally, individual neighborhoods should have some clearly delineted power to protect their special character.

REFERENCES

Aliaâ, H., Kassou, N., & Kacimi, I. (2017) Using the SLEUTH Model to Evaluated and Analyze the Impact of Urban Sprawl on Agricultural Land in the cities of Kenitra and Mehdia (Morocco). In: Proceedings of the 2nd International Conference on Computing and Wireless Communication Systems. ACM, Larache Morocco, pp 1–7 https://doi.org/DOI: 10.1145/3167486.31675

Benzakour, S. (1978). *Essai sur la politique urbaine au Maroc, 1912-1975: sur le rôle de l'État*. Éditions maghrébines.

Bertrand, J. M. (1977). *Pratique de la ville* (Masson, C. G., Ed.).

Chouki, M. (2011). *La Ville Marocaine, Essai de lecture synthétique*. Edition Dar Attaouhidi.

Delarue, J. M. (2008). La Politique Urbaine. Migrations Société, Regards croisés sur les politiques françaises d'immigration. Vol. 20. n. 117-118.

Dethier, J. (1970). *Soixante ans d'urbanisme au Maroc*. Princeton University.

Gervais-Lambony, P. (2001). « La ségrégation dans la grande ville, un essai de définition », in Gervais-Lambony Marie-Anne, Les très grandes villes dans le monde. p. 33-38.

Glaeser, E. (2011). Triumph of The City, How our greatest invention makes us richer, smarter, greener, healthier, and happier. Ed. Penguin Books.

Hensesns, J. (1971). La villa comme habitat urbain. *Bulletin Economique et Social du Maroc*, •••, 122.

Karmaoui, A., Moumane, A., El Jaafari, S., Menouni, A., Al Karkouri, J., Yacoubi, M., & Hajji, L. (2023). Thirty Years of Change in the Land Use and Land Cover of the Ziz Oases (Pre-Sahara of Morocco) Combining Remote Sensing, GIS, and Field Observations. *Land (Basel)*, 12(12), 2127. DOI: 10.3390/land12122127

Loulad, S., Nguyen, T. T., Simou, M. R., Rhinane, H., & Buerkert, A. (2023). Monitoring rural-urban transformation in the coastal region of Rabat-Sale-Kenitra, Morocco. *PLoS One*, 18(8), e0290829. DOI: 10.1371/journal.pone.0290829 PMID: 37651348

Merlin, P., & Choay, F. (2010). Dictionnaire de l'Urbanisme et de l'Aménagement. (Ed. Quadrige / Presses Universitaires de France (PUF), 3e édition.

Moumane, A., Al Karkouri, J., Benmansour, A., El Ghazali, F. E., Fico, J., Karmaoui, A., & Batchi, M. (2022). Monitoring long-term land use, land cover change, and desertification in the Ternata oasis, Middle Draa Valley, Morocco. *Remote Sensing Applications: Society and Environment*, 26, 100745. DOI: 10.1016/j.rsase.2022.100745

Moumane, A., El Ghazali, F. E., Al Karkouri, J., Delorme, J., Batchi, M., Chafiki, D., & Karmaoui, A. (2021). Monitoring spatiotemporal variation of groundwater level and salinity under land use change using integrated field measurements, GIS, geostatistical, and remote-sensing approach: Case study of the Feija aquifer, Middle Draa watershed, Moroccan Sahara. *Environmental Monitoring and Assessment*, 193(12), 769. DOI: 10.1007/s10661-021-09581-2 PMID: 34735624

Nasraoui, R., & Benzidiya, K. (2024) The Impact of Industrial Activities on the Land Surface Temperature (LST) in the Urban Environment of the City of Kenitra (Morocco). In: Kahime K, El Yamani M, Pouffary S (eds) Climate Change Effects and Sustainability Needs. Springer Nature Switzerland, Cham, pp 163–180https:// DOI: 10.1007/978-3-031-59603-2_11

Rieucau, J. (2004). Dualisme urbain et glissement du centre décisionnel, de la cité prétouristique à la ville touristique, dans l'Espagne méditerranéenne du début du XXIe siècle. *Cahiers de Geographie de Quebec*, 48(133), 47–70. DOI: 10.7202/009762ar

Sabrou, F., & Mazzine, H. (2020). La politique urbaine au Maroc: entre la ségrégation ethnique et socio-spatiale ». Ouvrage collectif: La gestion des espaces urbaines au Maroc dans un contexte mouvant. Centre d'Etudes et de recherches Takamul (D. Assalam, Ed.)., PP 187-204.

Van der Linden, S., Okujeni, A., Canters, F., Degerickx, J., Heiden, U., Hostert, P., Priem, F., Somers, B., & Thiel, F. (2018). Imaging spectroscopy of urban environments. *Surveys in Geophysics*, 40(3), 471–488. DOI: 10.1007/s10712-018-9486-y

Wellmann, T. (2020). *Remote sensing in urban planning: Contributions towards ecologically sound policies? Landscape and Urban Planning*. Elsevier., DOI: 10.1016/j.landurbplan.2020.103921

Chapter 7
Application of Cloud-Based Remote Sensing With Google Earth Engine in Land Cover Change Analysis

Liem Duy Nguyen
https://orcid.org/0000-0001-5151-4560
Nong Lam University, Ho Chi Minh City, Vietnam

Ngan Thi Thu Pham
https://orcid.org/0009-0008-7286-0356
Nong Lam University, Ho Chi Minh City, Vietnam

ABSTRACT

This chapter reviews research on the Google Earth Engine (GEE) platform and includes a case study in Vietnam. The literature highlights GEE's transformative role in geo-big data analytics, with its scalable and cost-effective processing of large datasets, despite some limitations in spatial functions and algorithms. The study employed GIS and remote sensing with GEE to generate land cover maps for 1985, 1995, 2005, 2015, and 2022, analyzing changes over this period. Using Landsat imagery and the random forest algorithm, the study achieved high classification accuracy, with overall accuracies above 96% and Kappa coefficients over 0.95. Findings show a shift from bare land and forest to increased crop areas and reduced forest cover due to socio-economic factors. Water bodies initially shrank but later expanded with irrigation development. The study provides insights for land management, while noting limitations like satellite resolution and weather

DOI: 10.4018/979-8-3693-6452-9.ch007

conditions, and suggests future research should incorporate high-resolution data and additional factors affecting land cover changes.

1. INTRODUCTION

1.1. Land cover change

Land cover (LC) and land use are two interconnected concepts. LC pertains to the physical attributes of the Earth's surface, such as natural vegetation or man-made structures. In contrast, land use refers to the various economic and social activities carried out on the land, reflecting how humans utilize these areas for purposes like recreation, education, or conservation (Soesbergen, 2016; Turner & Meyer, 1994). In simpler terms, LC represents the tangible, visible cover of an area, while land use denotes the socio-economic roles or purposes attributed to that land. A given area can have a single LC (e.g., forest) but can serve multiple land uses (e.g., protection, tourism) (Giri, 2012).

LC change (LCC) encompasses both the transformation, where one LC type is converted into another, and modification, which refers to changes in the internal conditions within a specific LC (Turner & Meyer, 1994). Two primary human-induced drivers of these changes are technological advancement and population growth (Lambin & Meyfroidt, 2011). Accurate information on the status and dynamics of LC is crucial for effective natural resource management, monitoring global environmental changes, and assessing their impacts (Loveland & Belward, 1997).

1.2. Land cover change analysis using remote sensing

Remote sensing is the scientific discipline focused on developing and applying techniques to collect, measure, and analyze information about objects or areas from a distance, without direct physical contact. The spectral reflectance corresponding to each type of LC shows differences due to the interaction between electromagnetic radiation and the object. This allows remote sensing to identify or analyze the characteristics of the LC type through the measurement of spectral reflectance.

The rapid progress in remote sensing, GIS, and computing technologies has greatly improved the capability to evaluate and track land use and LC across multiple spatial and temporal dimensions (Hansen & DeFries, 2004). Remote sensing offers several advantages for tracking LC on scales ranging from regional to global, including: (1) low cost and rapid information collection over large areas due to global coverage and frequent data updates; (2) digital format of remote sensing data, which is convenient for processing and analysis; (3) ability to collect data from spectral bands beyond human vision; (4) capability to gather data from remote, inaccessible areas; and (5) provision of an objective perspective on LC and land use (Giri, 2012). Remote sensing data, available since the 1970s and freely provided from various sources, supports the evaluation and monitoring of LC, thereby enhancing our understanding of how Earth's systems and ecosystems are interconnected. Due to its cost and labor efficiency, remote sensing technology is frequently used to monitor real-time changes in natural ecosystems and biodiversity.

1.3. Google Earth Engine

When mapping LC across extensive areas, two major challenges emerge: processing large volumes of data and obtaining cloud-free images over broad regions (Noi Phan et al., 2020). Traditional methods of data processing- such as searching, filtering, downloading, mosaicking, and preprocessing- are extremely labor-intensive. Processing these extensive datasets necessitates not only significant effort but also considerable storage capacity and access to advanced computing resources. Additionally, achieving clear imagery for a large area within a short period is difficult due to cloud cover. To minimize the time needed to create cloud-free mosaics, partially cloudy images must be downloaded and preprocessed, increasing the complexity and workload significantly (Carrasco et al., 2019).

As a cloud-based computing platform, Google Earth Engine (GEE) addresses significant challenges associated with large-scale LC mapping (Noi Phan et al., 2020). It allows users to process extensive remotely sensed imagery directly through a web-based Integrated Development Environment (IDE) code editor, thereby eliminating the need for data downloads to local machines. This facilitates easy access, selection, and processing of large datasets for extensive study areas (Gorelick et al., 2017). Besides its fast processing capabilities, GEE's growing popularity is also due to the availability of numerous packages containing various algorithms, which simplify the use of remote sensing tools for both experts and novices.

Since 2013, the number of publications utilizing GEE has steadily risen. Notably, the volume of articles published between 2020 and 2022 matches the total number of papers released from 2013 to 2019, following the launch of the GEE platform (Pérez-Cutillas et al., 2023). Landsat 8 imagery is the most commonly applied data in studies utilizing the GEE cloud computing system. The most commonly employed methods for analysis on the Google Earth Engine (GEE) platform are non-parametric classification techniques. Among these, the top five include random forest, support vector machines, classification and regression trees, decision trees, and artificial neural networks. GEE has been applied across a wide spectrum of uses, with water resources being the most common, followed by land use/LC and natural hazards applications.

1.4. Land cover change in Vietnam

Vietnam is located in the eastern region of the Indochinese Peninsula in Southeast Asia. It is bordered by the Gulf of Thailand, the Gulf of Tonkin, the South China Sea, as well as by China, Laos, and Cambodia. The country covers an area of 331,210 km^2 (The World Factbook, 2024). The terrain of Vietnam includes a low, flat delta in the south and north, central highlands, and hilly, mountainous regions in the far north and northwest. Vietnam experiences both tropical and temperate climate zones, with the entire country affected by the annual monsoon (The World Bank Group, 2024). The rainy seasons correspond with monsoon patterns, delivering substantial rainfall to the northern and southern regions from May to October, and to the central regions from September to January. Average temperatures in the northern regions vary from 22–27.5°C in summer to 15–20°C in winter, whereas the southern regions maintain a more stable temperature range, with 28–29°C in summer and 26–27°C in winter. The climate is further influenced by the El Niño Southern Oscillation (ENSO), which affects monsoonal circulation, leading to complex variations in rainfall and temperature patterns, with notable regional differences observed at a sub-national level.

Since the economic reforms of 1986, Vietnam has experienced significant changes in land use and LC. Recent years have seen rapid urbanization, leading to the conversion of many farmlands and green spaces into urban areas (Fan et al., 2019). Despite this trend, Vietnam has successfully shifted from a net loss of forests to a net gain, largely due to the expansion of plantation forests through reforestation efforts in recent decades (McElwee, 2009; Meyfroidt & Lambin, 2008). A comprehensive framework has been developed for the automated creation of Vietnam's first nationwide annual land use/LC datasets, covering the period from 1990 to 2020, using available remotely sensed and inventory data (Phan et al., 2021). The results indicate that forests were the predominant land use/LC, occupying roughly half of the country's area. This was followed by croplands (16.3%), rice fields (14.2%), and open water (including saltwater areas, 8.1%). Grasslands and scrublands each comprised about 2.8% of the total area, with residential areas being the smallest land use/LC at 1.3%. Despite some minor recoveries in 2000 and 2010, there was a substantial net loss of forested areas (19,940 km^2), which were largely converted to croplands over the 30-year period. At the same time, productive croplands were increasingly converted into urban areas, showing an approximate tenfold increase. The tripling of aquaculture activities was a major driver of the loss of wetlands (1,914 km^2). The spatial and temporal variations in these changes were significant, with the most dynamic regions being the northwest, south central, and southern parts of Vietnam.

1.5. Objectives

The aim of this chapter is to synthesize the current research on the GEE platform. Additionally, it presents a case study on the application of GIS and remote sensing with GEE in Ba Ria-Vung Tau (BRVT) Province, Vietnam, for creating LC maps for the years 1985, 1995, 2005, 2015, and 2022, and analyzing the decadal LCC from 1985 to 2022.

BRVT is situated in the Southeast region of Vietnam and spans an area of 1,982.56 km^2 (People's Committee of Ba Ria - Vung Tau Province, 2023). It is bordered to the west by Ho Chi Minh City, to the north by Dong Nai province, to the east by Binh Thuan province, and to the south by the East Sea. The province comprises eight administrative units, including two cities (Ba Ria, Vung Tau), one town (Phu My), and five districts (Chau Duc, Xuyen Moc, Dat Do, Long Dien, and Con Dao). The terrain of BRVT is relatively flat, with a common slope of less than 3°, and consists of four types of landforms: peninsulas, low mountainous and semi-mountainous areas, coastal plain valleys, and marine and island regions. BRVT falls within a tropical monsoon climate zone, which features two distinct seasons: the wet season from May to October and the dry season from November to April of the following year. In 2020, the province's Gross Regional Domestic Product (GRDP) reached approximately USD 13.661 billion, ranking 5th out of 63 provinces and cities in Vietnam, with a per capita GRDP of USD 11,375 (People's Committee of Ba Ria - Vung Tau Province, 2023). The economic structure has been shifting towards an increasing proportion of the industrial and service sectors while the proportion of the agricultural sector has been decreasing. The province's urbanization rate is high, standing at 57.56%, ranking 6th out of 63 provinces and cities in the country (People's Committee of Ba Ria - Vung Tau Province, 2023).

2. LITERATURE REVIEW

Remote sensing offers extensive data gathered globally via various satellite sensors. The continuous release of open-access datasets from remote sensing, along with rapid sensor technology advancements, has led to a substantial increase in available information (Lin et al., 2013). Consequently, the challenge in remote sensing-based research lies not in data scarcity but in managing and analyzing the vast quantities of data available (Yang et al., 2017).

Currently, significant advancements in geo-big data analytics are being made on cloud platforms due to their increased accessibility and cost-effectiveness. These platforms offer flexible options for processing power, memory, and storage capacity. Specifically, cloud computing offers scalable storage solutions that enable the efficient management of large datasets at a reduced cost. Furthermore, cloud systems provide on-demand services, including infrastructure, platform, storage, and software (Sun & Scanlon, 2019). A key example of such a service is Google Earth Engine (GEE). GEE is distinguished by its extensive, analysis-ready data catalog containing several petabytes, coupled with a high-performance, inherently parallel computational service. This service is accessible via an Application Programming Interface (API) over the Internet and is equipped with a web-based interactive development environment, facilitating rapid prototyping and presentation of results (Gorelick et al., 2017).

GEE has certainly established itself as the most prevalent cloud processing tool, significantly contributing to the analysis of large-scale geographical data in remote sensing (Tamiminia et al., 2020). GEE's architecture integrates various elements, with remote sensing information and operational functions being essential. Firstly, GEE provides an extensive array of datasets, encompassing both raw and pre-processed data, accessible at various scales (global, national, and local). It features a comprehensive library of satellite imagery from major sensors involved in land monitoring programs, including Landsat, MODIS, and Sentinel. Moreover, GEE offers daily-updated climate, meteorological, and geophysical datasets that are instantly available for analysis. Additionally, users can leverage both platform-provided data and their own data, with the capability to integrate and analyze both types simultaneously.

In terms of operational capabilities, GEE facilitates the application of both spectral and spatial analyses to large sets of imagery. This includes a spectrum of tasks, from basic mathematical operations to sophisticated image processing and machine learning techniques, enabled through a library of Application Programming Interfaces that support programming languages such as JavaScript and Python (Tamiminia et al., 2020). Nevertheless, GEE offers limited spatial functions due to difficulties in parallel processing, which complicates the use of certain filtering methods like Gaussian and Laplacian, as well as edge detection techniques such as Sobel and Hough transforms (Pérez-Cutillas et al., 2023). Moreover, GEE does not accommodate functions related to FFT Wavelet algorithms, hierarchical methods, or physics-based models (Amani et al., 2020).

Since its launch in 2010, GEE has garnered increasing research interest. The number of published research papers in the period from 2020 to 2022 surpassed that from 2013 to 2019 (Pérez-Cutillas et al., 2023). Landsat 8 emerged as the most frequently used satellite. Non-parametric classification methods, especially Random Forest, were the most frequently used algorithms. The primary methodological applications were in water resources assessment and prediction. The extensive number of countries where GEE has been utilized underscores its effectiveness as a robust tool for global-scale geospatial data analysis. The ability to leverage vast datasets and time series on this open cloud-based platform presents considerable research opportunities, particularly in developing nations.

Recent advancements in processing and characterizing Landsat archive data have greatly enhanced our capacity to map global LC and land use with increased precision, frequency, and thematic detail. Potapov et al. shared preliminary results from a project centered on annual multidecadal land monitoring using GEE (Potapov et al., 2022). This global dataset, offering a spatial resolution of 30 meters, records variations in forest cover and height, agricultural land, urban areas, surface water, and perennial snow and ice between 2000 and 2020. The foundation for LC and land use mapping was provided by pre-processed Landsat imagery, with each thematic product being independently created through machine learning methods specifically calibrated for local and regional conditions. Validation through statistical sampling of reference data has confirmed high accuracy, with user and producer accuracies exceeding 85% for all LC and land use categories, except built-up areas. The dataset highlights significant changes in global LC and land use over the past two decades. Publicly available, this bitemporal dataset serves as a critical resource for the global land monitoring system.

Unlike satellite images, which are usually captured and processed in near-real-time, global LC datasets have traditionally been produced annually, often experiencing substantial delays between image processing and dataset release. Brown et al. proposed a groundbreaking automated method for achieving globally consistent, high-resolution, near-real-time (NRT) land use and LC classification by utilizing deep learning techniques on 10 m Sentinel-2 imagery (Brown et al., 2022). This method employs a highly scalable cloud-based system via GEE to provide an ongoing stream of land use and LC predictions concurrently with Sentinel-2 data acquisitions. The innovative near-real-time product, Dynamic World, meets a broad spectrum of user needs, offering both the most current land use/LC information and custom global composites for specific date ranges chosen by users. Moreover, the continuous output of this product enables the refinement, expansion, and even redefinition of land use/LC classifications. These distinctive features offer exceptional flexibility for a diverse range of users across multiple disciplines.

3. METHOD

3.1. Implementation process

The implementation process of the study began with the collection of base map, Landsat satellite image series, field surveys, land use status maps, and built surface data (see Figure 1). The base map includes layers of population, administrative boundaries, transportation, hydrology, and topography. Landsat 5 TM satellite image series were collected for the years 1985, 1995, and 2005. Landsat 8 OLI/TIRS satellite image series were collected for the years 2015 and 2022. Field surveys on LC were conducted in two trips: the first trip from August 11 to 15, 2022, and the second trip from February 27 to March 3, 2023. Land use status maps were collected for the years 1995, 2005, 2015, and 2020. Built surface data GHS-BUILT-S R2023A was collected for the years 1985, 1995, 2005, 2015, and 2020.

Next, the multi-temporal Landsat satellite image series for each time point were cloud-filtered to improve image quality in segments with cloud cover, fog, or shadows. The filtered images were then overlaid to create composite images for each time point and clipped according to the administrative boundaries of BRVT province. Based on the characteristics of the study area and the potential of the sensors, the LC classification system for BRVT province was developed, consisting of five LC classes: forest, crops, built-up land, water bodies, and barren land.

Next, the interpretation key for the LC classes was established based on the false color composite and ground-truth images. Based on the interpretation key and field surveys, samples were collected for LC classification. The random forest classification algorithm was applied to classify the LC classes using 70% of the samples for each time point. The accuracy of the LC classification results for each time point was assessed using overall accuracy and the Kappa coefficient, utilizing the remaining 30% of the samples. Post-classification processing was conducted to generalize and refine the information using the majority filter and the GHS-BUILT-S R2023A built surface data. The entire process, from Landsat data collection to post-classification processing, was conducted on the GEE platform with the complete GEE JavaScript API script for the years 1985, 1995, 2005, 2015, and 2022, available respectively at the following links: https://code.earthengine.google.com/ea23b99c7b9ecdf2f16e0b802337b330, https://code.earthengine.google.com/b03a39a34df4d6d8919ca6cc9ef0a26f, https://code.earthengine.google.com/f94a2c87ec4ecbef866fd38faef44c41, https://code.earthengine.google.com/feba298bd8c6a479f483bdad51fc7a1f, https://code.earthengine.google.com/9e5076b5aed346a2413b5e0cd91bea97.

Subsequently, pairwise intersection of the LC status data for the years 1985, 1995, 2005, 2015, and 2022 was performed to create LCC data for the periods 1985-1995, 1995-2005, 2005-2015, and 2015-2022, with attributes containing the LC combinations of each time point. Finally, area and rate matrices of LCC for the periods 1985-1995, 1995-2005, 2005-2015, and 2015-2022 were constructed.

Figure 1. Implementation process of land cover change in Ba Ria – Vung Tau Province

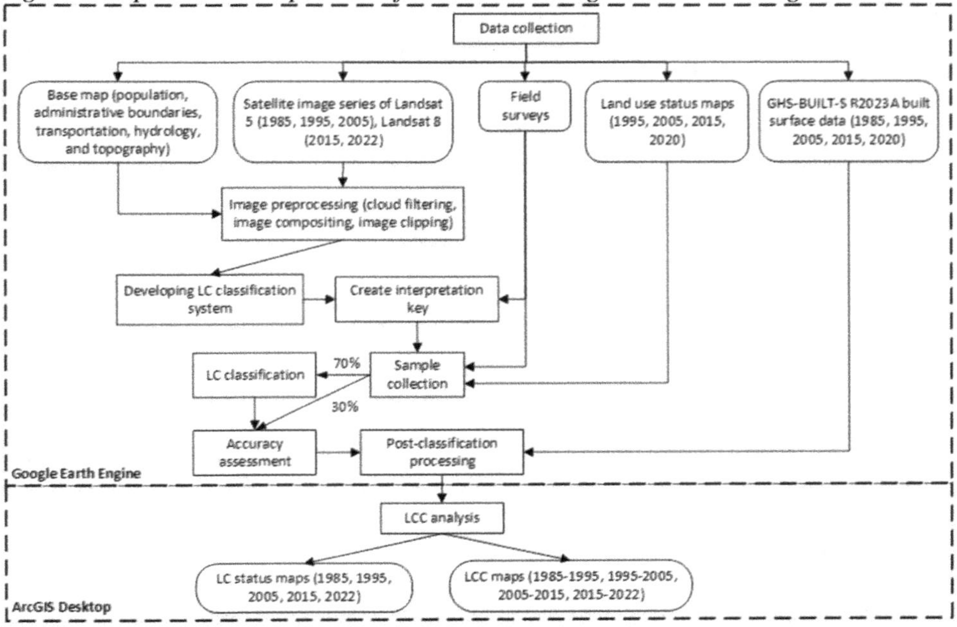

3.2. Data collection

The study collected primary and secondary data, including: base map of BRVT Province, Landsat satellite images (5 TM, 8 OLI/TIRS), field surveys, LC status maps of BRVT Province, and GHS-BUILT-S R2023A built surface data, as detailed in Table 1.

Table 1. Input data sources

No	Data type	Description	Resolution	Format	Source
1	Base map of BRVT Province	Layers: population, administrative boundaries, transportation, hydrology, topography	Map scale: 1:50,000	*.gdb	(Department of Natural Resources and Environment of Ba Ria-Vung Tau Province, 2022)
2	Landsat satellite images (5 TM, 8 OLI/TIRS)	Surface reflectance values	Spatial resolution: 30 m; Temporal frequency: 16 days; Time periods: 1982-1988, 1992-1998, 2002-2008, 2012-2018, 2019-2022;	*.tif	(U.S. Geological Survey, 2023a, 2023b)
3	Field surveys	LC data	Two trips: 1st trip (August 11-15, 2022), 2nd trip (February 27 - March 3, 2023)	*.csv	
4	Land use status maps of BRVT Province	Land use type	Map scale: 1:100,000 (1995), 1:50,000 (2005, 2015, 2020)	*.dgn (1995, 2015, 2020), *.tab (2005)	(Department of Natural Resources and Environment of Ba Ria-Vung Tau Province, 1995, 2005, 2015, 2020)
5	GHS-BUILT-S R2023A built surface data	Estimated built surface area	Spatial resolution: 100 m; Years: 1985, 1995, 2005, 2015, 2020	*.tif	(Martino & Panagiotis, 2023).

Note. Each data source includes specific details such as data type, description, resolution, format and source.

3.2.1. Field surveys

Field surveys were conducted at 92 sites within BRVT Province. The minimum sample size was 40 meters, corresponding to the Landsat image spatial resolution of 30 meters, with a GPS error of 5 meters (Justice & Townshend, 1981). The LC information collected was GPS-positioned, stored, and extracted using the Epicollect5 application, replacing paper forms to minimize data errors. The locations of the survey points are shown in Figure 2.

Figure 2. The locations of the survey points in Ba Ria – Vung Tau Province

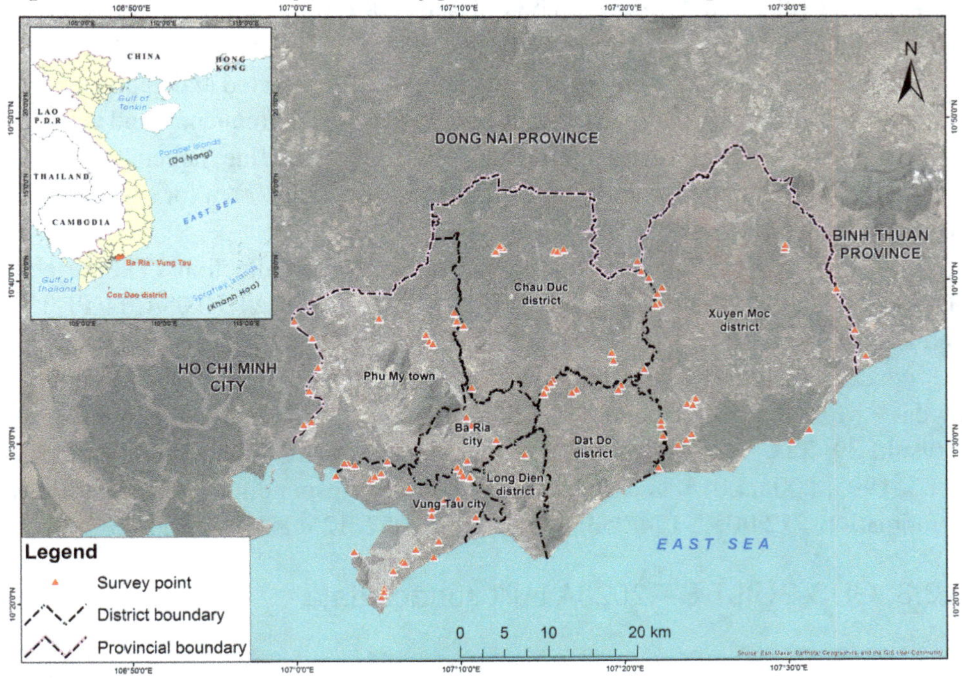

3.2.2. Landsat satellite images

Satellite imagery data were obtained from the Landsat program, which is managed by the U.S. Geological Survey (USGS) and the National Aeronautics and Space Administration (NASA). The Landsat program has been operational since 1972, providing continuous imagery. To date, nine generations of Landsat satellites have been launched, and their data have been widely used globally for resource management and environmental monitoring. The success of Landsat is attributed to its multi-spectral capabilities for earth observation, good spatial resolution, and relatively short revisit period.

For the years 1985, 1995, and 2005, the study utilized the Landsat 5 TM image series from Collection 2, Level 2 (USGS Landsat 5 Level 2, Collection 2, Tier 1), corresponding to the periods 1982-1988, 1992-1998, and 2002-2008. This series features surface reflectance values that have been atmospherically corrected and geometrically processed, with 8-bit radiometric resolution, 16-day revisit frequency, and includes 7 spectral bands with 30 m resolution for Blue, Green, Red, NIR, SWIR 1, and SWIR 2, and 60 m for Thermal Infrared.

For the years 2015 and 2022, the study used the Landsat 8 OLI/TIRS image series from Collection 2, Level 2 (USGS Landsat 8 Level 2, Collection 2, Tier 1), corresponding to the periods 2012-2018 and 2019-2022. This series includes surface reflectance values that have been atmospherically corrected and geometrically processed, with 12-bit radiometric resolution, 16-day revisit frequency, and includes 11 spectral bands with 30 m resolution for Coastal Aerosol, Blue, Green, Red, NIR, SWIR 1, SWIR 2, and Cirrus, 15 m for Panchromatic, and 100 m for two Thermal Infrared bands.

This study uses GEE platform to access Landsat image collections. The variable `landsatCollection` is defined as the Landsat 5 TM (LANDSAT/LT05/C02/T1_L2) and Landsat 8 OLI/TIRS (LANDSAT/LC08/C02/T1_L2) image collections using the `ee.ImageCollection` function. The `filterDate` function is used to filter images by date, while the `filterBounds` function filters by the study area (geometry). As a result, 39, 353, 323, 425, and 300 Landsat scenes were collected for the years 1985, 1995, 2005, 2015, and 2022, respectively, corresponding to the 6 path/row combinations: 124/052, 124/053, 124/054, 125/052, 125/053, and 125/054.

3.3.3. GHS-BUILT-S R2023A built surface data

Due to the limitations in distinguishing linear built-up features, such as road networks, from Landsat imagery, the study supplemented data with the GHS-BUILT-S R2023A dataset (Martino & Panagiotis, 2023). This dataset is part of the Global Human Settlement Layer (GHSL) project, which generates global spatial information based on evidence and knowledge about human presence on the planet. It utilizes input data from satellite imagery, census data, and geographic data provided by the community or volunteers.

The GHS-BUILT-S R2023A dataset is available in raster format and describes estimated built-up land areas from 1975 to 2030 at a 5-year interval, with spatial resolutions of 10 m (for 2018), 100 m, and 1 km. The data are generated through spatial-temporal interpolation or extrapolation based on a combination of Sentinel 2 and Landsat (MSS, TM, ETM) imagery (European Commission, 2023). In the context of this study, the GHS-BUILT-S R2023A data were collected for the years 1985, 1995, 2005, 2015, and 2020, at a spatial resolution of 100 m.

3.3. Image preprocessing

In remote sensing data analysis, image preprocessing refers to preliminary activities performed prior to the main analysis (Campbell et al., 2023). In this study, Landsat image preprocessing included: 1) cloud masking, 2) image compositing, and 3) image clipping. The purpose of these activities was to enhance image quality, providing a solid foundation for subsequent analysis and interpretation.

Cloud masking was applied to improve image quality in areas affected by cloud cover, fog, or shadows. From the downloaded Landsat image series, GEE platform was used to remove cloud pixels from the image series by utilizing the 'QA_PIXEL' and 'QA_RADSAT' bands, as well as the visible and thermal bands.

Following the removal of cloud-affected areas from the Landsat image series, composite images for each time period were generated by calculating the median value of all pixels across the bands in the series using the GEE platform. From the composite images for each time period, the images were clipped to the administrative boundaries of BRVT Province using the GEE platform, as shown in Figure 3.

Figure 3. The cloud-free Landsat composite images of the study area for a) 1985 (RGB = 321), b) 1995 (RGB = 321), c) 2005 (RGB = 321), d) 2015 (RGB = 432), e) 2022 (RGB = 432).

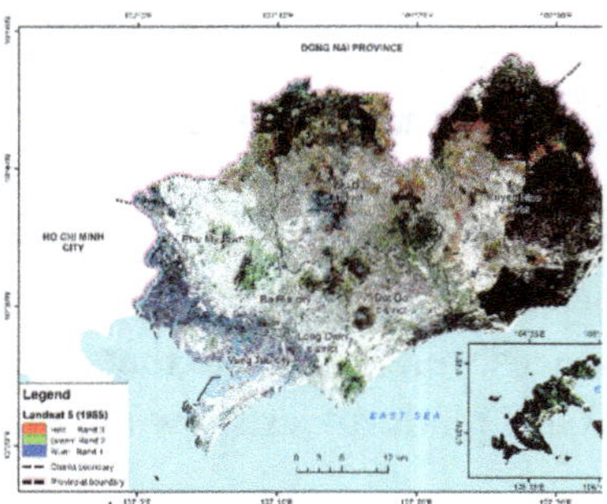

a)

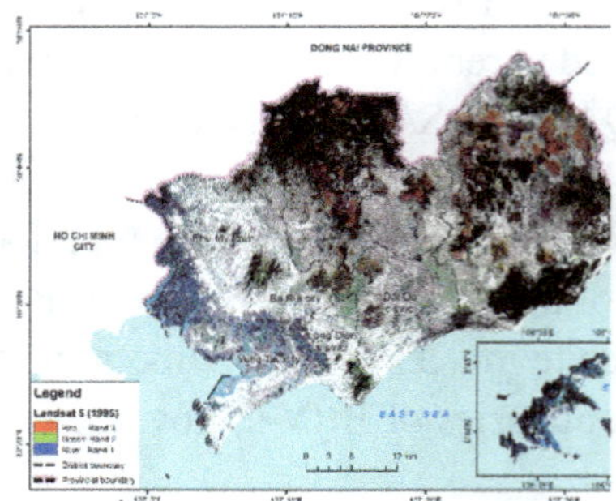

b)

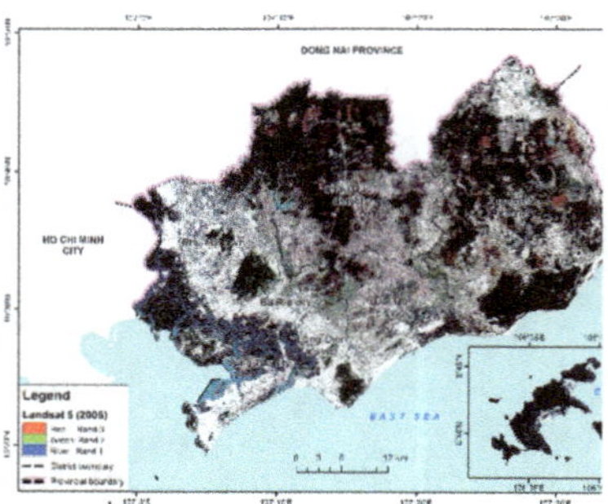

c)

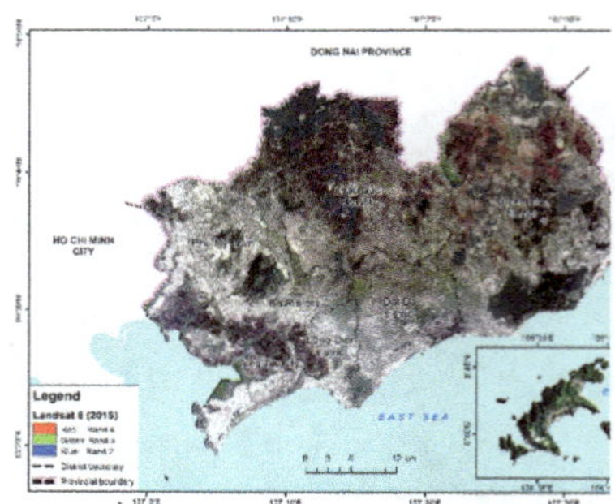

d)

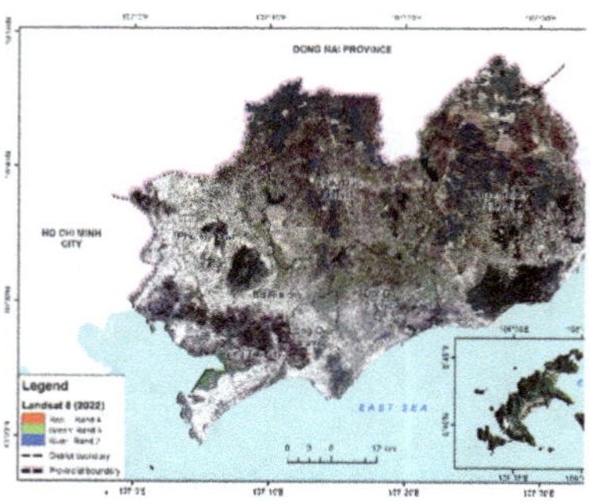

e)

3.4. Developing land cover classification system

The LC classification system is a standardized and comprehensive system designed to meet the specific requirements of users for creating LC maps, independent of the scale or mapping medium used (Di Gregorio & Jansen, 1998). One of the earliest and most widely used LC/ land use classification systems is the Anderson Classification System (Anderson et al., 1961). This system was specifically designed for use with remote sensing imagery. However, as better data sources become available, such as field surveys or high-resolution satellite imagery, the Anderson system remains flexible enough to incorporate detailed classifications extracted from these richer data sources. The Anderson classification system includes 9 main LC classes (built-up land; agricultural land; grasslands; forests; water bodies; wetlands; bare land; tundra; and permanent snow and ice) and 37 subclass categories (e.g., agricultural land includes croplands and pasture; orchards, nurseries, and gardens; restricted grazing; and other agricultural land).

The ecosystems in BRVT Province can be classified into 5 main types (Department of Natural Resources and Environment of Ba Ria - Vung Tau Province, 2019): 1) Upland Forests includes habitats such as tropical evergreen rainforests with dominant Dipterocarps, tropical evergreen secondary forests, seasonal deciduous forests, lowland broad-leaved forests, mixed wood-timber forests, and scattered tree savannas in areas like Dinh – Thi Vai Mountain, Lon – Nho Mountain, and Minh Dam Mountain; 2) Mangrove Forests, Estuaries, and Coastal Areas includes wetland areas (brackish-saline), mangrove forests, estuary systems, sand dunes, coastal plantations, and Casuarina forests; 3) Agricultural Areas includes deforested areas for agriculture, or river and stream corridors affected by agricultural expansion and urbanization, with habitats largely fragmented and mixed with shrubs, herbs, and tree species; 4) Urban Areas characterized by transportation corridors, cultural parks, urban canal corridors, home gardens, forest gardens, and vegetation recovering from land clearing; 5) Freshwater Bodies includes river and lake systems with habitats such as shrub savannas, wetlands, planted forests, home gardens, forest gardens, and croplands.

Based on the characteristics of the ecosystem and habitats in the study area, and referencing the Anderson classification system, this study developed a LC classification system for BRVT Province that includes 5 main LC classes: forest, cropland, built-up land, water bodies, and bare land (Table 2).

Table 2. Land cover classification system for Ba Ria – Vung Tau Province

Land cover	Description
Forest	- One or more species of woody trees, bamboo, rattan, or palm trees growing on land (terrestrial mountains, rocky mountains, freshwater wetlands, sandy land) with an area of 0.3 hectares or more. - Terrestrial land area covered by forests (including natural and planted forests) that meets the forest standards set by forestry legislation. - One or more species of woody trees, bamboo, rattan, or palm trees growing along the coastline (the average lowest mean sea level over many years) and major river mouths frequently or periodically inundated by saline tides. - Land with coastal water (beyond Mean Low Water Line) covered by mangrove forests.
Cropland	- Herbaceous plants that cover the ground, with low shading capability, belonging to the category of low-growing plants. - Land used for annual crops or perennial crops.
Built-up land	- Impervious surfaces that prevent or delay water from infiltrating into the soil, causing surface runoff, such as roofs and roads covered with impermeable materials. - Residential land, specialized land, religious sites, worship sites, cemeteries, funeral homes, and crematoriums.
Water bodies	- Water bodies of linear hydrological features without closed boundaries forming natural or artificial land parcels used for drainage, water conveyance, aquaculture, or tourism services, including rivers, streams, canals, brooks, and springs. - Water bodies of hydrological features such as ponds, lakes, and reservoirs with closed boundaries forming land parcels, used for purposes like hydropower, irrigation, aquaculture, or tourism services. - Coastal waters (beyond Mean Low Water Line) used for aquaculture, beach bathing, marine tourism, boat mooring, or marine mineral exploration, extraction, and processing.
Bare land	Areas with very little vegetation cover, including: abandoned land, sandy land, land awaiting construction, and bare rocky terrain.

Note. The land cover classification for Ba Ria – Vung Tau Province includes: Forest (woody trees, bamboo, mangroves, and coastal forest), Cropland (herbaceous plants for annual or perennial crops), Built-up land (impervious surfaces and urban areas), Water bodies (linear features and closed water bodies), and Bare land (areas with minimal vegetation, including sandy or rocky terrain).

3.5. Create interpretation key

A classification key is a standardized guide for identifying specific objects, consisting of a set of identification criteria established by the interpreter to facilitate rapid and accurate classification results for various objects. In this study, the development of a classification key for LC classes helps streamline the process of sampling for LC classification, ensuring consistency across different time points. The classification key is established for the 5 LC classes—forest, crops, built-up land, water bodies, and bare land—within BRVT Province (Table 3).

Table 3. Interpretation key for land cover classes

Land cover	Landsat 8 imagery (RGB = 543)	Ground-truth imagery	Identification factors
Forest	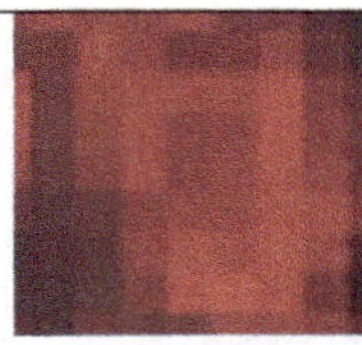		Dark red, speckled texture, indeterminate shape.
Cropland			Light red, smooth texture, with boundaries, indeterminate shape.
Built-up land			White, gray, blue, purple, indeterminate shape, speckled texture.
Water bodies	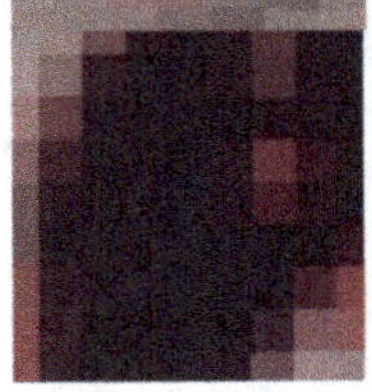		Dark blue, linear form, smooth texture.
Bare land			Brown-yellow, gray, with boundaries, indeterminate shape, smooth texture.

Note. For land cover classification: Forest appears dark red with a speckled texture; Cropland is light red with a smooth texture and clear boundaries; Built-up land is white, gray, blue, or purple with a speckled texture; Water bodies are dark blue with a linear form and smooth texture; Bare land is brown-yellow or gray with smooth texture and boundaries.

3.6. Sample collection

Based on the interpretation key combined with land use status maps and field survey data, samples for LC classification were collected for each time point. The number of samples in each LC class ensured that there were at least 10 times the number of image bands used in the classification (Jensen, 1996). The locations and quantity of the samples are shown in Figure 4 and Table 4.

Figure 4. Sample locations for different time points: a) 1985, b) 1995, c) 2005, d) 2015, e) 2022

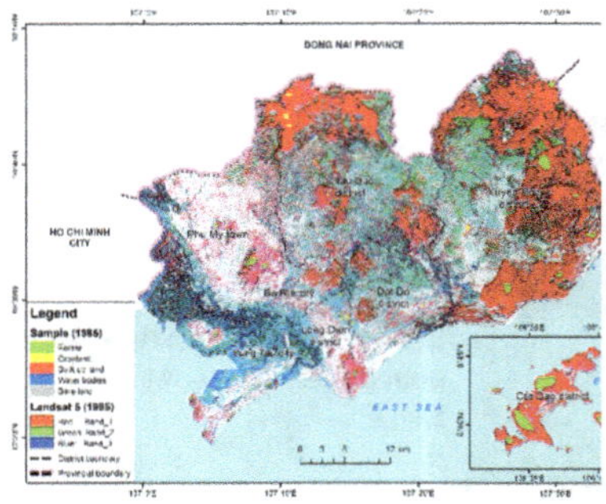

a)

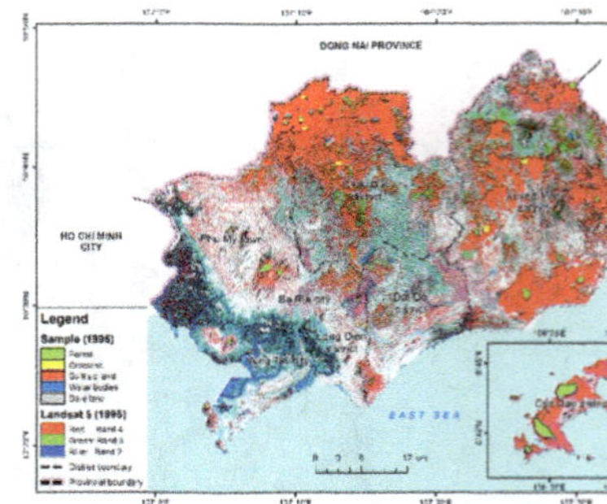

b)

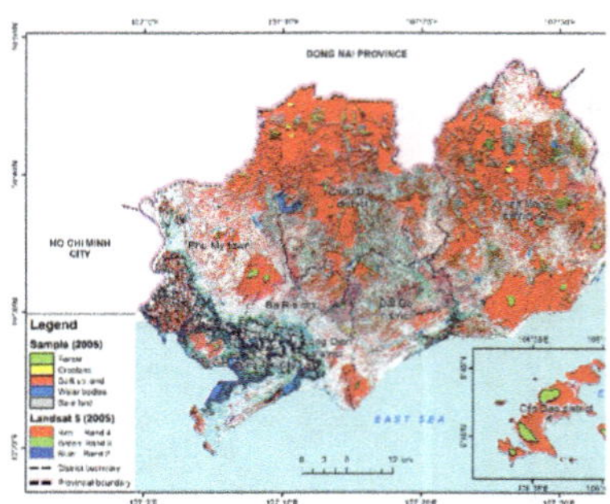

c)

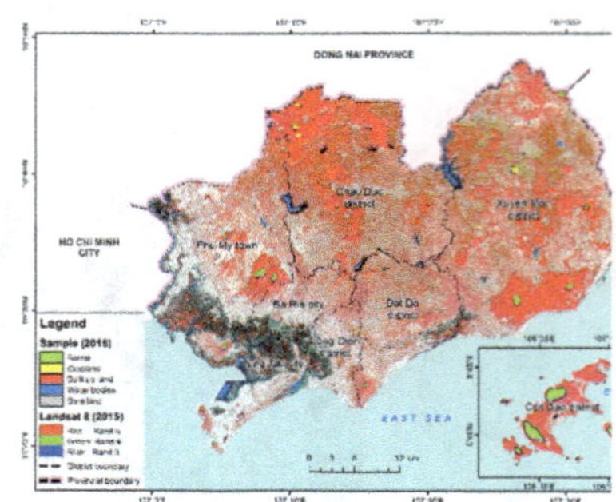

d)

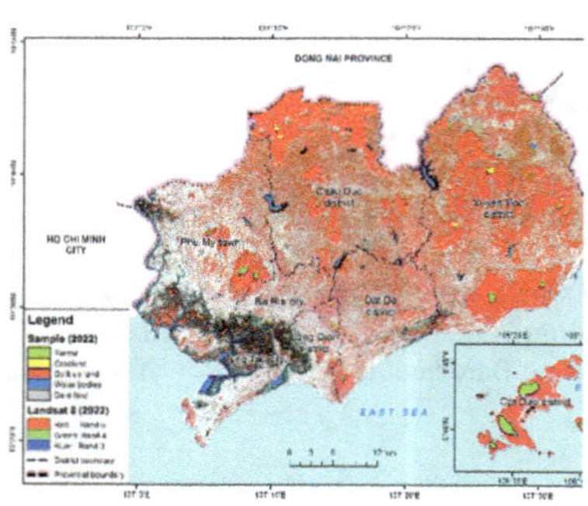

e)

Table 4. Number of samples for different time points

No.	Land cover	1985	1995	2005	2015	2022
1	Forest	2,506	2,447	2,554	2,138	2,389
2	Cropland	1,809	1,592	1,625	1,713	1,841
3	Built-up land	300	445	658	769	1,038
4	Water bodies	1,819	2,011	1,983	1,910	2,199
5	Bare land	2,819	2,464	1,871	2,067	2,292
Total		9,253	8,959	8,691	8,597	9,759

Note. The number of samples for land cover types across different years are: Forest (2,506 in 1985 to 2,389 in 2022), Cropland (1,809 in 1985 to 1,841 in 2022), Built-up land (300 in 1985 to 1,038 in 2022), Water bodies (1,819 in 1985 to 2,199 in 2022), and Bare land (2,819 in 1985 to 2,292 in 2022). The total number of samples increased from 9,253 in 1985 to 9,759 in 2022.

3.7. Land cover classification

Image classification is the process of segregating or aggregating information based on the spectral, spatial, and temporal characteristics provided by the image of the object of interest. The image classification method is performed by assigning a class name (information type) based on the brightness values (DN) of certain pixels (spectral characteristics) belonging to a group of objects that have relatively homogeneous spectral properties, in order to differentiate those groups within the image. Depending on the number of information types to be identified, the feature extraction of different object types is carried out based on training data through a predetermined decision rule (classification algorithm).

Various image classification techniques are employed to map LC using remote sensing data, including unsupervised algorithms, supervised parametric methods, and advanced machine learning approaches. Machine learning algorithms have increasingly replaced traditional parametric methods due to their greater accuracy and efficiency in handling complex, high-dimensional data, particularly for large-scale mapping (Rogan et al., 2003). These methods often outperform traditional techniques as they bypass the need for specific data distribution assumptions, such as normality, and typically provide improved accuracy (Friedl & Brodley, 1997). However, some machine learning approaches, such as neural networks and support vector machines, can be complex and prone to overfitting due to the large number of parameters needing adjustment (Breiman et al., 2017; Foody, 2004). A significant advancement is the use of ensemble classifiers, including neural network ensembles, random forests, bagging, and boosting, which generate multiple classifications of the same data through repeated application of the same base classifier or a combination of different classifiers (Breiman, 2001; Mountrakis et al., 2009). Ensemble learning techniques aggregate results from multiple classifiers using rule-based approaches or error-minimizing methods, achieving higher accuracy by combining the strengths of individual classifiers and reducing their weaknesses (Ghimire et al., 2010).

Random forest, among ensemble learning techniques, is being progressively used for LC classification with multispectral and hyperspectral satellite data imagery (Chan & Paelinckx, 2008; Ghimire et al., 2010) as well as with LiDAR and radar datasets (Guo et al., 2011; Latifi et al., 2010). It is developed based on decision trees, bagging, and enhanced bootstrapping techniques (Breiman, 2001). Random forest classification builds multiple decision trees from randomly selected training pixels. Key parameters are `ntree`, the number of trees (ranging from 1 to 500), and `mtry`, the number of variables used for splitting at each node. The model determines the optimal `ntree` for the lowest error rate and uses majority voting from all trees to produce the final classification result.

This study used bands 1-5 and 7 (Landsat 5 TM), and bands 1-7 (Landsat 8 OLI/TIRS) to perform LC classification using the random forest method with 10 trees developed on the GEE platform.

3.8. Accuracy assessment

Accuracy assessment is conducted to evaluate the quality of interpreted satellite imagery or to compare the classification accuracy of different methods. The comparison of the correspondence between ground truth and classified types by a classification algorithm is typically represented by an error matrix. The overall accuracy indicates the proportion of correctly classified pixels in the test dataset. Additionally, the Kappa coefficient is often used to provide information about the proportion of accurate classifications across the entire satellite image.

This study used 30% of the samples to perform the accuracy assessment of LC classification on the GEE platform.

3.9. Post-classification processing

The LC classification results were post-processed to generalize the information. Specifically, the majority filter on the GEE platform was used to assign the dominant LC value within a 3x3 pixel window to the central pixel.

Additionally, due to the limited capability of Landsat imagery to distinguish linear built-up features such as roads, the study used GHS-BUILT-S R2023A built-up surface data from the years 1985, 1995, 2005, 2015, and 2020 to update the LC classification results.

3.10. Land cover change analysis

The study overlaid with intersecting LC data for the years 1985, 1995, 2005, 2015, and 2022 using the Combine tool in ArcGIS Desktop. The results generated LCC for the periods: 1985-1995, 1995-2005, 2005-2015, and 2015-2022, with attributes containing the LC combinations for each period. Then, created an area matrix and calculated the LCC rate for the periods 1985-1995, 1995-2005, 2005-2015, and 2015-2022 using the Pivot Table tool in MS Excel.

4. RESULTS

4.1. Land cover status maps

4.1.1. Year 1985

For the year 1985, the LC classification results are very good with an overall accuracy of 96% and a Kappa coefficient of 0.95, as shown in Table 5. The LC map for 1985 is illustrated in Figure 5.

Table 5. Error matrix for the year 1985 (unit: pixel)

		Classified					
		Forest	**Cropland**	**Built-up land**	**Water bodies**	**Bare land**	**Total**
Reference	Forest	742	25	0	3	5	775
	Cropland	8	517	0	5	5	535
	Built-up land	1	3	72	0	11	87
	Water bodies	6	4	0	494	4	508
	Bare land	11	12	5	1	775	804
Total		768	561	77	503	800	2,709
		Overall accuracy = 96%		Kappa coefficient = 0.95			

Note. The error matrix for 1985 shows the number of pixels classified for each land cover type. For example, forest had 742 correct classifications and 33 misclassifications.

Figure 5. Land cover status map for the year 1985

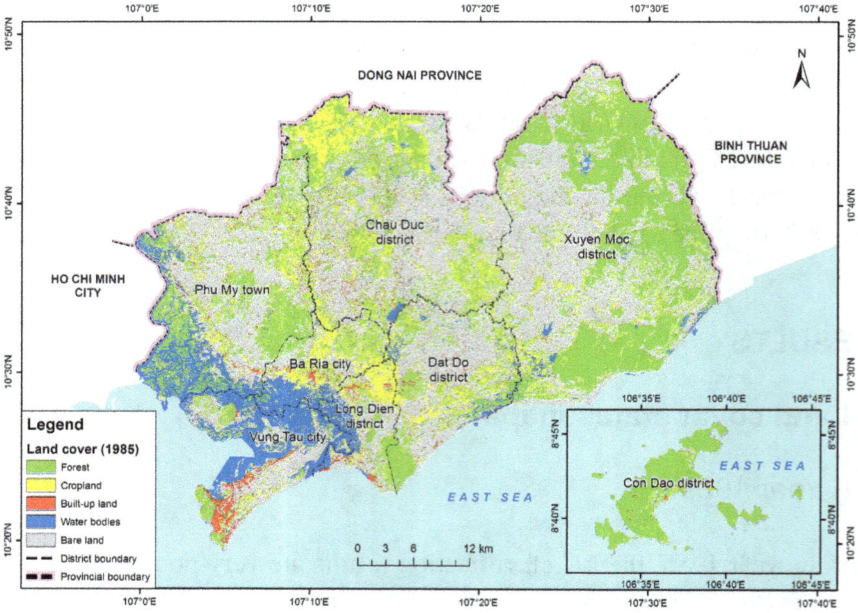

4.1.2. Year 1995

For the year 1995, the LC classification results are very good with an overall accuracy of 97% and a Kappa coefficient of 0.96, as shown in Table 6. The LC map for 1995 is illustrated in Figure 6.

Table 6. Error matrix for the year 1995 (unit: pixel)

		Classified					
		Forest	**Cropland**	**Built-up land**	**Water bodies**	**Bare land**	**Total**
Reference	Forest	687	6	2	0	9	704
	Cropland	9	476	0	0	2	487
	Built-up land	0	1	108	0	11	120
	Water bodies	2	0	0	588	5	595
	Bare land	18	2	6	2	709	737
Total		716	485	116	590	736	2,643
		Overall accuracy = 97%	Kappa coefficient = 0.96				

Note. The error matrix for 1995 shows the number of pixels classified for each land cover type. For example, forest had 687 correct classifications and 17 misclassifications.

Figure 6. Land cover status map for the year 1995

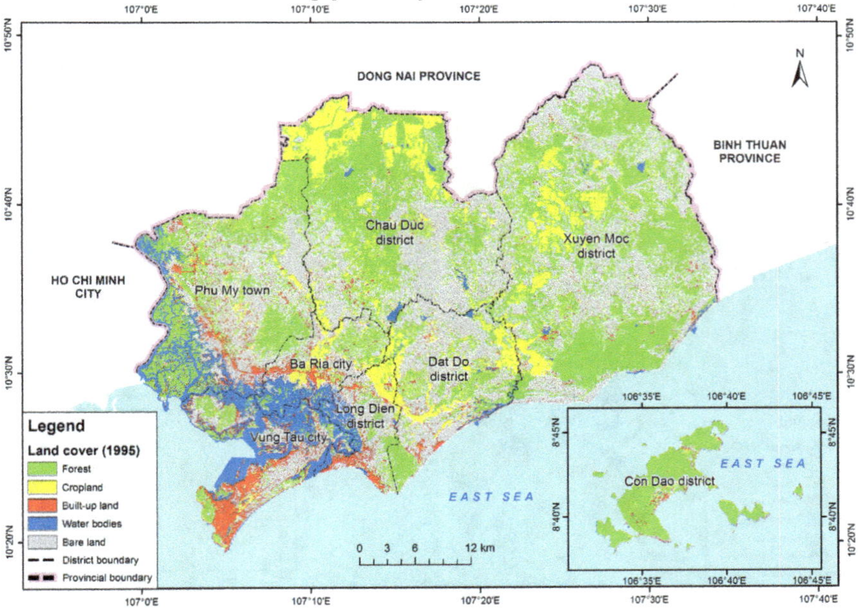

4.1.3. Year 2005

For the year 2005, the LC classification results are very good with an overall accuracy of 98% and a Kappa coefficient of 0.97, as shown in Table 7. The LC map for 2005 is illustrated in Figure 7.

Table 7. Error matrix for the year 2005 (unit: pixel)

		Classified					
		Forest	**Cropland**	**Built-up land**	**Water bodies**	**Bare land**	**Total**
Reference	Forest	705	6	0	0	5	716
	Cropland	8	444	0	0	1	453
	Built-up land	0	1	165	0	9	175
	Water bodies	0	2	0	572	10	584
	Bare land	7	3	2	6	558	576
Total		720	456	167	578	583	2,504
		Overall accuracy = 98%		Kappa coefficient = 0.97			

Note. The error matrix for 2005 shows the number of pixels classified for each land cover type. For example, forest had 705 correct classifications and 11 misclassifications.

Figure 7. Land cover status map for the year 2005

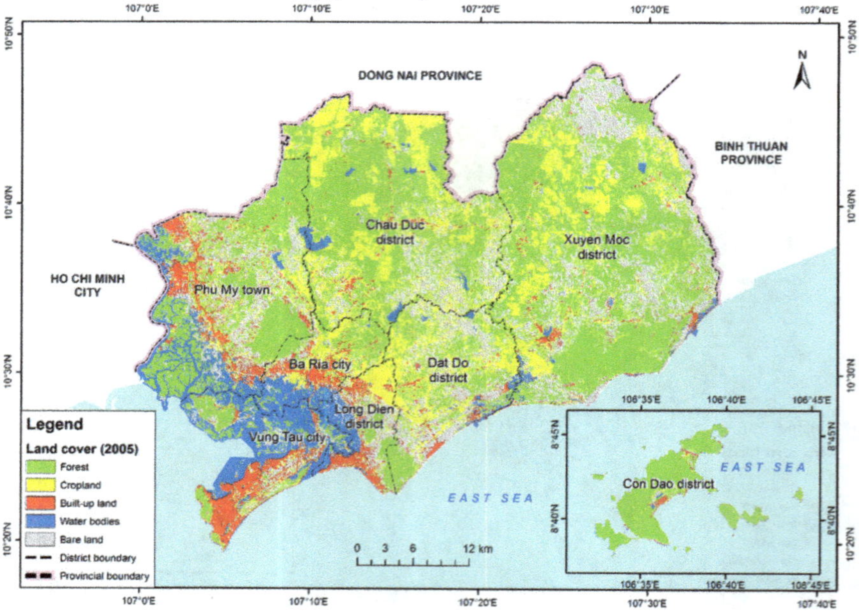

4.1.4. Year 2015

For the year 2015, the LC classification results are very good with an overall accuracy of 99% and a Kappa coefficient of 0.98, as shown in Table 8. The LC map for 2015 is illustrated in Figure 8.

Table 8. Error matrix for the year 2015 (unit: pixel)

		Classified					
		Forest	**Cropland**	**Built-up land**	**Water bodies**	**Bare land**	**Total**
Reference	Forest	660	3	2	0	1	666
	Cropland	0	529	0	0	4	533
	Built-up land	0	0	244	0	9	253
	Water bodies	1	0	0	608	3	612
	Bare land	1	1	2	5	607	616
Total		662	533	248	613	624	2,680
		Overall accuracy = 99%		Kappa coefficient = 0.98			

Note. The error matrix for 2015 shows the number of pixels classified for each land cover type. For example, forest had 660 correct classifications and 6 misclassifications.

Figure 8. Land cover status map for the year 2015

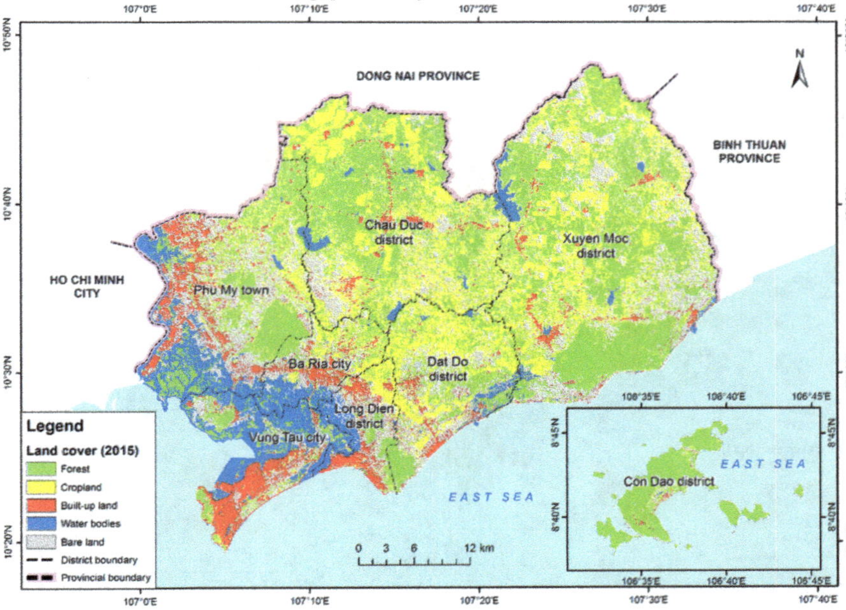

4.1.5. Year 2022

For the year 2022, the LC classification results are very good with an overall accuracy of 99% and a Kappa coefficient of 0.98, as shown in Table 9. The LC map for 2022 is illustrated in Figure 9.

Table 9. Error matrix for the year 2022 (unit: pixel)

		Classified					
		Forest	**Cropland**	**Built-up land**	**Water bodies**	**Bare land**	**Total**
Reference	Forest	713	3	1	0	0	717
	Cropland	4	547	0	0	1	552
	Built-up land	0	0	299	0	9	308
	Water bodies	1	1	0	631	4	637
	Bare land	0	2	9	2	673	686
Total		718	553	309	633	687	2,900
		Overall accuracy = 99%		Kappa coefficient = 0.98			

Note. The error matrix for 2022 shows the number of pixels classified for each land cover type. For example, forest had 713 correct classifications and 4 misclassifications.

Figure 9. Land cover status map for the year 2022

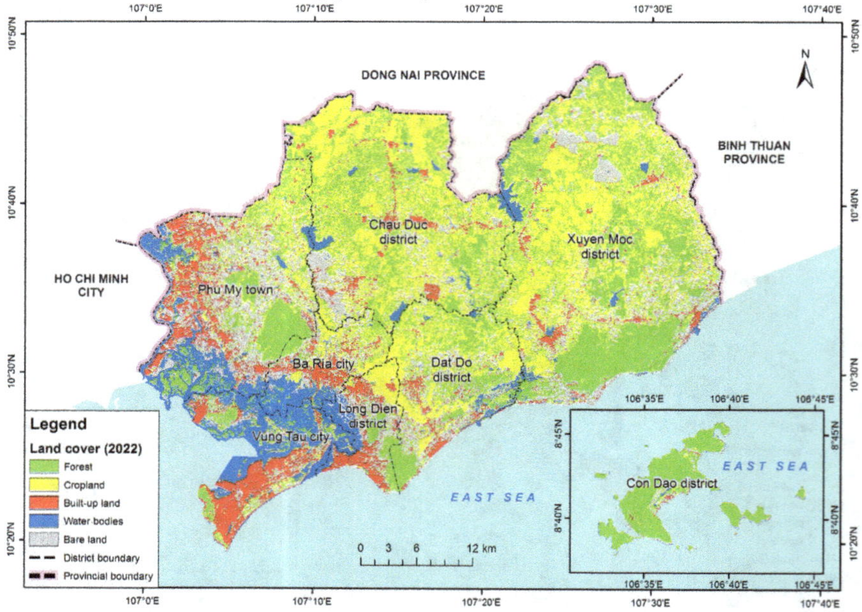

4.2. Land cover change maps

4.2.1. Period 1985-1995

The changes in LC during the period 1985-1995 are summarized in Table 10, Table 11, and illustrated in Figure 10. During this period, the majority of the forest area was preserved, with only 38.41% of it converting to other LC types, predominantly to bare land. Crop areas experienced significant transformation, with 70.73% changing mainly to bare land and forest. Built-up areas saw 43.36% of their space transitioning to other LC types, primarily to bare land. Water bodies had a transition rate of 32.10% to other LC types, mainly to forest. Bare land also underwent notable changes, with 42.29% of its area converting to other LC types, predominantly to forest.

Table 10. Land cover change matrix from 1985 to 1995 (unit: ha)

		1995					
		Forest	**Cropland**	**Built-up land**	**Water bodies**	**Bare land**	**Total**
1985	Forest	38,058.48	5,206.23	600.66	549.18	17,381.88	61,796.43
	Cropland	8,271.54	7,721.55	904.14	493.02	8,986.32	26,376.57
	Built-up land	603.36	235.44	1,508.04	17.46	1,113.30	3,477.60
	Water bodies	2,367.36	813.33	739.26	11,889.72	1,700.10	17,509.77
	Bare land	24,042.42	9,117.00	3,700.62	659.88	51,193.71	88,713.63
Total		73,343.16	23,093.55	7,452.72	13,609.26	80,375.31	197,874.00

Note. The land cover change matrix from 1985 to 1995 shows the area (in hectares) of land cover transitions. For example, forest areas retained 38,058.48 ha and converted to cropland (5,206.23 ha), built-up land (600.66 ha), water bodies (549.18 ha), and bare land (17,381.88 ha).

Table 11. Land cover change matrix from 1985 to 1995 (unit: %)

		1995					
		Forest	**Cropland**	**Built-up land**	**Water bodies**	**Bare land**	**Total**
1985	Forest	61.59	8.42	0.97	0.89	28.13	100.00
	Cropland	31.36	29.27	3.43	1.87	34.07	100.00
	Built-up land	17.35	6.77	43.36	0.50	32.01	100.00
	Water bodies	13.52	4.65	4.22	67.90	9.71	100.00
	Bare land	27.10	10.28	4.17	0.74	57.71	100.00

Note. The land cover change matrix from 1985 to 1995 shows percentage changes in land cover types. For example, forest areas retained 61.59% and converted to other uses: 8.42% to cropland, 0.97% to built-up land, 0.89% to water bodies, and 28.13% to bare land. Total percentages for each land cover type sum to 100%.

Figure 10. Land cover change map from 1985 to 1995

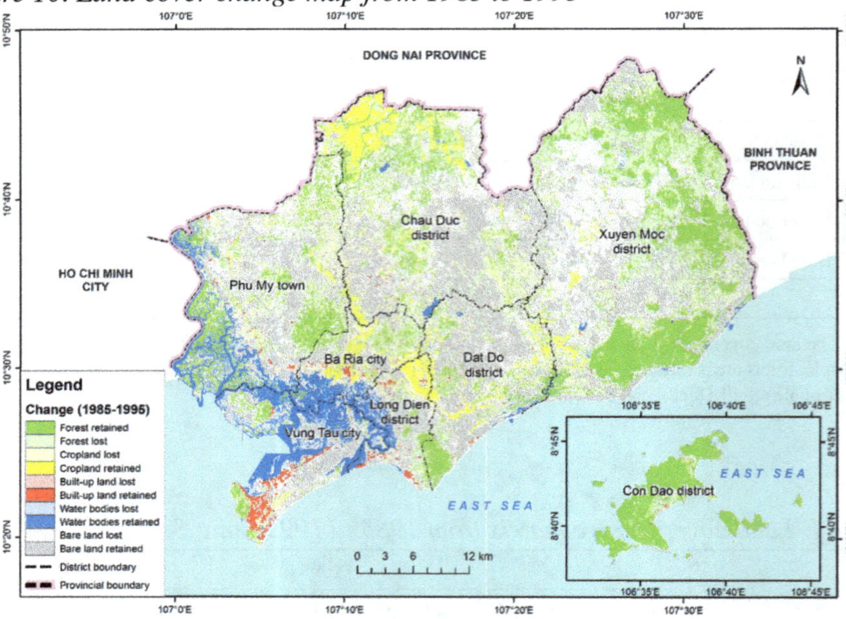

4.2.2. Period 1995-2005

The changes in LC during the period 1995-2005 are summarized in Table 12, Table 13, and illustrated in Figure 11. During this period, most of the forest area was preserved, with only 30.19% of the forest area transitioning to other LC types, predominantly crops and bare land. Cropland experienced significant transformation, with 42.40% of its area converting to other LC types, mainly forest. Built-up areas saw 45.23% of their space changing to other LC types, primarily to bare land and cropland. Water bodies had a transition rate of 13.51% to other LC types, mainly to forest and bare land. Bare land also underwent notable changes, with 58.86% of its area converting to other LC types, predominantly to forest and cropland.

Table 12. Land cover change matrix from 1995 to 2005 (unit: ha)

		2005					
		Forest	**Cropland**	**Built-up land**	**Water bodies**	**Bare land**	**Total**
1995	Forest	51,203.43	11,653.92	1,111.14	1,303.11	8,079.12	73,350.72
	Cropland	7,021.26	13,301.28	639.27	337.77	1,794.33	23,093.91
	Built-up land	514.53	874.44	4,082.13	695.34	1,286.46	7,452.90
	Water bodies	904.77	118.44	187.65	11,770.83	627.57	13,609.26
	Bare land	23,803.29	17,534.97	3,981.87	1,991.88	33,065.19	80,377.20
Total		83,447.28	43,483.05	10,002.06	16,098.93	44,852.67	197,883.99

Note. The land cover change matrix from 1995 to 2005 shows the area (in hectares) of land cover transitions. For example, forest areas retained 51,203.43 ha and converted to cropland (11,653.92 ha), built-up land (1,111.14 ha), water bodies (1,303.11 ha), and bare land (8,079.12 ha).

Table 13. Land cover change matrix from 1995 to 2005 (unit: %)

		2005					
		Forest	**Cropland**	**Built-up land**	**Water bodies**	**Bare land**	**Total**
1995	Forest	69.81	15.89	1.51	1.78	11.01	100.00
	Cropland	30.40	57.60	2.77	1.46	7.77	100.00
	Built-up land	6.90	11.73	54.77	9.33	17.26	100.00
	Water bodies	6.65	0.87	1.38	86.49	4.61	100.00
	Bare land	29.61	21.82	4.95	2.48	41.14	100.00

Note. The land cover change matrix from 1995 to 2005 shows percentage changes in land cover types. For example, forest areas retained 69.81% and converted to other uses: 15.89% to cropland, 1.51% to built-up land, 1.78% to water bodies, and 11.01% to bare land. Total percentages for each land cover type sum to 100%.

Figure 11. Land cover change map from 1995 to 2005

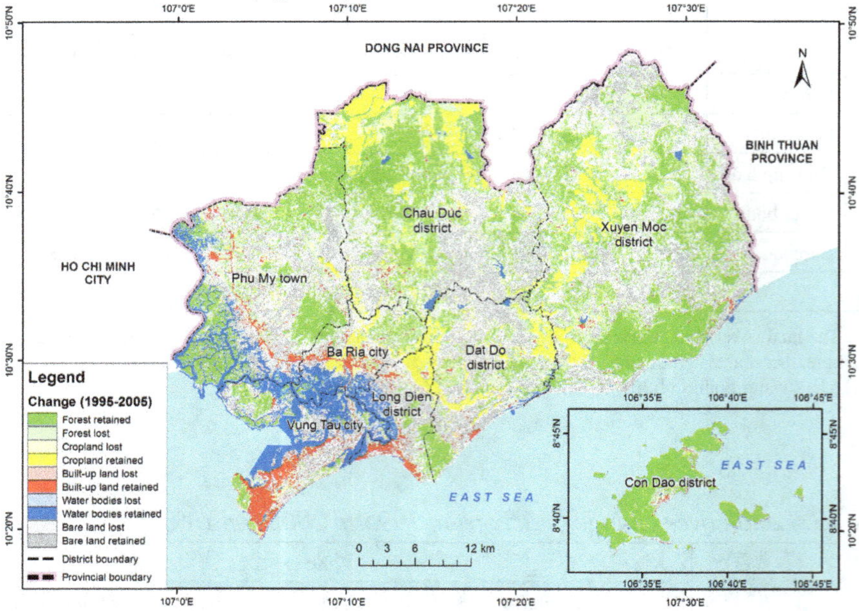

4.2.3. Period 2005-2015

The changes in LC during the period 2005-2015 are summarized in Table 14, Table 15, and illustrated in Figure 12. During this period, most of the forest area was preserved, with only 42.40% of the forest area transitioning to other LC types, primarily to cropland. Cropland experienced a significant change, with 55.09% of its area converting to other LC types, mainly to bare land and forest. Built-up areas saw 30.60% of their space changing to other LC types, primarily to bare land. Water bodies had a transition rate of 15.12% to other LC types, mainly to bare land. Bare land also underwent notable changes, with 53.00% of its area converting to other LC types, predominantly to cropland.

Table 14. Land cover change matrix from 2005 to 2015 (unit: ha)

		2015					
		Forest	**Cropland**	**Built-up land**	**Water bodies**	**Bare land**	**Total**
2005	Forest	48,069.63	18,519.57	1,256.40	2,812.77	12,788.91	83,447.28
	Cropland	9,678.78	19,527.57	1,669.14	430.56	12,177.00	43,483.05
	Built-up land	115.29	242.73	6,943.14	125.91	2,577.24	10,004.31
	Water bodies	464.22	467.64	339.75	13,664.88	1,162.44	16,098.93
	Bare land	6,928.20	12,842.19	2,455.92	1,544.22	21,082.14	44,852.67
Total		65,256.12	51,599.70	12,664.35	18,578.34	49,787.73	197,886.24

Note. The land cover change matrix from 2005 to 2015 shows the area (in hectares) of land cover transitions. For example, forest areas retained 48,069.63 ha and converted to cropland (18,519.57 ha), built-up land (1,256.40 ha), water bodies (2,812.77 ha), and bare land (12,788.91 ha).

Table 15. Land cover change matrix from 2005 to 2015 (unit: %)

		2015					
		Forest	**Cropland**	**Built-up land**	**Water bodies**	**Bare land**	**Total**
2005	Forest	57.60	22.19	1.51	3.37	15.33	100.00
	Cropland	22.26	44.91	3.84	0.99	28.00	100.00
	Built-up land	1.15	2.43	69.40	1.26	25.76	100.00
	Water bodies	2.88	2.90	2.11	84.88	7.22	100.00
	Bare land	15.45	28.63	5.48	3.44	47.00	100.00

Note. The land cover change matrix from 2005 to 2015 shows percentage changes in land cover types. For example, forest areas retained 57.60% and converted to other uses: 22.19% to cropland, 1.51% to built-up land, 3.37% to water bodies, and 15.33% to bare land. Total percentages for each land cover type sum to 100%.

Figure 12. Land cover change map from 2005 to 2015

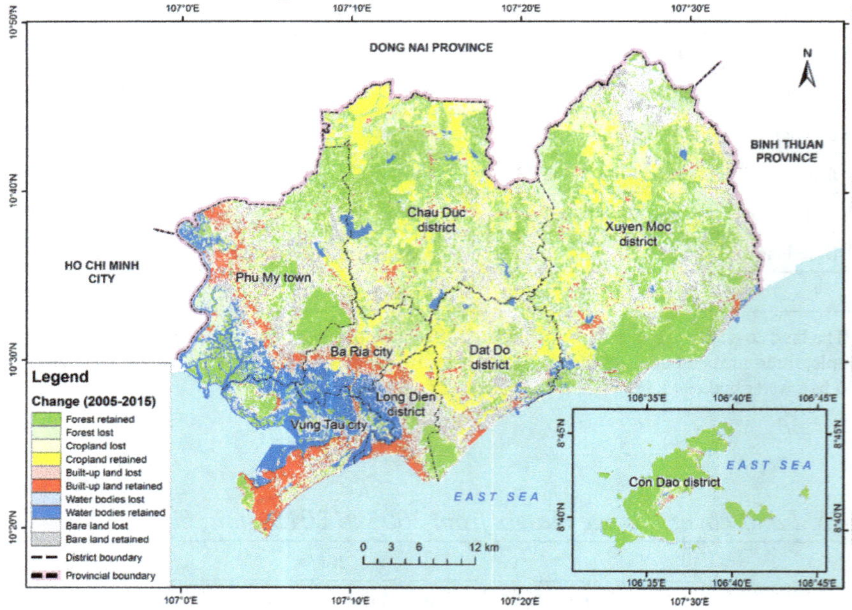

4.2.4. Period 2015-2022

The changes in LC during the period 2015-2022 are summarized in Table 16, Table 17, and illustrated in Figure 13. During this period, most of the forest area was preserved, with only 37.42% of the forest area transitioning to other LC types, primarily to cropland. Cropland experienced a notable change, with 33.81% of its area converting to other LC types, mainly to forest and bare land. Built-up areas saw 13.42% of their space changing to other LC types, primarily to bare land. Water bodies had a transition rate of 14.51% to other LC types, mainly to bare land. Bare land also underwent significant changes, with 43.08% of its area converting to other LC types, predominantly to cropland.

Table 16. Land cover change matrix from 2015 to 2022 (unit: ha)

		2022					
		Forest	**Cropland**	**Built-up land**	**Water bodies**	**Bare land**	**Total**
2015	Forest	40,840.38	20,204.55	201.78	713.25	3,296.16	65,256.12
	Cropland	8,368.92	34,153.38	347.31	561.69	8,168.40	51,599.70
	Built-up land	55.08	54.54	10,968.30	29.16	1,562.04	12,669.12
	Water bodies	691.47	709.38	277.47	15,882.21	1,017.81	18,578.34
	Bare land	4,455.72	13,170.24	3,004.83	819.36	28,337.58	49,787.73
Total		54,411.57	68,292.09	14,799.69	18,005.67	42,381.99	197,891.01

Note. The land cover change matrix from 2015 to 2022 shows the area (in hectares) of land cover transitions. For example, forest areas retained 40,840.38 ha and converted to cropland (20,204.55 ha), built-up land (201.78 ha), water bodies (713.25 ha), and bare land (3,296.16 ha).

Table 17. Land cover change matrix from 2015 to 2022 (unit: %)

		2022					
		Forest	**Cropland**	**Built-up land**	**Water bodies**	**Bare land**	**Total**
2015	Forest	62.58	30.96	0.31	1.09	5.05	100.00
	Cropland	16.22	66.19	0.67	1.09	15.83	100.00
	Built-up land	0.43	0.43	86.58	0.23	12.33	100.00
	Water bodies	3.72	3.82	1.49	85.49	5.48	100.00
	Bare land	8.95	26.45	6.04	1.65	56.92	100.00

Note. The land cover change matrix from 2015 to 2022 shows percentage changes in land cover types. For example, forest areas retained 62.58% and converted to other uses: 30.96% to cropland, 0.31% to built-up land, 1.09% to water bodies, and 5.05% to bare land. Total percentages for each land cover type sum to 100%.

Figure 13. Land cover change map from 2015 to 2022

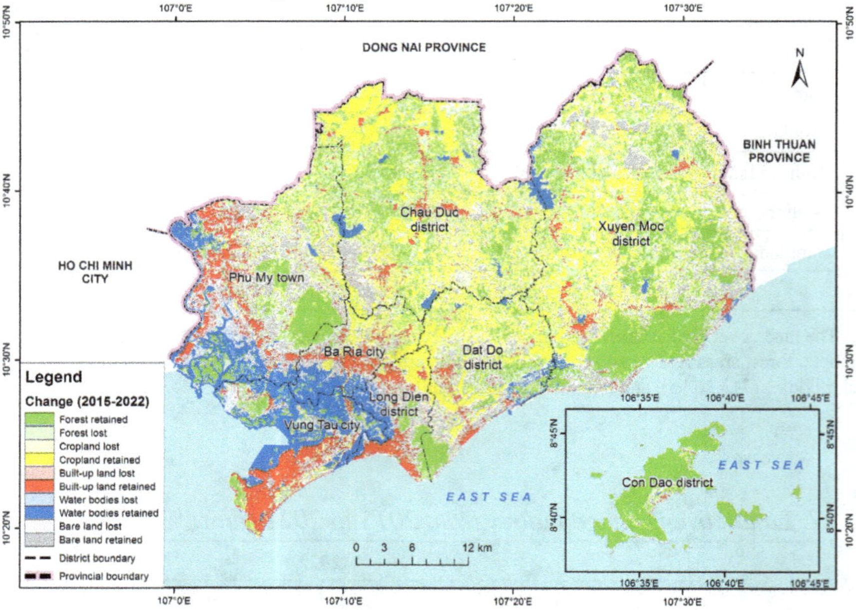

4.3. Discussion

In 1985, ten years after the reunification of Vietnam, the majority of the province's land was either bare land (45%) or forest (31%) (see Table 19). By 1995, nine years into the renovation period starting from 1986, there was an expansion of built-up areas (increasing by 3,975 hectares compared to 1985) alongside a reduction in bare land (decreasing by 8,337 hectares compared to 1985). Despite this, bare land still dominated the landscape, comprising 41% of the total area. By 2005, forest area continued to expand and became the predominant LC type, covering 42% of the area. The area of crops has significantly increased, nearly doubling the crop area from 1995, reaching 10,004 ha. By 2015, forest area declined due to the expansion of cropland. Although forest still was dominant, its area dropped to 33%, while cropland occupied 26% of the land. By 2022, agricultural production was further intensified, leading to an additional 16,692 hectares of cropland, which became the primary LC type. Forest area continued to shrink, accounting for only 27% of the total area.

Table 19. Statistics on land cover change from 1985 to 2022

LC	1985		1995		2005		2015		2022	
	(ha)	**(%)**	**(ha)**	**(%)**	**(ha)**	**(%)**	**(ha)**	**(%)**	**(ha)**	**(%)**
Forest	61,796	31	73,351	37	83,447	42	65,256	33	54,412	27
Cropland	26,377	13	23,094	12	43,483	22	51,600	26	68,292	35
Built-up land	3,478	2	7,453	4	10,004	5	12,669	6	14,800	7
Water bodies	17,510	9	13,609	7	16,099	8	18,578	9	18,006	9
Bare land	88,714	45	80,377	41	44,853	23	49,788	25	42,382	21
Total	197,874	100	197,884	100	197,886	100	197,891	100	197,891	100

Note. From 1985 to 2022, the land cover statistics show notable changes. Forested areas decreased from 61,796 hectares (31%) in 1985 to 54,412 hectares (27%) in 2022. Cropland expanded significantly, increasing from 26,377 hectares (13%) to 68,292 hectares (35%). Built-up land grew from 3,478 hectares (2%) to 14,800 hectares (7%). Water bodies remained relatively stable, around 18,000 hectares (9%). Bare land decreased from 88,714 hectares (45%) to 42,382 hectares (21%).

In terms of trends, forest area increased during the period 1985-2005 due to policies promoting forest protection and development. However, it decreased in the subsequent period 2005-2022 due to rising pressures from population growth and socio-economic development. For crops and built-up areas, a nearly continuous increase was observed from 1985 to 2022, reflecting the pressures of growing population and economic development. Water bodies decreased during 1985-2005 but increased in the period 2005-2022 due to the development of irrigation systems. Bare land predominantly decreased continuously from 1985 to 2022.

In terms of spatial analysis, the total area of LC disturbance across the province for the periods 1985-1995, 1995-2005, 2005-2015, and 2015-2022 were 44%, 43%, 45%, and 34%, respectively. During 1985-1995, the primary disturbance states were forest converting to bare land (17,381.88 ha) and vice versa (24,042.42 ha). In the 1995-2005 period, the main disturbance states involved bare land converting to forest and crops, with areas of 23,803.29 ha and 17,534.97 ha, respectively. For the 2005-2015 period, key disturbances included forest converting to crops and bare land, with areas of 18,519.57 ha and 12,788.91 ha, respectively; crops converting to bare land (12,177.00 ha); and bare land converting to crops (12,842.19 ha). In the 2015-2022 period, the main disturbance states involved crops, with forest and bare land converting to crops, at areas of 20,204.55 ha and 13,170.24 ha, respectively.

5. CONCLUSION

The literature review highlighted the significant role of remote sensing in global data acquisition, emphasizing the shift from data scarcity to the challenge of managing and analyzing vast datasets due to advancements in sensor technology and open-access data. The emergence of cloud platforms, particularly GEE, has revolutionized geo-big data analytics by providing scalable, cost-effective solutions for processing large datasets. GEE stands out for its extensive, analysis-ready data catalogue and high-performance computational capabilities, making it the most widely-used tool in remote sensing for geographical big data analysis. Despite some limitations in spatial functions and algorithmic support, GEE's versatility in handling both platform and user data, coupled with its ability to perform complex spectral and spatial analyses, has led to its widespread adoption in research. The surge in GEE-related publications, especially in the last few years, underscores its importance as a global-scale tool, offering substantial research opportunities, particularly in water resources assessment and in developing countries.

The study completed the classification of LC for the years 1985, 1995, 2005, 2015, and 2022 based on multi-temporal Landsat optical satellite images (5 TM, 8 OLI/TIRS) for BRVT province, Vietnam using the random forest algorithm. The classification accuracy for all periods was excellent, with overall accuracy above 96% and Kappa coefficients above 0.95. In 1985, the majority of the province's area was bare land (45%) and forest (31%). By 1995, there was an expansion in built-up areas accompanied by a reduction in bare land, though bare land remained the predominant LC (41%). By 2005, forest area continued to expand, becoming the dominant LC (42%), while crop area increased significantly. By 2015, forest area declined due to the expansion of crop areas but still remained predominant (33% of the area), whereas crops covered 26% of the area. By 2022, agricultural production intensified, making crops the primary LC, while forested areas continued to shrink, covering only 27% of the area.

The changes in LC during the periods 1985-1995, 1995-2005, 2005-2015, and 2015-2022 were analyzed using post-classification comparison methods based on GIS technology. The results indicated that forest area expanded from 1985 to 2005 as a result of forest protection and development policies. However, this trend reversed in the following period (2005-2022), with a decrease in forest area attributed to population growth and socio-economic development pressures. For crops and built-up areas, there was a nearly continuous increase from 1985 to 2022 due to the same pressures. Water bodies decreased from 1985 to 2005 but increased from 2005 to 2022 due to the development of irrigation systems. Bare land continuously decreased throughout the period from 1985 to 2022. Spatially, the total area of LC disturbance across the province for the periods 1985-1995, 1995-2005, 2005-2015, and 2015-2022 were 44%, 43%, 45%, and 34%, respectively.

This research holds significant importance both theoretically and practically. Theoretically, it offers valuable insights into LCC over various periods from 1985 to 2022, utilizing satellite data and advanced classification algorithms such as random forest. This contributes to reinforcing current knowledge about LC dynamics concerning economic growth, social changes, and environmental conservation policies. Practically, the study offers valuable information for land management, urban planning, and environmental policy-making, supporting decision-makers with precise and up-to-date data analysis.

Despite the positive outcomes, the study has certain limitations. Firstly, the resolution of satellite imagery may affect classification accuracy, especially in distinguishing between similar LC types. Lastly, factors such as cloud cover and inconsistent weather conditions could impact the quality of satellite data. To address these limitations and broaden the scope of research, future studies should integrate data from various sources, such as high-resolution satellite imagery, LiDAR, and radar, to enhance classification accuracy. Furthermore, investigating the factors influencing LCC—such as climate change, land management practices, and development policies—will yield a more thorough understanding of the dynamics involved. This approach will enhance the accuracy of LC assessments and provide valuable insights into the underlying drivers of change, thereby supporting more informed decision-making in land management and environmental policy.

REFERENCES

Amani, M., Ghorbanian, A., Ahmadi, S. A., Kakooei, M., Moghimi, A., Mirmazloumi, S. M., Moghaddam, S. H. A., Mahdavi, S., Ghahremanloo, M., Parsian, S., Wu, Q., & Brisco, B. (2020). Google Earth Engine Cloud Computing Platform for Remote Sensing Big Data Applications: A Comprehensive Review. *IEEE Journal of Selected Topics in Applied Earth Observations and Remote Sensing*, 13, 5326–5350. DOI: 10.1109/JSTARS.2020.3021052

Anderson, J. R., Hardy, E. E., Roach, J. T., & Witmer, R. E. (1961). *A land use and land cover classification system for use with remote sensor data.*

Breiman, L. (2001). Random Forests. *Machine Learning*, 45(1), 5–32. https://doi.org/http://dx.doi.org/10.1023/A:1010933404324. DOI: 10.1023/A:1010933404324

Breiman, L., Friedman, J. H., Olshen, R. A., & Stone, C. J. (2017). *Classification and Regression Trees*. Routledge., DOI: 10.1201/9781315139470

Brown, C. F., Brumby, S. P., Guzder-Williams, B., Birch, T., Hyde, S. B., Mazzariello, J., Czerwinski, W., Pasquarella, V. J., Haertel, R., Ilyushchenko, S., Schwehr, K., Weisse, M., Stolle, F., Hanson, C., Guinan, O., Moore, R., & Tait, A. M. (2022). Dynamic World, Near real-time global 10 m land use land cover mapping. *Scientific Data*, 9(1), 1–17. DOI: 10.1038/s41597-022-01307-4 PMID: 35013360

Campbell, J. B., Wynne, R. H., & Thomas, V. A. (2023). *Introduction to Remote Sensing* (6th ed.). The Guilford Press.

Carrasco, L., O'Neil, A. W., Morton, R. D., & Rowland, C. S. (2019). Evaluating Combinations of Temporally Aggregated Sentinel-1, Sentinel-2 and Landsat 8 for Land Cover Mapping with Google Earth Engine. *Remote Sensing (Basel)*, 11(3), 288. DOI: 10.3390/rs11030288

Chan, J. C.-W., & Paelinckx, D. (2008). Evaluation of Random Forest and Adaboost tree-based ensemble classification and spectral band selection for ecotope mapping using airborne hyperspectral imagery. *Remote Sensing of Environment*, 112(6), 2999–3011. DOI: 10.1016/j.rse.2008.02.011

Department of Natural Resources and Environment of Ba Ria-Vung Tau Province. (1995). *Land Use Status Map of 1995.*

Department of Natural Resources and Environment of Ba Ria-Vung Tau Province. (2005). *Land Use Status Map of 2005.*

Department of Natural Resources and Environment of Ba Ria-Vung Tau Province. (2015). *Land Use Status Map of 2015.*

Department of Natural Resources and Environment of Ba Ria - Vung Tau Province. (2019). *Project "Action Plan for Biodiversity in Ba Ria - Vung Tau Province to 2020 and Vision to 2030."*

Department of Natural Resources and Environment of Ba Ria-Vung Tau Province. (2020). *Land Use Status Map of 2020.*

Department of Natural Resources and Environment of Ba Ria-Vung Tau Province. (2022). *Electronic Atlas of Ba Ria-Vung Tau Province.* http://atlas.baria-vungtau .gov.vn

Di Gregorio, A., & Jansen, L. J. M. (1998). *Land Cover Classification System (LCCS).* Classification Concepts and User Manual.

European Commission. (2023). *GHSL Data Package 2023.* Publications Office of the European Union.

Fan, P., Ouyang, Z., Nguyen, D. D., Nguyen, T. T. H., Park, H., & Chen, J. (2019). Urbanization, economic development, environmental and social changes in transitional economies: Vietnam after Doimoi. *Landscape and Urban Planning*, 187, 145–155. DOI: 10.1016/j.landurbplan.2018.10.014

Foody, G. M. (2004). Thematic Map Comparison. *Photogrammetric Engineering and Remote Sensing*, 70(5), 627–633. DOI: 10.14358/PERS.70.5.627

Friedl, M. A., & Brodley, C. E. (1997). Decision tree classification of land cover from remotely sensed data. *Remote Sensing of Environment*, 61(3), 399–409. DOI: 10.1016/S0034-4257(97)00049-7

Ghimire, B., Rogan, J., & Miller, J. (2010). Contextual land-cover classification: Incorporating spatial dependence in land-cover classification models using random forests and the Getis statistic. *Remote Sensing Letters*, 1(1), 45–54. DOI: 10.1080/01431160903252327

Giri, C. P. (Ed.). (2012). *Remote Sensing of Land Use and Land Cover - Principles and Applications* (1st ed.). CRC Press., DOI: 10.1201/b11964

Gorelick, N., Hancher, M., Dixon, M., Ilyushchenko, S., Thau, D., & Moore, R. (2017). Google Earth Engine: Planetary-scale geospatial analysis for everyone. *Remote Sensing of Environment*, 202, 18–27. DOI: 10.1016/j.rse.2017.06.031

Guo, L., Chehata, N., Mallet, C., & Boukir, S. (2011). Relevance of airborne lidar and multispectral image data for urban scene classification using Random Forests. *ISPRS Journal of Photogrammetry and Remote Sensing*, 66(1), 56–66. DOI: 10.1016/j.isprsjprs.2010.08.007

Hansen, M. C., & DeFries, R. S. (2004). Detecting Long-term Global Forest Change Using Continuous Fields of Tree-Cover Maps from 8-km Advanced Very High Resolution Radiometer (AVHRR) Data for the Years 1982?99. *Ecosystems (New York, N.Y.)*, 7(7), 695–716. DOI: 10.1007/s10021-004-0243-3

Jensen, J. R. (1996). *Introductory Digital Image Processing: A Remote Sensing Perspective* (2nd ed.). Prentice Hall.

Justice, C. O., & Townshend, J. R. G. (1981). Integrating ground data with remote sensing. In Townshend, J. R. G. (Ed.), *Terrain Analysis and Remote Sensing* (pp. 38–58). Allen & Unwin.

Lambin, E. F., & Meyfroidt, P. (2011). Global land use change, economic globalization, and the looming land scarcity. *Proceedings of the National Academy of Sciences of the United States of America*, 108(9), 3465–3472. DOI: 10.1073/pnas.1100480108 PMID: 21321211

Latifi, H., Nothdurft, A., & Koch, B. (2010). Non-parametric prediction and mapping of standing timber volume and biomass in a temperate forest: Application of multiple optical/LiDAR-derived predictors. *Forestry*, 83(4), 395–407. DOI: 10.1093/forestry/cpq022

Lin, F.-C., Chung, L.-K., Wang, C.-J., Ku, W.-Y., & Chou, T.-Y. (2013). Storage and processing of massive remote sensing images using a novel cloud computing platform. *GIScience & Remote Sensing*, 50(3), 322–336. DOI: 10.1080/15481603.2013.810976

Loveland, T. R., & Belward, A. S. (1997). The International Geosphere Biosphere Programme Data and Information System global land cover data set (DISCover). *Acta Astronautica, 41*(4–10), 681–689. DOI: 10.1016/S0094-5765(98)00050-2

Martino, P., & Panagiotis, P. (2023). *GHS-BUILT-S R2023A - GHS built-up surface grid, derived from Sentinel2 composite and Landsat, multitemporal (1975-2030)*. European Commission, Joint Research Centre. https://doi.org/DOI: 10.2905/9F06F36F-4B11-47EC-ABB0-4F8B7B1D7

McElwee, P. (2009). Reforesting "Bare Hills" in Vietnam: Social and Environmental Consequences of the 5 Million Hectare Reforestation Program. *Ambio*, 38(6), 325–333. DOI: 10.1579/08-R-520.1 PMID: 19860156

Meyfroidt, P., & Lambin, E. F. (2008). The causes of the reforestation in Vietnam. *Land Use Policy*, 25(2), 182–197. DOI: 10.1016/j.landusepol.2007.06.001

Mountrakis, G., Watts, R., Luo, L., & Wang, J. (2009). Developing Collaborative Classifiers using an Expert-based Model. *Photogrammetric Engineering and Remote Sensing*, 75(7), 831–843. DOI: 10.14358/PERS.75.7.831

Noi Phan, T., Kuch, V., & Lehnert, L. W. (2020). Land cover classification using google earth engine and random forest classifier-the role of image composition. *Remote Sensing (Basel)*, 12(15), 2411. Advance online publication. DOI: 10.3390/rs12152411

People's Committee of Ba Ria - Vung Tau Province. (2023). *Comprehensive Report on the Provincial Planning of Ba Ria - Vung Tau for the Period 2021-2030, Vision to 2050.*

Pérez-Cutillas, P., Pérez-Navarro, A., Conesa-García, C., Zema, D. A., & Amado-Álvarez, J. P. (2023). What is going on within google earth engine? A systematic review and meta-analysis. *Remote Sensing Applications: Society and Environment, 29*(September 2022), 100907. DOI: 10.1016/j.rsase.2022.100907

Phan, D. C., Trung, T. H., Truong, V. T., Sasagawa, T., Vu, T. P. T., Bui, D. T., Hayashi, M., Tadono, T., & Nasahara, K. N. (2021). First comprehensive quantification of annual land use/cover from 1990 to 2020 across mainland Vietnam. *Scientific Reports*, 11(1), 1–20. DOI: 10.1038/s41598-021-89034-5 PMID: 33976255

Potapov, P., Hansen, M. C., Pickens, A., Hernandez-Serna, A., Tyukavina, A., Turubanova, S., Zalles, V., Li, X., Khan, A., Stolle, F., Harris, N., Song, X.-P., Baggett, A., Kommareddy, I., & Kommareddy, A. (2022). The Global 2000-2020 Land Cover and Land Use Change Dataset Derived From the Landsat Archive: First Results. *Frontiers in Remote Sensing*, 3(April), 1–22. DOI: 10.3389/frsen.2022.856903

Rogan, J., Miller, J., Stow, D., Franklin, J., Levien, L., & Fischer, C. (2003). Land-Cover Change Monitoring with Classification Trees Using Landsat TM and Ancillary Data. *Photogrammetric Engineering and Remote Sensing*, 69(7), 793–804. DOI: 10.14358/PERS.69.7.793

Soesbergen, A. van. (2016). *A review of land use change models.*

Sun, A. Y., & Scanlon, B. R. (2019). How can Big Data and machine learning benefit environment and water management: A survey of methods, applications, and future directions. *Environmental Research Letters*, 14(7), 073001. DOI: 10.1088/1748-9326/ab1b7d

Tamiminia, H., Salehi, B., Mahdianpari, M., Quackenbush, L., Adeli, S., & Brisco, B. (2020). Google Earth Engine for geo-big data applications: A meta-analysis and systematic review. *ISPRS Journal of Photogrammetry and Remote Sensing*, 164, 152–170. DOI: 10.1016/j.isprsjprs.2020.04.001

The World Bank Group. (2024). *Vietnam - Climatology*. https://climateknowledgeportal.worldbank.org/country/vietnam/climate-data-historical

The World Factbook. (2024). *Vietnam*. https://www.cia.gov/the-world-factbook/countries/vietnam/

Turner, B., & Meyer, W. (1994). Global Land-Use and Land-Cover Change: An Overview. In Meyer, W., & Turner, B. (Eds.), *Changes in Land Use and Land Cover: A Global Perspective* (pp. 3–10). Cambridge University Press.

U.S. Geological Survey. (2023a). *Landsat 5*. https://www.usgs.gov/landsat-missions/landsat-5

U.S. Geological Survey. (2023b). *Landsat 8*. https://www.usgs.gov/landsat-missions/landsat-8

Yang, C., Huang, Q., Li, Z., Liu, K., & Hu, F. (2017). Big Data and cloud computing: Innovation opportunities and challenges. *International Journal of Digital Earth*, 10(1), 13–53. DOI: 10.1080/17538947.2016.1239771

KEY TERMS AND DEFINITIONS

Accuracy Assessment: The process of evaluating the correctness of classification results using metrics such as overall accuracy and Kappa coefficient. It involves comparing classified results with ground-truth data.

Change Matrix: A table that shows the transition of land cover types from one period to another, typically presented in absolute values (hectares) and percentages. It helps in understanding the magnitude and direction of land cover changes.

Cloud Masking: The process of removing or correcting areas in satellite imagery affected by clouds, fog, or shadows to improve image quality for analysis.

Cloud-Free Imagery: Refers to satellite images without cloud cover, which is crucial for accurate land cover mapping over large areas.

Error Matrix: A table used to summarize the performance of a classification algorithm by comparing the classified results against reference data.

Field Surveys: On-the-ground data collection involving visits to specific locations to gather information about land cover and land use. This data is used to validate and refine remote sensing classifications.

Geographic Information System (GIS): A system for managing, analyzing, and visualizing spatial and geographic data, often used in conjunction with remote sensing technology.

Google Earth Engine (GEE): A cloud-based computing platform that allows users to analyze large volumes of remotely sensed images using a web-based Integrated Development Environment (IDE) code editor, without the need for local data downloads.

Image Clipping: The process of extracting a specific area of interest from a larger image based on defined boundaries, such as administrative boundaries.

Image Compositing: Combining multiple images taken over time or from different sensors into a single composite image to reduce noise and enhance the quality of the data.

Interpretation Key: A standardized guide used for identifying and classifying objects in imagery based on specific criteria. It ensures consistency and accuracy in classification.

Kappa Coefficient: A statistical measure that assesses the agreement between classified and reference data, correcting for chance agreement. It ranges from 0 to 1, where 1 indicates perfect agreement.

Land Cover (LC): Refers to the physical state of the Earth's surface, such as vegetation or constructed structures. It describes the observable physical cover of the land.

Land Cover Change (LCC): Involves the transformation or modification of land cover. Transformation refers to changing one type of LC to another, while modification refers to altering internal conditions within an LC.

Landsat Satellite Images: Satellite imagery provided by the Landsat program, which has been operational since 1972. These images are used for earth observation and come in different versions, such as Landsat 5 TM and Landsat 8 OLI/TIRS, each with specific spectral bands and spatial resolutions.

Overall Accuracy: A measure of the correctness of land cover classification, calculated as the ratio of correctly classified pixels to the total number of pixels. It indicates the percentage of accurate classifications across the entire image.

Random Forest Classification: A machine learning algorithm that uses an ensemble of decision trees to classify data. It is known for its accuracy and robustness in handling large datasets with high-dimensional features.

Remote Sensing: The science of collecting, measuring, and analyzing information about observed objects without direct contact. It uses spectral reflectance to identify or analyze characteristics of land cover types.

Chapter 8
Remote Sensing and GIS for Weed Detection Using Enhanced Spatial Insights: A Review

Faran Masood Peerzada
https://orcid.org/0000-0003-3238-1330
College of E&ME, National University of Sciences and Technology, Islamabad, Pakistan

Zeeshan Ali Shah
https://orcid.org/0000-0002-1606-6181
Department of Electrical Engineering, Wah Engineering College, University of Wah, Pakistan

ABSTRACT

Remote sensing (RS) and Geographic Information Systems (GIS) technologies are receiving significant attention due to their combined potential to revolutionize spatial analysis and enhance understanding of complex environmental and geographic phenomena. This integration has vast applications, spanning from urban planning to engineering technologies. Remote sensing and GIS play essential roles in detecting and mapping weed infestations in agricultural fields, facilitating precise weed management and reducing herbicide use. Advancements include high-resolution imagery and machine learning algorithms, which enhance detection accuracy and efficiency. This chapter reviews and discusses the framework of weed detection

DOI: 10.4018/979-8-3693-6452-9.ch008

based on different techniques, modes of data collection, and categories of weed features, including spectral and spatial features of weeds. Additionally, it discusses the different challenges encountered in the detection of weed.

1. INTRODUCTION

Remote Sensing (RS) and Geographic Information Systems (GIS) are two technologies without which modern farming would not have been possible. They have lifted the industry to the level of spatial analysis, monitoring the environment as well as the process automation through the wide range of technologies. Remote sensing is the art of observing the surface of the Earth without physical contact. It is usually done by means of satellites, drones, or aircraft that have been loaded with sensors. These sensors capture the signals or images or any other form of data by then, these sensors can pick up distinct wavelengths like visible light, infrared, or radar to discover the changes even in physical properties like size and time across different media. Geographic Information Systems (GIS) systems mostly include tools for recording, storing, explaining and displaying location data on a map. GIS brings together the spatial data (the data that says the location) and the attribute data (the data that describes what it is) to make detailed maps and execute sophisticated spatial analyses (generally differentiation of spatial variables related to each other) (Bibhuti Bhusan & Giri, 2023)

In urban planning, the integration of RS and GIS can be very helpful in monitoring urban sprawl, managing land use, and assessing the environmental impacts (Kumar et al., 2023). In case of agriculture, these technologies can be utilized to monitor the health of crop, map the different properties of soil, and detect different type of weed infestations, leading to more efficient and precise farm management practices. This integration's capacity to track and evaluate changes over time also makes it indispensable for research on climate change, natural resource management, and catastrophe management. Using RS technologies, which offer high-resolution images and spectral data essential for precisely identifying weed species and separating them from crops, improves detection accuracy. In addition to monitoring weed growth patterns and identifying infestations at different phases of growth, temporal analysis of RS data can be used to plan timely treatments and gain insight into the lifecycle of weeds (Xu et al., 2023)

The ultimate result of combining RS and GIS technologies is precision weed management, which lessens the need for broad herbicide treatments and has a smaller negative impact on the environment. Farmers can minimize the impact of herbicide applications on non-target species and the environment while lowering chemical use and expenses by properly mapping weed infestations (Ghatrehsamani et al., 2023)

This review's main goal is to review the framework and developments in crop management especially weed detection through the combined use Remote Sensing and GIS technologies. This review aims to consolidate existing research, highlight technological advancements, and discuss the practical applications and challenges of RS and GIS in this domain.

This review will address several key questions and focus areas to provide a comprehensive understanding of the topic:

1. What are the current techniques and methodologies used in weed detection through remote sensing and GIS?
2. How do high-resolution imagery and machine learning algorithms enhance the accuracy and efficiency of weed detection?
3. What are the different modes of data collection used in these technologies, and how do they impact the quality of weed detection?
4. What are the spectral and spatial features of weeds that are most commonly analyzed, and how do they contribute to precise identification?
5. What challenges are encountered in the detection of weeds using RS and GIS, and how can they be mitigated?

2. TECHNIQUES USED IN CROP MANAGEMENT

Weed can typically detected through three methods;

1. Remote Sensing
2. Geographic Information System
3. Integration of Remote Sensing (RS) and Geographic Information System (GIS)

Remote sensing is becoming more and more popular as the technology is advancing and it has been proven to be a very useful tool in the explanation and the translation of weed infestations in the fields (Victor et al., 2024). The most popular RS techniques that are used to different weed are given below along with recent trends in Table 1:

- Multispectral Imaging: The method captures data related to different wavelengths throughout the entire electromagnetic spectrum (for example, infrared or ultraviolet).
- Hyperspectral Imaging: The sensors capture the needed data across a large number of the specific spectral bands, thereby revealing the spectral properties of the vegetation.

- Thermal Imaging: Thermal sensors are used to monitor the temperature changes on the surface of the earth.
- LiDAR: Light Detection and Ranging (LiDAR) technology enables the use of laser pulses to produce high-resolution 3D models of the vegetation.

Table 1. Different Remote Sensing Techniques in Weed Management

Methods	Objective	Feature	References
Multispectral Imaging	Reducing computational time and effort	Invasive Weed Optimization with Capsule Network for Multispectral Image Classification (IWOCN-MSIC) technique	(M & Rj, 2023)
	Increasing Hemp Production through selective hemp variety	Utilization of digital multispectral phenotyping system with amalgamation of machine learning	(Singh et al., 2024)
	Precision farming	Data collection through U-net and classification through index-based threshold approach	(Seiche et al., 2024)
Hyperspectral Imaging	Detection of fully grown trifoliate	Machine vision algorithm implemented on Robotic system to grasp based on Hyperspectral data	(Chen et al., 2023)
	To identify different diseases on leaf	Using Hyperspectral reflectance spectra	(Yuan et al., 2014)
	Identify parasitic plant at early stage	Classification between infected and non-infected flower through regression technique using Hyperspectral imaging	(Atsmon et al., 2022)
Thermal Imaging	Identification of water level in winter wheat	Thermal imaging of data collected through experimentation in order to calculate crop water stress index	(Ma et al., 2024)
	Detection of irrigations malfunctions in drip-irrigated orchards	Classification on the basis of temperature difference through UAV	(Rozenfeld et al., 2024)
	Assess temperature distribution across crops	Machine Learning Thermal Imaging	(Marais-Sicre et al., 2024)
LiDAR	Assess weed management in almond orchard	UAV-LiDAR data combined with manual measurements	(Caras et al., 2024)
	Weed Control in Orchids	LiDAR data collection	(Sampurno & Ahamed, 2024)

Geographic Information Systems (GIS) are one of the most important modern technologies in agriculture especially in weed control as shown in Figure 1. Die spatial approach to weed control is enhanced by the GIS; it means that the use of these tools to analyze, visualize, and interpret data helps the data- driven and more efficient weed irrigation process.

Figure 1. GIS Application in Agriculture

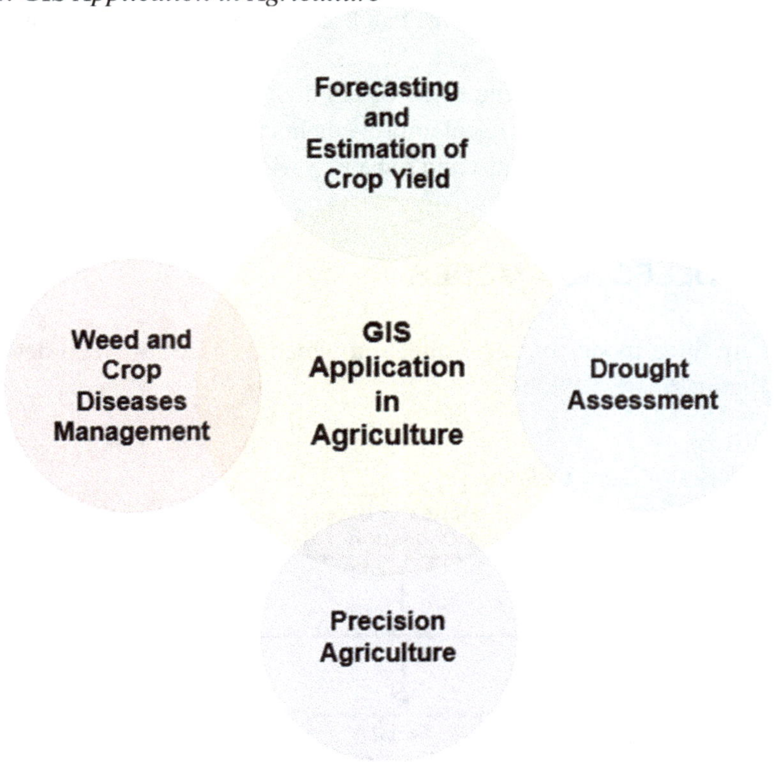

Software was developed in order to enable farmers to manage fertilizer's units (Xie et al., 2012). An irrigation simulation was proposed in (Wang et al., 2011) in order to maximize crop production by assessing crop patters. The soil properties were quantified on the basis of soil fertility index through GIS (Tunçay et al., 2021). A GIS based approach is used in (Tsegaye & Bharti, 2021) in order to estimate soil erosion. A novel crop monitoring system is proposed in (Kumar K & Babu Db, 2016) in order to avoid weed.

Integrating RS with GIS (Geographic Information System) can weigh in the revolution of weed detection via space analysis, therefore the accuracy and consequently higher efficiency and effectiveness in the control of weed infestations. Through the application of the synergetic benefits of both the technologies, we will have a more in-depth understanding of weed distribution and its lifespan. As a result, a reduction in the application of herbicides and the utilization of has been achieved (Lakra et al., 2024).

RS and GIS methods are used in (Roslim et al., 2021) in order to enhance weed management practices. In (Kulwant & Patel, 2024), application of RS and GIS is used in agriculture.

By combining the various functions of RS and GIS, these examples show off the potential of current technological improvements for weed detection and control, drawing us closer to the healthiest and most eco-friendly plant agriculture.

3. DATA COLLECTION MODES

There are three modes of data collection which us used for weed detection as given in Figure 2

Figure 2. Modes of Data Collection

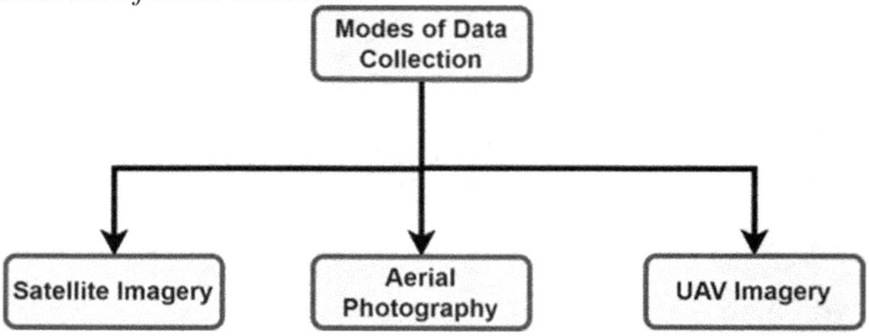

Satellite imagery is a major source of remote sensing data in telemetry weeds. Satellites with a myriad of sensors that orbit the Earth will take images with different spectral bands. The satellite imagery has the properties as follows:

- Wide Coverage: Because of satellites' ability to cater to large agricultural areas, the monitoring of such areas is easier.
- Multispectral and Hyperspectral Data: Satellites gather data at various portions of the light spectrum in their flights, which helps them to distinguish between the different crops and the various weeds based on their spectral signatures.
- Frequent Revisit Times: The fact that a lot of the satellites make regular sweeps of the territory thanks to the revisit times has a net effect on cropland monitoring and the nipping in the bud of weed invasions that are properly and timely done (Segarra, 2024).

In aerial photography, various activities were performed with the help of manned planes that were equipped with the highest resolution of lenses. This method deals mainly with small-size agricultural mapping and detailed mapping, able to create an accurate map of a small place. Aerial photography contains the characteristics such as:

- High Spatial Resolution: Aerial photography can also manage better very high spatial resolution, such as a few centimeters per pixel, due to which it may come with very precise weed identification.
- Flexibility: Airplanes that are not in flight for the whole day enables such aircraft to be very flexible when it comes to acquiring the necessary data which are precious to some specificity crop leveling (Khan et al., 2024).

Unmanned Aerial Vehicles (UAVs), which are also known as drones, are safe for farming because it allows for acquiring agriculture-grade. Key elements are:

- Ultra-High Spatial Resolution: This setup allows the drones to catch the images of a few centimeters of the spots per pixel, which helps in finding the small weed patches
- Flexibility and Accessibility: To acquire the necessary data drones are used for both one-time and regular inspections. Consequently, both real-time and delayed data can be obtained for analysis.
- Multispectral and Thermal Sensors: The drones are equipped with a variety of sensors such as multispectral and thermal cameras that are used to gather detailed data on the crop and weed conditions (Zhang et al., 2024).

4. CATEGORIES OF WEED FEATURES

There are two features of weed that is used in image processing in order to detect it; Spectral feature and spatial feature.

Spectral signatures are, in fact, specific patterns that materials have, such as vegetation, that control the reflection or absorption of the sun's energy. Weeds like any other plant either reflect or absorb light with different wavelengths across the electromagnetic spectrum. Once spectral characteristics have been analyzed, vegetation can be distinguished from weeds and other plant life.

- Visible Spectrum (400-700 nm): Typically, the color of healthy vegetation is the strongest at around 550 nm which is the green light. Only planted cultures, on the infected carry, the virus behavior pesticides treatment or the vegetation warm from water can only give different reflectance patterns.

- Near-Infrared (NIR) Spectrum (700-1300 nm): NIR light is provided from the leaves to the plant due to its internal leaf structure. Weed and other plants that take from nitrogen store in this spectrum can have high reflectance compared to those that can be stressed or not living properly.
- Shortwave Infrared (SWIR) Spectrum (1300-2500 nm): These are the region where you can see water in plants. Different SWIR reflectance might be an indication of differences in plant moisture levels, which can help in identifying stressed or water-deficient weeds (Lin et al., 2017).

Spectral indices are mathematical combinations of reflectance values at specific wavelengths designed to highlight particular features of vegetation. Commonly used spectral indices for weed detection include:

- Normalized Difference Vegetation Index (NDVI): NDVI is calculated using the formula

$$\frac{(\mathrm{NIR} - \mathrm{Red})}{(\mathrm{NIR} + \mathrm{Red})} \tag{1}$$

It is widely used to assess plant health and vigor. Weeds and crops typically have distinct NDVI values, allowing for differentiation.

- Green Normalized Difference Vegetation Index (GNDVI): Similar to NDVI, GNDVI uses green light instead of red light and is calculated as

$$\frac{(NIR - GREEN)}{(NIR + GREEN)} \tag{2}$$

It is particularly useful for detecting chlorophyll content and plant health.

- Enhanced Vegetation Index (EVI): EVI improves upon NDVI by reducing the effects of atmospheric conditions and soil background, calculated as

$$2.5 * \frac{(NIR - RED)}{((NIR + 6 * RED - 7.5 * BLUE) + 1)} \tag{3}$$

It provides more detailed information about vegetation structure and condition.

Spatial features correspond to the distribution of weeds in an explorable area, that being said, the agricultural field. The spatial arrangement of these patterns is very important to the management of weeds (Vioix et al., 2002). The elements below are important:

- Density and Clustering: The weeds usually grow in some clusters and do not form a uniform distribution. The practice of studying the density of the weeds and locating the clusters of outbreaks directs the isolated environmental activities, for example pesticide application.
- Patch Size and Shape: The distribution of jungle patches is the main source of data on the infectious power and the underlying ecological context influencing the growth of these plants. For example, the strips that are elongated may lead us to believe that wind or water inflow has caused the seed to be transported to other places.

The combination of spectral and spatial features of the plants and the input of remote sensing and GIS technologies may provide a thorough and correct report on weed infestation. As a result, better weed management skills and thus the sustainable approach of agricultural activities are feasible.

5. MACHINE LEARNING ALGORITHMS

The connection of machine learning (ML) algorithms with data from distant sensors is a huge step forward in weed detection technologies. Machine learning is implemented using computational models all over the remote sensing data, thereby significantly facilitating the identification and classification of weeds with more precision. This paragraph presents ways in which machine learning algorithms are employed to improve weed detection, showing various techniques of algorithms, their applications, and the advantages they bring. The application of machine learning in weed detection consists of the following:

- Automated Analysis: Machine learning models automatically process and analyze high-resolution imagery, which reduces the need for manual interpretation. This ensures the detection of weeds is done in a timely manner and the highest possible efficiency is obtained (Rani et al., 2022).
- Pattern Recognition: Application of machine learning in the form of the latest algorithms can see through the seemingly chaotic array of lines within the image and find complex patterns that are indicative of weeds of different

species. This new technology precisely locates those problematic plants and prevents them from pretending to be other plants (Tenhunen et al., 2019).

- Predictive Modeling: Machine learning models are used to be able to forecast the spread and growth of weed outbreaks by using historical data and current environmental conditions. This proactiveness helps concerned parties with data-driven decisions (Abd Manaf et al., 2022).

Many forms of machine learning algorithms are making weed detection possible by analyzing remote sensing data, each having their features and strengths. The development of machine learning is typically seen as a major reason for the improvement of the weed detection accuracy and the overall operational efficiency. Besides processing big data volumes, these algorithms can find and classify repeating data configurations and also make predictions which are based on data.

- Supervised Learning Algorithms: These intelligent programs deal with validation of labeled datasets where it is given that the input data (e.g., remote sensing images) and the corresponding output (e.g., weed presence or absence) are already known. The most popular supervised learning algorithms include:
  - o Support Vector Machines (SVM): This is a technology which is being effective for classification problems using remote sensing images like where SVMs classify the pixels based on their special signatures. This feature among SVMs permits them to separate the weeds from crops by determining the optimal boundary that distinguishes various categories in the feature space (Kok et al., 2021).
  - o Random Forests (RF): A good compromise technique combining decision trees is random forest, thereby improving the success rate in classification. Random forests prevent overfitting and are therefore better in handling large, high-dimensions of data sets, which means they are performing well in the agricultural data, specifically weed detection (Basha et al., 2020).
  - o Neural Networks: With neural networks, chiefly, the weapons of choice are convolutional neural networks (CNNs), on the other hand, they prove to be the best for visual tasks. Talking about weed detection, CNNs may be used to inspect high-resolution pictures for weed clusters and order different weed species according to their spectral and spatial features. Deep learning models such as CNNs master the ability to learn from convoluted patterns in remote sensing images, and subsequently, make a correct identification of weed patch (Barbosa et al., 2020).

- K-Nearest Neighbors (KNN): KNN is a simple algorithm for classifying data in respect of their proximity to the known examples which still works well though. KNN can likewise point out and classify new data points that are close to an already labeled one given a chance (Bhambri et al., 2020).
  - Deep Learning: Deep learning methods that are implemented in models of neural networks, particularly those that rely on deep learning architectures, have illustrated excellent performance in image analysis. These models are discovered to perceive multiple hierarchical features in raw data, thus producing highly correct weed detection and classification results (Dharani et al., 2021).
- Unsupervised Learning Algorithms: These algorithms are trained without any labeled data and are used for the identification of the patterns and clusters in the data. Examples of unsupervised learning algorithms are:
  - K-means Clustering: This technique groups alike data arguments into different clusters as per their mentioned features. It can be used to split a set of pixels in remote sensed images into different areas with similar vegetation, such as weeds (Utkarsha P. Narkhede & K.P.Adhiya, 2014).
  - Principal Component Analysis (PCA): With PCA, the number of the parameters is reduced, while the data are kept at the same level of representation. It gives ways to pull out characteristics of the spectrum that belong only to the weeds as compared to the rest of the vegetation (Liu et al., 2023).
- **Semi-supervised Learning Algorithms:** These algorithms combine the labeled and unlabeled data, therefore, they are perfect for the situation when the labeled data is rare. They can increase the detection of the presence of weeds by using all of the unlabeled remote sensing images available (Li et al., 2024).

Using supervised, unsupervised or semi-supervised learning approaches, machine learning models can use smart strategy for classification, detection, and monitoring of weed infestations with a high degree of precision. Remote sensing technologies have become even more advanced by the integration of machine learning, as it makes more weed management processes more effective and sustainable practices in agriculture. As the research and development in this area keep on, the capabilities and applications of machine learning in weed detection are apprehended to be more and more widespread, thus, agriculture through a completely new prism is being established.

6. CHALLENGES IN WEED DETECTION

Despite the marked technological strides in the remote sensing (RS) technology and the Geographic Information Systems (GIS) technology for weed detection, there are a number of critical issues which one needs to tackle in order to be able to actually use them to the full. Aside from technical issues, the environmental factors and the practical constraints- altogether they integrate and interact to damp the efficacy (success rate), the scalppsity (quality and quantity of labor force) and the acceptance of these technologies in agriculture. The component deals with the major difficulties associated with Rs and geographic information systems (GIS) techniques in solving the issue of weed detection and thus ensure that all of the necessary aspects of the challenge are correctly defined and they could be improved.

The data quality, processing, and integration issues are the technical bottlenecks. Weeds and field crops can be accurately distinguished with high-resolution images and advanced machine learning algorithms. One way to solve the challenges raised by these data is the use of sensors that provide so much spatial and spectral data that each location is distinguished by it, and data are retrieved from different sources at the same time.

A major technological problem in weed detection is the production of high-quality and high-resolution images that pave the way for better results. The accuracy of remote sensing data can be affected by a number of kinds of objects:

- Spatial Resolution: The use of images with high spatial resolution is a necessity in recognizing and mapping weeds accurately, especially in an agricultural space filled with different crops. However, to get such precise data may be very expensive, and it might not be obtained in all parts or as frequently as one would desire. Moreover, high-resolution image data contains more details, which complicates the processing and analysis of such large datasets. Effectively balancing the need for high-quality data with the desire for data sets of reasonable size continues to be a problem.
- Spectral Resolution: Whereas spectrally-based multispectral and hyperspectral imaging can distinguish plant species, they may also require appropriate sensors as well as intensive data sets tee too complex for management and analysis (Xu et al., 2023).
- Sensor Calibration: Sensor calibration disparities potentially can cause the captured data to be inconsistent and thus affect the accuracy of weed detection. Regular sensor calibration as well as maintenance are necessary to keep the data quality.
- Atmospheric Interference: Different factors such as the presence of such as clouds, haze, and aerosols, can scatter the spectral signals recorded by the re-

mote sensors. The interference reduces the clarity and precision of the images and hence the distinction of weed types becomes more of a challenge (Xu et al., 2023).

7. LIMITATIONS OF CURRENT TECHNOLOGIES

Now that we have come across the technical revolution, we still find that there are difficulties that prevent us from applying these techniques to the full extent of weed detecting. These issues are:

- Complex Algorithms: Above all else, the analysis of remote sensing data for weed detection is carried out by means of complicated machine learning algorithms and models that necessitate loads of computation power and theoretical know-how to design and install them.
- Data Volume: The amount of data that is collected with high-resolution imaging and continuous monitoring is often so much measurement that it's not simply a question of too fast. Ecosystem data storage, process and analysis are important but can usually be too much time spent on human resource support (Müllerová et al., 2023).
- Spectral Overlap: The spectral signatures of plants of varying types and of different colors can be so close that it's almost impossible to tell them apart. A sort of an algorithm that is to be combined with multispectral and hyperspectral sensors are needed to boost the accuracy of the species classification.
- Temporal Resolution: Regularly obtaining data on the increase and spread of algae through drone checks and computer remotely acquired images is an example. The situation with cloud cover, satellite revisit intervals, and the working conditions of drones and aerial platforms has never been stable. The changes over time are so short, too. High temporal resolution is essentially required for a proper weeding process.
- Data Processing and Analysis: The large amount of data received by high-resolution remote sensing techniques means that powerful processing and analysis tools are required. The design of the best algorithms and the procurement of the most powerful machines for processing and storage should be a top priority in managing these data sets. Moreover, data collection from various sources such as drones, satellites, and ground sensors all with differing resolutions and formats presents a large difficulty. The basic requirement for the success of this lies on the compatibility and consistency of the data that will be needed for the analysis (Müllerová et al., 2023).

8. A CASE STUDY

In Lahore, a major agricultural and urban center in Pakistan, the lockdown resulted in notable changes in agricultural practices and environmental conditions. Reduced industrial activity and lower vehicular traffic likely led to changes in Land Surface Temperature (LST), while restrictions on movement and reduced human intervention in agricultural activities impacted vegetation and soil moisture levels.

Figure 3. Study Area

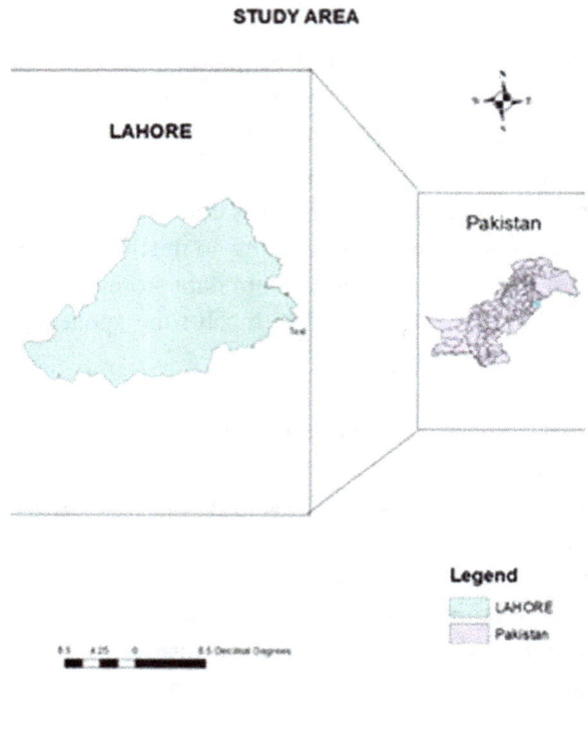

In order to analyze the effects of the lockdown on agricultural conditions in the Lahore region, this study relied on Landsat satellite data from 2019 and 2020. The study wishes to estimate the effects of the lockdown on the agricultural system of the local area, by looking at Land Surface Temperature (LST), Normalized Difference Vegetation Index (NDVI), and Normalized Difference Moisture Index (NDMI) as given in Figure 3.

The study utilized Landsat 8 OLI/TIRS imagery from 2019 and 2020, focusing on data from the same months each year to ensure a valid comparison as given in Table 2.

Table 2. Information about Dataset Used

Data Type	Source	Description	Date
Lahore Shapefile	https://data.humdata.org/dataset/cod-ab-pak	Administrative boundaries	***
Satellite Images	earthexplorer.usgs.gov	Landsat 8 OLI/TIRS (30×30 meters)	2019 and 2020 Images

In order to assess the impact of the COVID-19 lockdown on agricultural conditions, the following methodology was employed as given in Figure 4:

Figure 4. Methodology

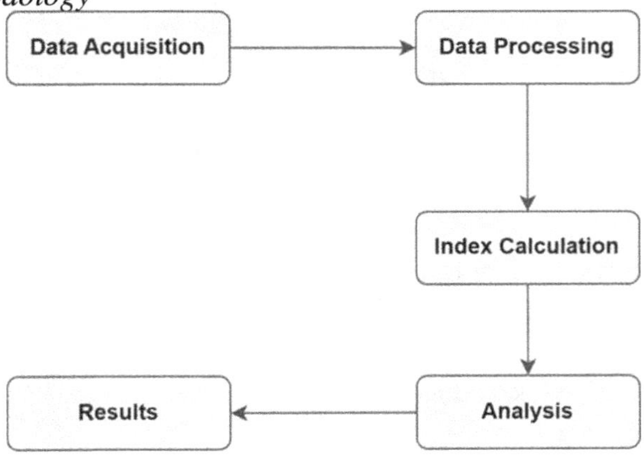

1. Data Acquisition: Landsat satellite imagery from 2019 and 2020 was obtained to capture pre-lockdown and lockdown periods. This data provided essential information on surface temperature, vegetation health, and moisture content.
2. Data Processing: The imagery was processed using QGIS, an open-source geographic information system. QGIS was used to apply atmospheric corrections and extract relevant indices for analysis.
3. Index Calculation: Land Surface Temperature (LST): LST was derived from the thermal infrared bands of the Landsat images, reflecting changes in surface temperature due to reduced human activities.

Figure 5. LST Map Before and During Lockdown

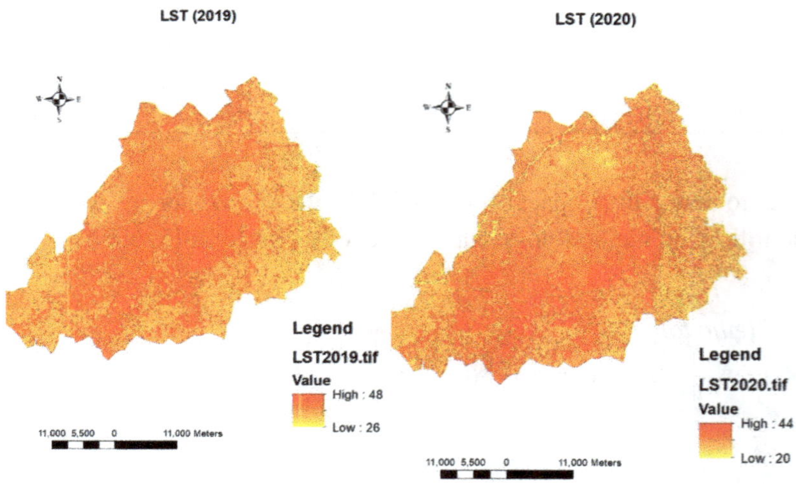

Figure 5(a) shows the Land Surface Temperature of Lahore before and during the COVID-19 lockdown. It highlights the baseline temperature patterns across the region. Figure 5(b) illustrates the Land Surface Temperature during the lockdown period, showing noticeable reductions in temperature due to decreased industrial and vehicular activities.

Figure 6. NDVI Map Before and During Lockdown

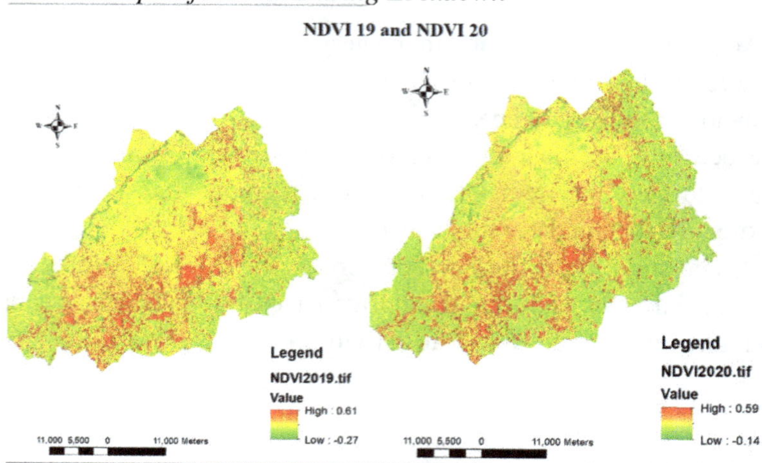

Normalized Difference Vegetation Index (NDVI): NDVI was calculated using the red and near-infrared bands of the Landsat images to assess changes in vegetation health

Figure 7. NDMI Map Before and During Lockdown

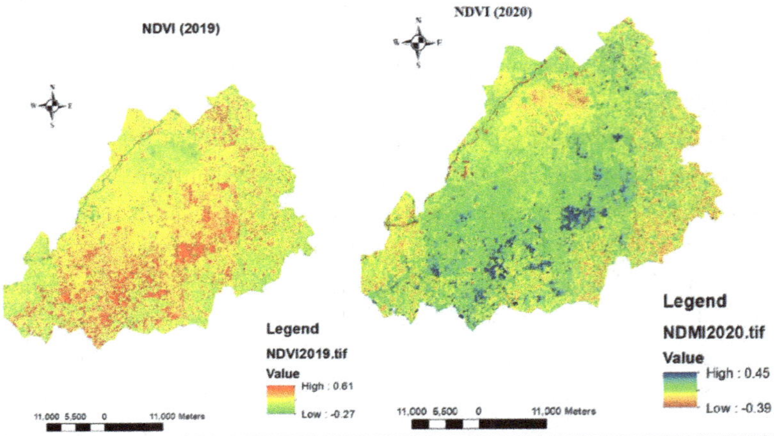

Figure 6 (a) provides an overview of vegetation health prior to the lockdown, indicating baseline vegetation coverage and health. While Figure 6 (b) demonstrates improvements in vegetation health after the lockdown, likely due to reduced environmental stress.

Normalized Difference Moisture Index (NDMI): NDMI was calculated using the near-infrared and shortwave infrared bands to evaluate variations in moisture content.

Figure 7 (a) shows soil moisture levels before the lockdown, providing a baseline for comparison While Figure 7 (b) reflects increased moisture content during the lockdown, suggesting improved soil and vegetation moisture levels.

4. Analysis: The indices were analyzed to detect changes between 2019 and 2020. This included comparing pre-lockdown and lockdown periods to identify trends and impacts on agricultural conditions.
5. Results Interpretation: The results were interpreted to understand the effects of reduced human activity on LST, NDVI, and NDMI, providing insights into the changes in agricultural environments during the lockdown.

9. RESULTS AND DISCUSSION

Technological advancements in Remote Sensing and Geographical Information Systems have resulted and will probably continue to shape the way we monitor agriculture and conduct environmental analysis, with limitless advantages in that realm. Whereas Remote Sensing (RS) gets a detailed snapshot of coverage of land and vegetation through imagery, which is a data source providing all the necessary information such as spatial and temporal data in a comprehensive way. GIS, a spatial tool, acts as a gateway to the integration and analysis of data of various formats that in turn give us insights into spatial relations and environmental dynamics.

Remote sensing is such a huge advantage for the environment in that it can map large areas without being so much of disturbance. This system makes it possible to perform the monitoring of crop health, land use, and environmental conditions. However, this method has its disadvantages as well. The high resolutions of the imaging might be very costly and will need highly dependent processing capabilities to handle huge datasets. Furthermore, the cloudiness and haze conditions can be a problem as they can interfere with the data transmission from a satellite, which in turn results in reduced data accuracy. The dearth of sensor calibration remains a tough issue, because inconsistencies can lead to inaccuracies in the data collected.

GIS advances data analysis by illustrating the different types of spatial data integrated together, and it gives an opportunity to gain insight from the most complex spatial factors. Creativity with advanced analytical techniques and spatial modeling is also possible, which are critical for knowing patterns and trends. However, GIS also requires software engineering knowledge and management of data, hence vexing interactions among diverse data sources can complicate the process. In the same way, when working with large datasets, it might require a lot of computational resources, which in some cases can become an issue.

Applying remote sensing (RS) and geographic information system (GIS) techniques has been effective in monitoring crop health, managing pests and diseases, detecting external environmental causes, and managing measures at agro-ecosystems. Remote-sensing technology, such as multispectral and Hyperspectral imaging, enable a precise assessment of plant health and stress levels by collecting and analyzing data over a variety of wavelengths. This can result in farmers being able to spot diseases and pests early or to forecast weather issues in addition to crop assessment.

The utilization of indices such as the Normalized Difference Vegetation Index (NDVI) and the Normalized Difference Moisture Index (NDMI) is normal in the applications of Remote Sensing. The NDMI is surreal to the mode in which vegetation changes with time as well as the impact of environmental indicators, while NDVI evaluates the health condition of the vegetation by checking the contrast between the near-infrared and red light reflectance. Similarly, NDMI interprets soil moisture by comparing the near-infrared and the shortwave infrared bands. These indices play a vital role for people who would like to know the effects of the changes in the environment as well as those who need to practice agriculture.

The analysis of Landsat 8 OLI/TIRS imagery for Lahore before and during the COVID-19 lockdown brought about some very important changes in the agricultural and the environmental conditions. The Land Surface Temperature (LST) maps indicated that, prior to the lockdown, Lahore experienced typical urban heat island effects, with higher temperatures in industrial and densely populated areas. On the other hand, during the lockdown period, there was a perceptible fall in surface temperatures. This reduction could have resulted from the less industrial activity and lower vehicle emissions-cutting down the heat generation in the process and thus accompanied with better thermal conditions.

The Normalized Difference Vegetation Index (NDVI) in Figure 6 showed a marked improvement in vegetation health during the lockdown. NDVI values before the lockdown reflected the standard vegetation health and coverage for that time of year, while post-lockdown values demonstrated enhanced vegetation health. This improvement can be attributed to reduced environmental stress resulting from decreased pollution and less human interference.

Similarly, the Normalized Difference Moisture Index (NDMI) maps in Figure 7 indicated an increase in soil moisture content during the lockdown. NDMI values before the lockdown represented typical soil moisture levels, whereas the lockdown period showed improved moisture content. This improvement suggests that reduced human activity alleviated environmental stress, allowing natural processes to recover and enhancing agricultural conditions.

The case study highlights the potential of RS and GIS to assess the impacts of human activities on the environment. It demonstrated how remote sensing can provide insights into environmental changes over time and how GIS can integrate and analyze these changes to support effective management practices.

While the RS and GIS technologies are progressing, a number of problems and difficulties are remaining. Quality of the data and its processing are namely hard points, as high-resolution images used in mapping can be costly and are difficult to work with. The imperative for the regular sensor calibration and the atmospheric interference that could kill the accuracy of the data are the other matters to deal with. Furthermore, different plant species are not separated into different space regions with their own spectral characteristics in images. Therefore, the accurate classification and detection of plants require the development of more precise algorithms.

Temporal and spatial variability in data also presents challenges. The frequency of data acquisition can be affected by satellite revisit intervals and weather conditions, while the patchy distribution of crops and weeds requires sophisticated detection methods. Integrating data from various sources, such as drones, satellites, and ground sensors, adds complexity and demands compatibility and consistency.

Overall, while RS and GIS offer powerful tools for environmental monitoring and agricultural management, addressing these challenges through continued research and technological innovation is crucial. By overcoming these hurdles, we can enhance the effectiveness of these techniques and improve agricultural practices and environmental stewardship

10. CONCLUSION

The integration of Remote Sensing (RS) and Geographic Information Systems (GIS) has profoundly transformed agricultural practices, particularly in the realms of weed detection and environmental monitoring. This chapter has illustrated how these technologies provide accurate, timely, and spatially detailed data essential for effective land management. Through advanced analytics and real-time monitoring, RS and GIS enable precise mapping of weed occurrences, which in turn facilitates more efficient herbicide use and improved land management practices. The case study presented in this chapter further highlights the significant impact of reduced human activities on agricultural environments, as seen during the COVID-19 lockdown. By utilizing Landsat satellite imagery and GIS tools, we were able to analyze changes in Land Surface Temperature (LST), Normalized Difference Vegetation Index (NDVI), and Normalized Difference Moisture Index (NDMI). This analysis offered valuable insights into how decreased industrial and vehicular activities influenced surface temperatures, vegetation health, and soil moisture levels. The use of RS and GIS in this context underscored their critical role in monitoring and assessing environmental changes, demonstrating their effectiveness in capturing and quantifying the impacts of global disruptions on agricultural systems.

REFERENCES

Abd Manaf, M. N. H., Juraimi, A. S., Che'Ya, N. N., Su, A. S. M., Roslim, M. H. M., Ahmad, A., & Noor, N. M. (2020). Predictive Modelling for Rice Weeds in Climate Change: A Review. Advances in Agricultural and Food Research Journal.

Atsmon, G., Nehurai, O., Kizel, F., Eizenberg, H., & Nisim Lati, R. (2022). Hyperspectral imaging facilitates early detection of Orobanche cumana below-ground parasitism on sunflower under field conditions. *Computers and Electronics in Agriculture*, 196, 106881. DOI: 10.1016/j.compag.2022.106881

Barbosa, A., Trevisan, R., Hovakimyan, N., & Martin, N. F. (2020). Modeling yield response to crop management using convolutional neural networks. *Computers and Electronics in Agriculture*, 170, 105197. DOI: 10.1016/j.compag.2019.105197

Basha, S. M., Rajput, D. S., J, J., Somula, R. S., & Ram, S. (2020). Principles and Practices of Making Agriculture Sustainable: Crop Yield prediction using Random Forest. *Scalable Computing: Practice and Experience*, 21(4), 591–599. DOI: 10.12694/scpe.v21i4.1714

Bhambri, D., Dhanoa, D. I. S., Sinha, D. V. K., & Kaur, E. J. (2020). Paddy Crop Production Analysis Based on SVM and KNN Classifier. [IJRTE]. *International Journal of Recent Technology and Engineering*, 8(5), 2790–2793. DOI: 10.35940/ijrte.D8650.018520

Bibhuti Bhusan, S., & Giri, R. (2023). Application of Remote Sensing & Geographic Information Systems for Agriculture-A Review. In *Advances in Agricultural Technology*.

Caras, T., Lati, R. N., Holland, D., Dubinin, V. M., Hatib, K., Shulner, I., Keiesar, O., Liddor, G., & Paz-Kagan, T. (2024). Monitoring the effects of weed management strategies on tree canopy structure and growth using UAV-LiDAR in a young almond orchard. *Computers and Electronics in Agriculture*, 216, 108467. DOI: 10.1016/j.compag.2023.108467

Chen, Z., Wang, J., & Jin, J. (2023). Fully automated proximal hyperspectral imaging system for high-resolution and high-quality in vivo soybean phenotyping. *Precision Agriculture*, 24(6), 2395–2415. DOI: 10.1007/s11119-023-10045-5

Dharani, M. K., Thamilselvan, R., Natesan, P., Kalaivaani, P., & Santhoshkumar, S. (2021). Review on Crop Prediction Using Deep Learning Techniques. *Journal of Physics: Conference Series*, 1767(1), 012026. DOI: 10.1088/1742-6596/1767/1/012026

Ghatrehsamani, S., Jha, G., Dutta, W., Molaei, F., Nazrul, F., Fortin, M., Bansal, S., Debangshi, U., & Neupane, J. (2023). Artificial Intelligence Tools and Techniques to Combat Herbicide Resistant Weeds—A Review. *Sustainability (Basel)*, 15(3), 1843. DOI: 10.3390/su15031843

Khan, U., Khan, M. K., Latif, M. A., Naveed, M., Alam, M. M., Khan, S. A., & Su'ud, M. M. (2024). A Systematic Literature Review of Machine Learning and Deep Learning Approaches for Spectral Image Classification in Agricultural Applications Using Aerial Photography. *Computers, Materials & Continua*, 78(3), 2967–3000. DOI: 10.32604/cmc.2024.045101

Kok, Z. H., Mohamed Shariff, A. R., Alfatni, M. S. M., & Khairunniza-Bejo, S. (2021). Support Vector Machine in Precision Agriculture: A review. *Computers and Electronics in Agriculture*, 191, 106546. DOI: 10.1016/j.compag.2021.106546

Kulwant, M., & Patel, D. (2024). Application of Remote Sensing, GIS, and AI Techniques in the Agricultural Sector. In S. Gupta, S. K. Himanshu, & P. K. Gupta, *Agri-Tech Approaches for Nutrients and Irrigation Water Management* (1st ed., pp. 248–271). CRC Press. DOI: 10.1201/9781003441175-15

Kumar, S. K., & Babu, S. D. B. (2016). A web GIS based decision support system for agriculture crop monitoring system-A case study from part of Medak District. *J Remote Sensing & GIS*, 5(4), 177–197.

Lakra, D., Singh, S. K., Gupta, S. K., & Kanga, S. (2024). Enhancing integrated resource management through remote sensing and GIS. *Journal of Geography and Cartography*, 7(1), 4265. DOI: 10.24294/jgc.v7i1.4265

Li, J., Chen, D., Yin, X., & Li, Z. (2024). Performance evaluation of semi-supervised learning frameworks for multi-class weed detection. *Frontiers in Plant Science*, 15, 1396568. DOI: 10.3389/fpls.2024.1396568 PMID: 39228840

Lin, F., Zhang, D., Huang, Y., Wang, X., & Chen, X. (2017). Detection of Corn and Weed Species by the Combination of Spectral, Shape and Textural Features. *Sustainability (Basel)*, 9(8), 1335. DOI: 10.3390/su9081335

Liu, J., Kang, H., Tao, W., Li, H., He, D., Ma, L., Tang, H., Wu, S., Yang, K., & Li, X. (2023). A spatial distribution – Principal component analysis (SD-PCA) model to assess pollution of heavy metals in soil. *The Science of the Total Environment*, 859, 160112. DOI: 10.1016/j.scitotenv.2022.160112 PMID: 36375553

Ma, S., Liu, S., Gao, Z., Wang, X., Ma, S., & Wang, S. (2024). Water Deficit Diagnosis of Winter Wheat Based on Thermal Infrared Imaging. *Plants*, 13(3), 361. DOI: 10.3390/plants13030361 PMID: 38337894

Marais-Sicre, C., Queguiner, S., Bustillo, V., Lesage, L., Barcet, H., Pelle, N., Breil, N., & Coudert, B. (2024). Sun/Shade Separation in Optical and Thermal UAV Images for Assessing the Impact of Agricultural Practices. *Remote Sensing (Basel)*, 16(8), 1436. DOI: 10.3390/rs16081436

Müllerová, J., Brundu, G., Große-Stoltenberg, A., Kattenborn, T., & Richardson, D. M. (2023). Pattern to process, research to practice: Remote sensing of plant invasions. *Biological Invasions*, 25(12), 3651–3676. DOI: 10.1007/s10530-023-03150-z

Narkhede, U. P., & Adhiya, K. P. (2014). Evaluation of Modified K-Means Clustering Algorithm in Crop Prediction. *International Journal of Advanced Computer Research*, 4(3), 799–807.

Rani, S. V. J., Kumar, P. S., Priyadharsini, R., Srividya, S. J., & Harshana, S. (2022). Automated weed detection system in smart farming for developing sustainable agriculture. *International Journal of Environmental Science and Technology*, 19(9), 9083–9094. DOI: 10.1007/s13762-021-03606-6

RJ, K. (2024). Invasive weed optimization with deep transfer learning for multispectral image classification model. Multimedia Tools and Applications, 83(15), 45519-45534.

Roslim, M. H. M., Juraimi, A. S., Che'Ya, N. N., Sulaiman, N., Manaf, M. N. H. A., Ramli, Z., & Motmainna, M. (1809). Che'Ya, N. N., Sulaiman, N., Manaf, M. N. H. A., Ramli, Z., & Motmainna, Mst. (2021). Using Remote Sensing and an Unmanned Aerial System for Weed Management in Agricultural Crops: A Review. *Agronomy (Basel)*, 11(9), 1809. Advance online publication. DOI: 10.3390/agronomy11091809

Rozenfeld, S., Kalo, N., Naor, A., Dag, A., Edan, Y., & Alchanatis, V. (2024). Thermal imaging for identification of malfunctions in subsurface drip irrigation in orchards. *Precision Agriculture*, 25(2), 1038–1066. DOI: 10.1007/s11119-023-10104-x

Sampurno, R. M., & Ahamed, T. (2024). Challenges in Orchard Weed Management: Perspectives on the Use of 3D Cameras and LiDAR to Develop a Low-Cost Small-Scale Robotic Weeder. In Ahamed, T. (Ed.), *IoT and AI in Agriculture* (pp. 229–247). Springer Nature Singapore., DOI: 10.1007/978-981-97-1263-2_14

Segarra, J. (2024). Satellite Imagery in Precision Agriculture. In P. M. Priyadarshan, S. M. Jain, S. Penna, & J. M. Al-Khayri (Eds.), *Digital Agriculture* (pp. 325–340). Springer International Publishing. DOI: 10.1007/978-3-031-43548-5_10

Seiche, A. T., Wittstruck, L., & Jarmer, T. (2024). Weed Detection from Unmanned Aerial Vehicle Imagery Using Deep Learning—A Comparison between High-End and Low-Cost Multispectral Sensors. *Sensors (Basel)*, 24(5), 1544. DOI: 10.3390/s24051544 PMID: 38475081

Singh, G., Slonecki, T., Wadl, P., Flessner, M., Sosnoskie, L., Hatterman-Valenti, H., Gage, K., & Cutulle, M. (2024). Implementing Digital Multispectral 3D Scanning Technology for Rapid Assessment of Hemp (Cannabis sativa L.) Weed Competitive Traits. *Remote Sensing (Basel)*, 16(13), 2375. DOI: 10.3390/rs16132375

Tenhunen, H., Pahikkala, T., Nevalainen, O., Teuhola, J., Mattila, H., & Tyystjärvi, E. (2019). Automatic detection of cereal rows by means of pattern recognition techniques. *Computers and Electronics in Agriculture*, 162, 677–688. DOI: 10.1016/j.compag.2019.05.002

Tsegaye, L., & Bharti, R. (2021). Soil erosion and sediment yield assessment using RUSLE and GIS-based approach in Anjeb watershed, Northwest Ethiopia. *SN Applied Sciences*, 3(5), 582. DOI: 10.1007/s42452-021-04564-x

Tunçay, T., Kılıç, Ş., Dedeoğlu, M., Dengiz, O., Başkan, O., & Bayramin, İ. (2021). Assessing soil fertility index based on remote sensing and gis techniques with field validation in a semiarid agricultural ecosystem. *Journal of Arid Environments*, 190, 104525. DOI: 10.1016/j.jaridenv.2021.104525

Victor, N., Maddikunta, P. K. R., Mary, D. R. K., Murugan, R., Chengoden, R., Gadekallu, T. R., Rakesh, N., Zhu, Y., & Paek, J. (2024). Remote Sensing for Agriculture in the Era of Industry 5.0—A Survey. *IEEE Journal of Selected Topics in Applied Earth Observations and Remote Sensing*, 17, 5920–5945. DOI: 10.1109/JSTARS.2024.3370508

Vioix, J.-B., Douzals, J.-P., Truchetet, F., Assémat, L., & Guillemin, J.-P. (2002). Spatial and Spectral Methods for Weed Detection and Localization. *EURASIP Journal on Advances in Signal Processing*, 2002(7), 793080. DOI: 10.1155/S1110865702204072

Wang, Y., Chen, Y., & Peng, S. (2011). A GIS Framework for Changing Cropping Pattern Under Different Climate Conditions and Irrigation Availability Scenarios. *Water Resources Management*, 25(13), 3073–3090. DOI: 10.1007/s11269-011-9846-6

Xie, Y. W., Yang, J. Y., Du, S. L., Zhao, J., Li, Y., & Huffman, E. C. (2012). A GIS-based fertilizer decision support system for farmers in Northeast China: A case study at Tong-le village. *Nutrient Cycling in Agroecosystems*, 93(3), 323–336. DOI: 10.1007/s10705-012-9519-8

Xu, K., Shu, L., Xie, Q., Song, M., Zhu, Y., Cao, W., & Ni, J. (2023). Precision weed detection in wheat fields for agriculture 4.0: A survey of enabling technologies, methods, and research challenges. *Computers and Electronics in Agriculture*, 212, 108106. DOI: 10.1016/j.compag.2023.108106

Yuan, L., Huang, Y., Loraamm, R. W., Nie, C., Wang, J., & Zhang, J. (2014). Spectral analysis of winter wheat leaves for detection and differentiation of diseases and insects. *Field Crops Research*, 156, 199–207. DOI: 10.1016/j.fcr.2013.11.012

Zhang, J., Yu, F., Zhang, Q., Wang, M., Yu, J., & Tan, Y. (2024). Advancements of UAV and Deep Learning Technologies for Weed Management in Farmland. *Agronomy (Basel)*, 14(3), 494. DOI: 10.3390/agronomy14030494

Chapter 9
Precision Farming With Automated Weed Detection Using Machine Learning

Garima Mathur
Sagar Institute of Science and Technology, India

Harsha Pandey
Madhya Pradesh State Open School Education Board, India

ABSTRACT

An artificial intelligence-based weed detection system is a computerized system designed to automatically identify and classify different types of weeds in agricultural fields. The system utilizes advanced computer vision techniques and machine learning algorithms to accurately detect and differentiate weeds from crops or other elements in the field. The weed detection system typically consists of hardware components such as cameras or drones that capture high-resolution images or videos of the agricultural area. These images are then analyzed by the artificial intelligence algorithms which have been trained on large datasets of weed images to recognize and distinguish various weed species. This paper explores the application of AI in weed detection and offers a promising solution for automating weed detection in crops. Furthermore, the work addresses the potential benefits of using automated weed detection systems such as reduced labor costs decreased herbicide usage, and improved environmental sustainability.

DOI: 10.4018/979-8-3693-6452-9.ch009

INTRODUCTION

Overview: Weeds are a significant factor that can negatively impact crop yields. Site-specific weed control is becoming a highly effective approach, especially when integrated with machine learning and image processing techniques. With the global population rapidly increasing and projected to surpass nine billion soon, agricultural production must increase by approximately 70% to meet future demands (Toğaçar, Mesut et al., 2022). However, during this period, the agricultural sector faced several problems, like shrinking arable surfaces and the need for intensified production. Water scarcity and climate change will also have an impact on output. Precision or technological agriculture presents viable answers to these problems (Jawed M. K et al., 2022).

Weeds impact crop productivity and quality because they spread swiftly and uncontrollably (S. Abouzahir, M. Sadik, and E. Sabir et al., 2022). For essential resources including nutrients, sunlight, water, and space, they compete with crops (X. Jin, J. Che, and Y. Chen et al., 2021). Farmers must thus invest time and resources in weed management. Numerous issues plague crop protection and yield enhancement methods, including cost, time, and labor intensiveness. Additionally, these practices may be harmful to people's health, plants, animals, and the environment (Hameed, S. et al., 2018; Amin et al., 2019).

Automated weed control methods offer both economic and environmental benefits. By using machines to remove weeds, labor costs can be reduced, and selective spraying can decrease herbicide use (M.J.. et al., 2003). Accurate weed detection and identification are crucial for developing autonomous weed management systems (X. Li, R. Zeng, and H. Liao, et al., 2016).

According to S. I. Moazzam et al. (2021), there are four primary steps in a conventional weed identification system: image capture, preliminary processing, feature extraction, weed detection, and classification. These procedures have been enhanced by several contemporary technologies. The classification and identification of weeds are the two most important aspects of this approach.

When it comes to picture classification, identifying objects, and recognition, deep learning is an essential portion of neural networks (ML) that outperforms conventional ML techniques in various ways. Differentiating between crops and weeds can be challenging with ML techniques due to their similar appearances. However, deep learning methods are more proficient at extracting features from images, making them more effective in addressing these challenges.

Weeds, or unwanted plants, impede crop growth by taking up space with vital resources like sunlight, water, and nutrients. Crop productivity is lowered as a result. Farmers may find it labor- and time-intensive to manually identify and remove weeds (Ying, B.; Xu, Y.; Zhang, S.; Shi, Y.; Liu, L. et al., 2021). But detecting weeds in agricultural contexts may now be automated because to advances in artificial intelligence (AI).

BACKGROUND DETAILS

Plant weed identification is the technique of identifying and distinguishing between weeds and crops in agricultural settings. Weeds need to be found and eliminated from agricultural regions to guarantee the best possible crop development and output.

Labor is needed for all economic activity, but agriculture requires the most. The livelihoods of the 7.7 billion people on Earth are dependent on it, either directly or indirectly. A lot of resources are used in traditional farming, such as soil, water, fertilizers, pesticides, and herbicides. Modern farmers make use of cutting-edge technology, including computers, iPads, robots, drones, and other equipment.

Herbicides are used by farmers to keep weeds under control and shield crops from harm. Persicaria perfoliata, also referred to as Mile-a-Minute weed, is one example of an invasive weed that can spread swiftly and displace native plants across large areas. Herbicide use is crucial to lowering these risks and increasing crop yields.

Herbicides, on the other hand, poison the land and water, having detrimental effects on human health, wildlife, and the environment (Toğaçar, Mesut, et al., 2022; Y. Du, G. Zhang, et al., 2022).

Using smart agricultural technologies to increase farming production and environmental health is known as precision agriculture. The primary aim of precision farming is to apply the right amount of inputs at the right time and location. Identifying and classifying weeds is one of the main issues with precision farming.

An essential component of farm management is weed detection, which includes data segmentation, anomaly detection, and crop and weed identification. For these jobs, conventional machine learning techniques are applied. Deep learning (DL) is a crucial area of machine learning that has many benefits over traditional machine learning (ML) methods for object identification, picture categorization, and recognition.

Given both weeds and crops can have similar appearances, it may be difficult to identify and choose distinctive traits using machine learning. However, when working with larger datasets, deep learning techniques perform better than conventional ML algorithms. Deep learning, is utilized for several tasks, such as segmentation, detection, recognition, and classification. Figure 1 displays several deep-learning weed detection techniques.

Figure 1. Weed detection using deep learning.

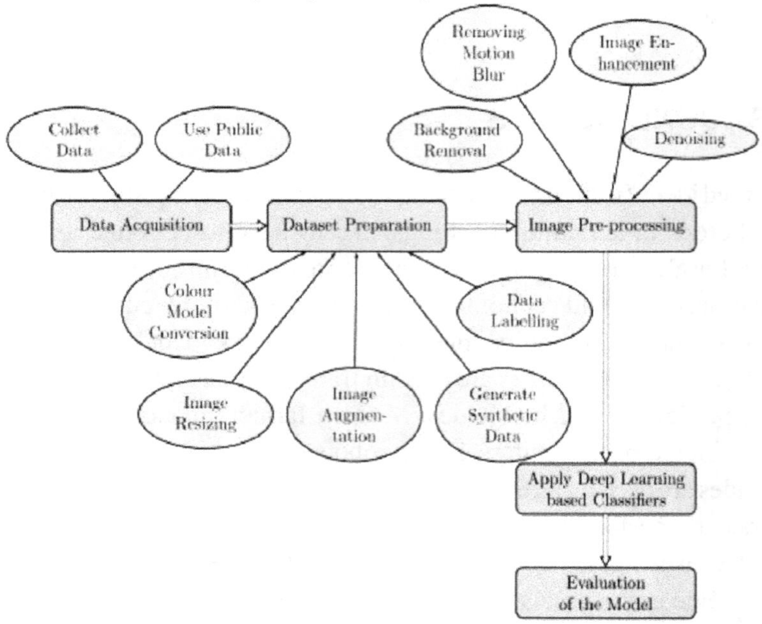

Various methods and technologies are available for detecting and managing weeds in crops. Some of them are

1. **Visual Inspection**
2. **Satellite Imagery and Remote Sensing**
3. Machine Learning and Computer Vision
4. Sensor-based Technologies
5. **Weed Control Strategies**

CLASSIFICATION OF WEED

The weed is classified into different categories such as based on life cycle, based on habitat, and based on morphology. Furthermore, the life cycle is also categorized as annual, biennial, and perennial weeds. The habitat weed is classified as terrestrial and aquatic. The morphological weeds are divided into monocot and dicot weeds. Furthermore, ecological weeds have been categorized based on ecology, soil type, specificity, toxic, and grassland. The detailed classification is given in Figure 2.

Figure 2. Classification of weeds

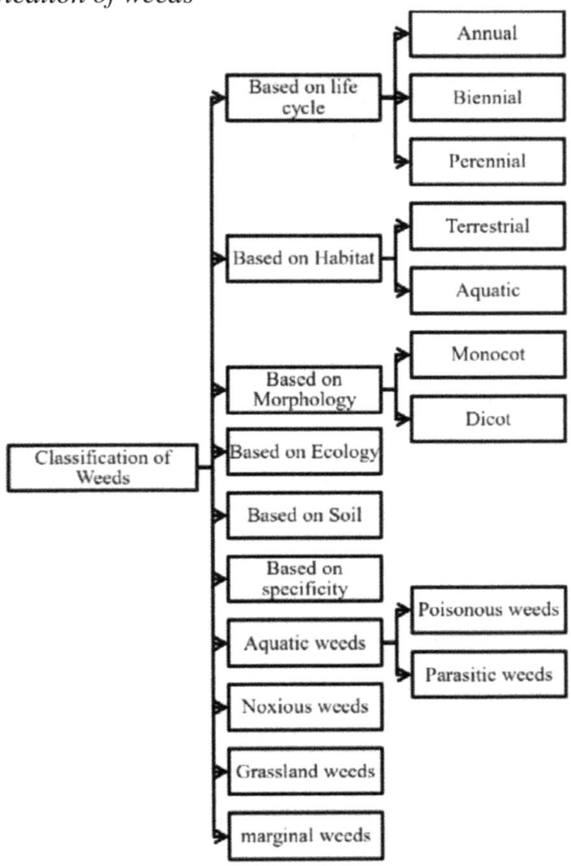

1. **Based on the Life Cycle:** The life cycle of weeds grows and reproduce in a seasonal pattern. The weeds are classified as Annual, Biennial, and Perennial. The complete descriptions are given in the next subsections.
   a. Annual Weeds: The annual weed has a one-year lifespan. These weeds, which resemble herbs and have thin stems and shallow roots, spread via seeds. Following sowing, annual weeds emerge, wither away, and then begin to produce for the following season's crop (Hameed, S et al., 2018; Amin et al., 2019). Furthermore, the annual weeds are classified into two subcategories winter and summer annual weeds. It is shown in Figure 3 a.

Figure 3a. Weeds are based on the Monsoon and the Annual season.

b. Biennial Weeds: The lifespan of biennial weeds is two years. They only produce leaves and flowers the following year; in the first year, they are purely vegetative. Weeds like "Alternanthera echinata," "Daucus Carota," and others are considered biennials (Aitkenhead, M.J.; Dalgetty, I.A.; Mullins, C.E.; McDonald, A.J.S.; Strachan, N.J.C. et al., 2003). Figure 3 b depicts a handful of typical weeds.

Figure 3b. A few examples of Biennial weeds

c. Perennial Weeds: The life cycle of certain weed varieties is more than two years. Annual weeds sprout from the same plant each year and are challenging to eradicate, though not as challenging as they are frequently portrayed to be. Minor growth can be handled rather easily if deep tap roots have been destroyed, as most have. The life span of certain weeds, such as "Commelina Benghalensis" and "Boerhavia Erecta," is only four months (Hameed, S et al., 2018; Amin et al., 2019). Additionally, it has been divided into three categories: corm, bulbous, and simple weeds. Simple weeds, such as "Sonchus Arvensis," "Bulbous," and "Allium sp.", are propagated via seeds. The Corn weed plant breeds through cream and seeds such as „Phleum pretense". There are a few common weeds as depicted in Figure 3 c.

Figure 3c. Bermuda grass Curly Dock

2. **Based on Habitat:** These weeds can spread throughout any location, site, and climate. Also included in the classification of weeds are aquatic and terrestrial. It is explained in the subsection that follows (X. Li, R. Zeng and H. Liao, et al., 2016).
    a. Terrestrial Weeds: Because terrestrial weeds have such a negative impact on ecosystems, including urban areas, agriculture, and forests, these invasive species must be managed properly (S. I. Moazzam et al., 2021). Restricting their spread is difficult because of their rapid adaptation and morphological improvement (Michalski, R.S., Carbonell, J.G. and Mitchell, T.M. eds. et al., 2013).
    b. Aquatic Weeds: Aquatic plants have adapted to living in aquatic environments. They are also called hydrophytes or saprophytes to distinguish them from algae and other microphytes.
3. **Based on Morphological:** The morphological features of the plants are categorized based on insignificance, and they have been categorized into two types: monocot and dicot weeds. Furthermore, it has been classified into three categories: Grasses, Sedges, and Broadleaves weeds (Celebi, M.E., Kingravi, H.A. and Vela, P.A., et al., 2013). The grasses are belonging to the "Poaceae" family. The edge is the cyperaceous family also called "Graminoid". 5,500 species of weeds are described, but 2,000 are identified in sedge plants. The weeds come under dicotyledonous families, i.e., "Flavaria Australacica", "Digera arvensis", "Tridax procumbens", "Rubus spp", "Bramble", "Butterfly" weed, etc.
4. **Based on Ecology:** These types of weeds are directly or indirectly associated with the environment. These types of weeds are part of ecology, there are some common weeds such as wetlands, Irrigation Lands, and drylands weeds are categorized (Tettamanzi, A.G. and Tomassini, M. et al., 2013). Wetland weeds are semi-aquatic plants that grow in two different sorts of ecological situations (Gagan Kumar, Rohit Kalra. et al., 2016). The first type of condition is one in which the plants are somewhat dry. Additionally known as irrigation land weeds, these weeds that no longer need more water will also grow on dry soil. Both

dry land and dehydrated land weeds can adapt by developing hairy and sticky root systems.

5. **Based on soil (Edaphic):** The few common weeds that germinate based on types of soil. Soil has been categorized into four sub-categories Regular soil, Red Clay soil, Loamy soil, and Lateritic soil. In Regular soils, the weeds are grown in a dehydrated situation, and special kinds of plants are grown in Clay soil (V. Vapnik. et al., 1995). The loamy soils of weeds produce sewage-like conditions e.g., "Leucas Aspera", "Lantana camera", "Spergula arvensis" etc.
6. **Based on the specificity:** Specificity is used to identify certain weeds; these weeds are classified as either parasitic or poisonous. Moreover, there are four categories of parasitic weeds: total root parasites, partial root parasites, total stem parasites, and partial stem parasites (V. Vapnik. et al., 1995). The following subsection contains thorough explanations.
   a. Poisonous weeds: These lead to cattle eating hazardous plant life, such as "Fastuosa (L.)", "Danert", and "Stramonium fastuosum (L.)", which is harmful to all living things. Alternatively, these lead to livestock being accumulated alongside barley and kept as farm animals. "Withania somnifera" berries and "Abrus precatorius" seeds are poisonous weeds (N. Cruz-Ramírez, H. G. Acosta-Mesa, H. CarrilloCalvet, L. Alonso Nava-Fernández, and R. E. Barrientos-Martínez, et al., 2007).
   b. Parasitic weeds: Most likely a combination of both the parasitic parasites and the host plant, the weeds are parasitic. Some parasites include partial, total, and total stem parasites as well as total root and partial stem parasites. Total root parasite plants, such as "Dendrophthoe," "Orobanche," "Viscum," "Santalum," "Aeginetia," "Lathrea," "Cistanche," and so on, are dependent on and provide nourishment for one another.
7. **Aquatic weeds:** There are four varieties of aquatic weed plants, namely submerged, emerging, marginal, and floating weeds, and they all grow in water throughout their life cycle: years (S. Lavania, P.S. Matey, et al., 2015).
8. **Noxious Weeds:** A toxic or poisonous weed plant is exceptionally unwelcome, difficult to manage, and inconvenient. As new weed management techniques are developed, a plant's classification as a toxic weed will also change following the legal translation of a country or state (S. Badhan, K. Desai, M. Dsilva, R. Sonkusare, and S. Weakey, et al., 2021). The poisonous weeds are extremely adept at mimicking and dispersing, and they use risky strategies to thwart human attempts to eradicate them. Sometimes the terms "exceptional weeds" and "offensive weeds" are also used to refer to poisonous weeds.
9. **Grassland Weeds:** Given that these areas provide a different biological environment than harvest lands, weeds indeed, as their name suggests, belong in these classes and assault prairies, rangelands, and evergreen fields (Boser, B.

E., I. Guyon, and V. Vapnik et al., 1992). Known as grasses, these weeds of the grasslands belong to the Poaceae family and are distinguished by their tall, narrow, prickly leaves.

10. **Marginal weeds:** On the wet shore, this can develop deeply. The water covers the leaves, while the root lies below it (V. Vapnik et al., 1995). Weed anchored in water, having most of its leaves above the surface. For example, "Typha," "Polygonum," "Cephalanthus," "Scirpus," and "Nilumbium speciosum." Based on the classifications mentioned earlier, a few common weeds are prevalent in Kharif and Rabi crops.

Mostly all weeds have common features and it's necessary to remove these weeds from cultivated cropland. Farmers are experiencing the issue of discrimination against weed plants from cropland and use the traditional method for weed plant removal. Traditional weed control methods, including manual cultivation using hoes, mechanical cultivation with powered tools, smothering with mulch, killing weeds through high heat, burning, and applying chemical herbicides, are often labor-intensive and time-consuming (Celebi, M.E., Kingravi, H.A. and Vela, P.A. et al., 2013). For this work, farmers must identify and detect the weeds in a short time.

AI IN WEED DETECTION

In machine learning algorithms for classifying, prior knowledge is utilized to educate the system to learn new. The machine learning method receives a training dataset containing feature set pairs in addition to tags as input to create the classification framework, the steps of the training and testing phases are depicted in Figure 4.

Figure 4. Test & Train Stage

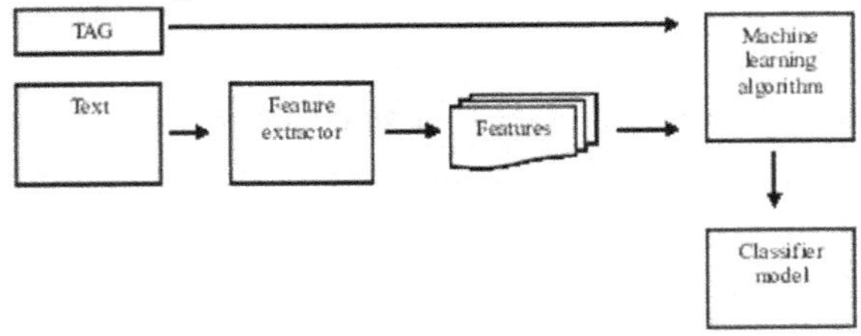

Once the training phase of the machine learning model is completed with various datasets then it can be used for prediction purposes. Feature extractors can now easily transform the uncovered data to their equivalent feature set which can be further given as input for making an accurate prediction in the classification model. The broad categories of ML approaches are unsupervised learning and supervised Learning as shown in Figure 5.

Figure 5. Classification of Machine Learning Technique

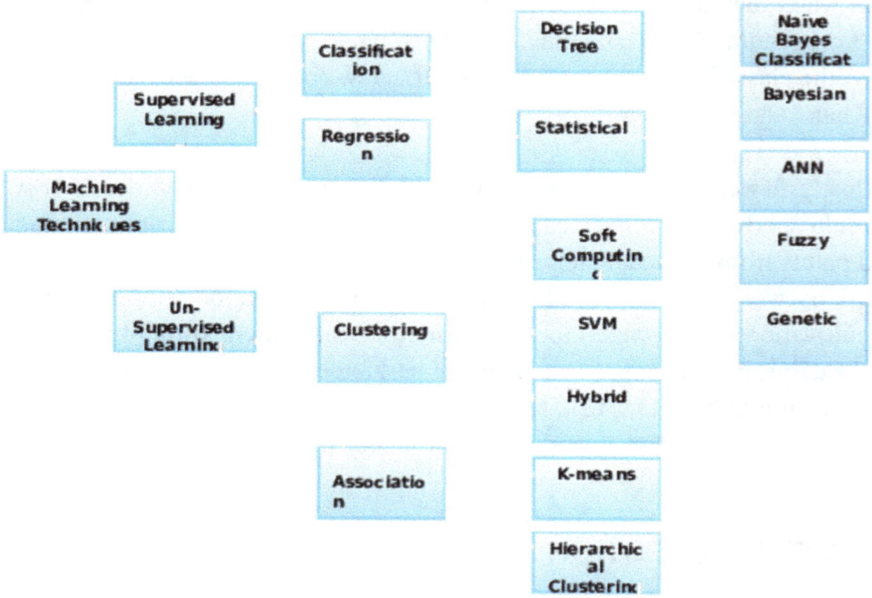

To relate an input dataset to its corresponding output in supervised learning, a mapping function is created using the training data that is provided. There are instances of input-output pairs in the training dataset.

Some most popular techniques used for weed detection are:

Decision Trees: A 'decision tree' is a hierarchical model in which the branches indicate potential attribute values and the main node indicates a test on a predictive feature (Tettamanzi & Tomassini, 2013). The class distributions are shown by the leaf nodes. Classifying an unlabeled object involves beginning from the root and working its way outward based on the values of its attributes.

Approximate computing: Deep Learning, genetic algorithms, and fuzzy logic algorithms are all used in conjunction with soft computing (Gagan Kumar, Rohit Kalra, et al., 2016). These methods are intended to resolve practical issues that were challenging to handle in the past. Neural networks are distributed systems made up of several parallel-arranged, interconnected nodes. According to Vapnik et al. (1995), every node is a processing unit that has a single output connection that can branch out to multiple others.

Artificial Neural Network: In machine learning (ML), a neural network is a system that is intended to replicate the neurological system of a human. It functions similarly to the human brain and excels in situations where other computers could falter (N. Cruz-Ramírez, et al., 2007).

Convolutional layers: Convolutional layers represent a type of deep learning model that convolutional neural networks (CNNs) employ for feature extraction from input data. This approach is motivated by the way biological neurons work, wherein increasingly complicated feature representations are developed by leveraging features from previous convolutional layers (S. Lavania, P.S. Matey, et al., 2015; S. Badhan, K. Desai, M. Dsilva, R. Sonkusare, and S. Weakey, et al., 2021). The weighted total of the inputs is calculated by several artificial neurons working in separate layers to produce activation levels for the output.

Statistical Techniques: To determine whether an instance belongs to a specific class or not, the statistical approach uses the probability theory to determine whether p is larger than 0 or less than 1, or equal to 0 or 1. Bayesian networks and the Naive Bayes classifier are examples of statistical techniques (V. Vapnik et al., 1995).

Support Vector Machines (SVM): Support Vector Machines, or SVMs, are recognized for their ability to produce results that are more accurate and frequently faster than those of some other machine learning models—all without the need for extra training. Nevertheless, it requires more resources to run its calculations. In essence, it creates a hyperplane that divides the space into two regions: one that contains vectors from a particular class and the other that contains vectors from a different class. The largest margin between the several classes is the definition of the hyperplane (Boser, B. E., I. Guyon, and V. Vapnik et al., 1992).

The training and tagging texts are represented using vectors and tags respectively. In case of high data complexity, the categorization of tags and vectors into limited subgroups becomes difficult. Even when the data complexity increases, this classification technique will always give the most precise as well as more accurate results. That makes SVM more special than other classification methods. To transfer the data from the original data to hyperspace and to separate the data into classes, a non-linear function is utilized.

Unsupervised Learning Techniques: To identify the hidden patterns and correlations from the provided dataset, the technique of unsupervised learning employs a mapping function. The dataset involves several examples in input and there is an output value in it or we can say no idea about the output. Clustering is used to find correlations as no teacher or feedback is there for making corrections. One such example is finding close friends on social media. The K-mean and hierarchical clustering methods can be used in this case (Celebi, M.E., Kingravi, H.A. and Vela, P.A. et al., 2013).

DETECTION AND CLASSIFICATION PROCESS

Identifying and classifying crops and weeds can be difficult in situations where there are similarities in color and texture, plants that are shaded in natural light, variations in color and texture caused by lighting, and weed species that look similar known as occlusion. During different stages of growth, similar agricultural plants or weeds may exhibit differences. Plant classification is made more challenging by motion blur and picture noise. In addition, different weed species can occur based on factors such as soil type, weather, crop variety, and geographic location as shown in Figure 6.

Figure 6. Weeds in different crops (weeds are indicated by red boxes and crops are indicated by green boxes)

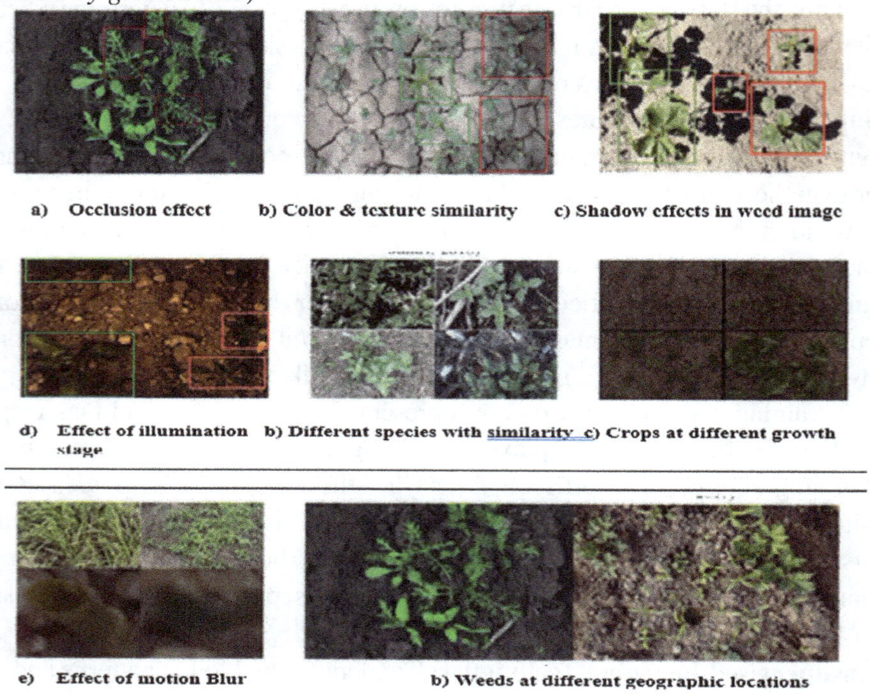

A typical weed identification system consists of the following steps acquisition and preliminary processing, feature extraction, and weed identification and categorization using images. To detect weeds in crop images it is important to preprocess the image data to enhance the quality and extract relevant features. Preprocessing steps can help remove noise standardize the images and highlight important information that will aid in weed detection. The process involves four key steps: data preprocessing, weed detection, feature extraction, and classification.

1. **Preprocessing-based methods:** Data preprocessing aims to enhance the image quality by improving its contrast. Images captured by digital cameras often contain noise, and this step helps facilitate further classification. It includes operations like enhancing images, converting color spaces, cropping, and flattening. The photographs may need to be trimmed and compressed for efficient analysis because they are usually taken in real-time and contain unnecessary backdrops. Below is an explanation of these strategies' specifics:
    a. **Enhancing the Image**: This step involves various methods such as histogram equalization and conversion of color and. The conversion of color is done to change RGB images into grayscale, while histogram equalization enhances the image's transparency. Additionally, techniques like image clipping and thresholding are employed for noise reduction and image preprocessing (M. Dyrmann, R.N. Jørgensen, H.S. Midtiby, et al., 2017). Examples of image enhancement methods are shown in Figure 7.

Figure 7a. Preprocessing Based Techniques

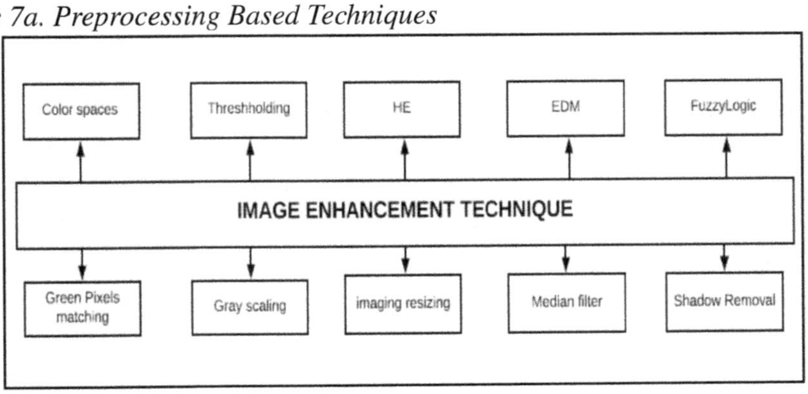

b. **Segmentation-based techniques:** Segmentation (J. Gui, L. Hao, Q. Zhang, and X. Bao, et al., 2015) is a process that divides an image into multiple parts to identify specific regions of interest. Some of the techniques used for segmentation are Otsu thresholding, K-means clustering, and color segmentation, which are widely used (depicted in Figure 19). K-means clustering, for instance, is applied for classification by forming different clusters of infected areas within the image. Each cluster consists of a set of pixels (S. Phadikar, J. Sil, and A.

D. Kumar, et al., 2013) that are similar to one another but distinct from pixels in other clusters. Otsu's method is often utilized for image compression and cropping. Another common segmentation method is thresholding, which is achieved by analyzing the histogram of the input image (A. Johannes, A. Picon, A. A. Gila, J. Echazarra, S. R. Vaamonde, and A. D. Navajas, et al., 2017).

Figure 7b. Segmentation Based Techniques

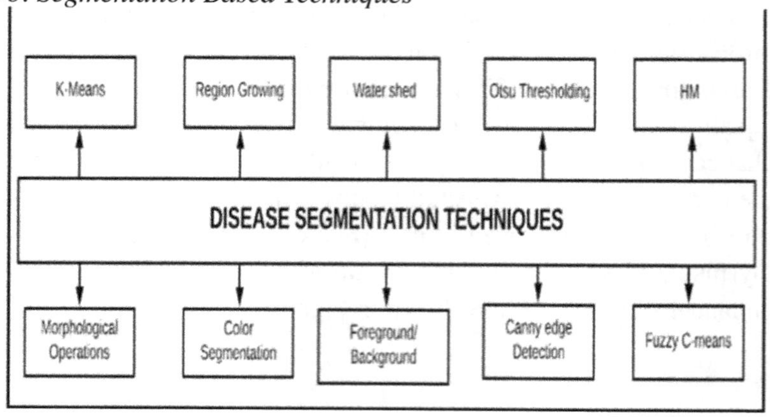

d. **Feature Extraction:** Features refer to the attributes of an object that differentiate it from others. These features can include aspects such as color, shape, size, and more (S. Goel, S. Pandey, and S. Kaur, et al., 2019). The selection of features like color, shape, and texture is critical since many weeds share similar characteristics with crops. Texture analysis primarily focuses on examining the surface properties of the object.

Once the images are preprocessed relevant features need to be extracted for weed detection. This could include color-based features texture features or shape features. These features will serve as input to the weed detection algorithm. Table 1 shows the types of feature extraction methods.

Table 1. Feature extraction methods

S.No.	Feature extractor type	Description
1.	Color-based features	• By extracting color-based features such as color histograms or color moments you can differentiate the color characteristics of weeds and crops. These features capture the variations in color distribution and can be used in classification algorithms.
2.	Texture-based features	• Texture refers to the spatial arrangement of pixels in an image. Weeds and crops often have different levels of texture due to variations in leaf shapes surface roughness or patterns.
3.	Shape-based features	• The shape of weed leaves or plant structures can be distinctive compared to crop leaves. Extracting shape-based features such as contour-based descriptors or geometric moments can help differentiate between weeds and crops based on their unique shape characteristics.
4.	Spatial features	• Spatial features capture the spatial relationships between different regions or objects within an image. By considering the spatial distribution of pixels or objects you can create features that represent the spatial layout of weeds relative to the crops.
5.	Gradient-based features	• The gradients in an image can provide information about the edges and boundaries between different objects. Extracting gradient-based features such as histograms of oriented gradients (HOG) or edge-based descriptors can help distinguish weeds from crops based on the variations in edge strength or direction.

1. Classification-Based Techniques

The final step in weed detection involves classifying the identified weeds. This step distinguishes between different classes, such as diseased leaves and healthy, non-infected leaves. The classifier is initially trained using a set of training images and then applied to test images. Several methods are used for plant leaf classification, including artificial neural networks (ANN), support vector machines (SVM), and K-means clustering. ANN is a self-learning technique (T. T. Dong, D. Song, C. Xiu, S. Geng, and J. Luo, et al., 2015) that adapts weights through backpropagation and efficiently handles noisy data. SVM, a widely-used non-linear classifier, operates in two stages: an offline training phase followed by an online decision-making phase for new images. The Radial Basis Function (RBF) classifier is known for its fast training and operates on Euclidean distance. Another classification method, probabilistic neural network (PNN) (S. Nandhini, R. Hemalatha, S. Radha, and K. Indumathi, et al., 2018), derived from RBF, is used for pattern classification. K-nearest neighbors (KNN) is a geometric classifier that calculates the Euclidean distance between positions and is simple to implement.

2. Machine Learning: A Modern Approach

Convolutional neural networks and machine learning have made major advancements in accuracy possible in recent years (CNNs). Deep learning has improved the ability to detect plant diseases from photos (A. E. Hassanien, et al., 2017). Even while conventional approaches have drawbacks—discussed previously—they might nevertheless produce fruitful outcomes. According to A. F. Aji et al. (2013), segmentation can be useful, for example, if the dataset's leaves are set against a black background. But when there are leaves or other background objects, these techniques become less accurate. Although there are numerous segmentation approaches available, their efficacy remains unknown. Histograms, SIFT, color-based approaches, and thresholding are a few examples of computationally demanding and potentially undergeneralized techniques (D. Majumdar, et al., 2015). As a result, machine learning has become more popular than manually designed features and segmentation problems. Large datasets can be handled effectively by techniques like CNNs (K. Wahid, A. Dinh, and M. Islam, et al., 2017; L. Wang et al., 2015; P. Sanyal et al., 2007).

The training stage and the verification phase are the two main stages of supervised learning. In the training phase, a machine learning algorithm is used to teach a model (F) (P. Sanyal, et al., 2007) based on the training data. In the testing phase, the model is given new input data, and it predicts the most appropriate labels (y) for the samples (X).

$$y_i=F(x_i)\ldots\ldots\ldots\ldots\ldots\ldots\ldots \quad (1)$$

Two subsets of data are typically created: one for testing and one for training. To get accurate and useful findings, a bigger percentage of the data is usually set aside for training. Neural network-based techniques like CNN (Rashid M., et al., 2021), NN, and BPNN are applied in machine learning; these techniques all adhere to the supervised learning paradigm. Convolution and pooling layers extract features; the fully connected layers use predetermined functions to classify these features (Picon A. et al., 2017).

LITERATURE SURVEY

Survey based on techniques used for weed identification: The author offered a strategy based on deep learning models that were refined using meta-heuristic optimization techniques by choosing the activations based on species that are transmitted to Softmax, the deep learning model's top layer (Toğaçar, Mesut, et al.,

2022). This activation that had been enhanced by the optimization techniques was then classed using Softmax.

In this study, the author suggested a pre-trained deep learning model, Mobile NetV2 (Y. Du, G. Zhang, D. Tsang, and M. K. Jawed et al., 2022), for a real-time weed detection tool. The 90% test accuracy achieved with weeds and crops as well as distortion, blur, and shadows is a significant step towards precision weed management in the real world. The authors of this research (S. Abouzahir, M. Sadik, and E. Sabir, et al., 2022) presented a reliable and simple detection method to produce weed maps under dense soybean fields. The findings show that leaf alignments compromise useful characteristics for weed-crop cluster discrimination.

Contrarily, the author of this research developed a new approach which is the integration of image processing and deep learning techniques (X. Jin, J. Che, and Y. Chen et al., 2021). The first step was to use a trained model for the identification of crops and the next is bounding boxe generation. Weeds were labeled as the leftover green objects that continued to fall out of the surrounding boxes. As a result, the model avoids controlling various weed species and instead focuses on recognizing only the vegetables.

The author's research focuses on using airborne multispectral camera sensors to identify weeds in the Sugar beetroot crop in Pakistan (S. I. Moazzam et al., 2021). They developed a unique weed detection model, a patch-based classification method that works better for aerial spraying systems designed for real-time intelligence than semantic segmentation.

The possibility of machine learning techniques for classifying weeds and crops from UAV photos is explored by the paper's author (N. Islam, M. Rashid, S. Wibowo, C. Xu, and A. Morshed, et al., 2021). To identify weeds using UAV photos, the performances of multiple machine learning methods, including random forest (RF), support vector machine (SVM), and k-nearest neighbors (KNN), are studied.

To identify weeds in crops, S. Badhan et al. (M. Dyrmann, R.N. Jørgensen, H.S. Midtiby, et al., 2017) have proposed a real-time weed detection strategy that combines stereo vision and machine learning for three-dimensional crop reconstruction. Two manual datasets of the cucumber and onion harvests are used to train the machine-learning model. The machine learning models are trained using the ResNet-50 method and Convolutional Neural Networks (CNN). ResNet-50's efficacy is comparable to that of traditional CNN.

Yanjee Li et al.(Y. Li et al., 2021) investigated the prospect of fast, accurate, and automated weed discrimination in pastures sown with ryegrass and clovers using hyperspectral imagery data and machine learning. The most dependable and solid prediction result was given by MLP built on the Sp technique (89.1% accuracy).

To create a system that can recognize and categorize plants as either crops or weeds, the author of this work applies machine learning to computer vision, enabling robotic appliances to perform plant cultivation and monitoring (D. R. Armstrong et al., 2021). The results show that larger photographs can yield results that are significantly more accurate, but at a major speed expense; low-resolution photos can yield results with relatively high-speed predictions.

An image is used as an input in a deep learning-based model developed by L. Boyina et al.(L. Boyina, G. Sandhya, S. Vasavi, L. Koneru, and V. Koushik, et al., 2021) to detect the presence of weeds in an agricultural field using image processing. For extracting invariant features, a convolutional neural network (CNN) is used. To diagnose faults in induction motors, A.Ibrahim et al.(A. Ibrahim, F. Anayi, M. Packianather, and O. Al-Omari, et al., 2021) describe a novel hybrid approach combining invasive weed optimization with machine learning. The motor's working state can be learned a lot from the vibration signal. The hybrid model has successfully demonstrated that it is reliable for defect diagnosis under various load circumstances.

The author's goal in this study is to present an automatic system for identifying plants and weeds, which might be very helpful when applying pesticides, fertilizers, and weedicides (S. A. Siddiqui, N. Fatima, and A. Ahmad et al., 2021). The model created in this paper uses a CNN to extract visual data and offers more accurate early weed detection in crop fields.

In this study, the author researched and proposed a YoloV5-based real-time system for weed identification (L. C. M. Junior and J. Alfredo C. Ulson, et al., 2021). The performance of the architecture was examined without and with transfer learning on a customized dataset consisting of 5 glyphosate-resistant weed species. The outcome clearly shows that the system can accurately identify the resistant weeds at 62 FPS. In this paper, the author discusses some of the significant studies done on weed and crop images for categorization and identification (R. Kamath, M. Balachandra, and S. Prabhu, et al., 2021).

In order to classify crops and weeds, the author of this study suggests a framework that enhances the original CNN with additional deep layers. To achieve this classification more accurately, the convolution neural network structure based on deep learning was developed (C. T. Selvi et al., 2021). In their paper, M. T. Khan et al. (2020) present a system for classifying crops and weeds in order to enable variable-rate pesticide spraying. Crops and weeds were identified and categorized using the RF classifier. The proposed vision-based pesticide spraying architecture has demonstrated real-time effectiveness in multiple field tests.

In this research, the author proposes a new method for weed and crop classification where the process of segmentation is done pixel-wise (A. Wang, Y. Xu, X. Wei, and B. Cui, et al., 2020). According to the findings in an encoder-decoder-based deep learning network, the transform using color space, and the index of vegetation without the near-infrared data will not improve segmentation outcomes for photos without augmentation, yet the addition of NIR information greatly increased the precision of segmentation. The best accuracy(average) achieved for this approach was 96.1%.

Using UAV-based photos, Y.Beehary et al.(Y. Beeharry and V. Bassoo, et al., 2020) investigated the effectiveness of the ANN and Alex Net weed detection methods. 15330 segments (approx) make up a picture dataset where 3247 data is of soil, 7373 of soybean, 3519 of grass, and 1190 of broadleaf weeds.

The results of a method for analyzing images introduced in the paper (M. G. Forero, S. Herrera-Rivera, J. Ávila-Navarro, C. A. Franco, J. Rasmussen, and J. Nielsen, et al., 2019) are superior to those of ML algorithms in several ways, including the speed and accuracy of weed detection in cereal crops. The best accuracy attained using machine learning algorithms was 97.07% with execution times of more than 2 minutes with 200 200-pixel sub-pictures.

The authors of the paper (S. T., S. T., S. G. G.S., S. S., and R. Kumaraswamy, et al., 2019) suggested a way to separate weeds from crops using ML. When a complete dataset is obtained for two types of weeds (Para grass and Nuts edge) and four different commercial crops, the Open CV and Keras framework are used to compare the performance of weed recognition algorithms.

In the study (M. D. Bah, A. Hafiane, R. Canals, and B. Emile, et al., 2019), the author suggests using 1-class classification and deep features on unsupervised data for the detection of weeds in UAV photos. The findings demonstrate that this new classification works well with both existing research and deep learning models built via supervised training data labeling. Deep convolutional neural networks, which are ideally suited for weed detection in turfgrass, are used by J Yu et al. (J. Yu, S. M. Sharpe, A. W. Schumann, and N. S. Boyd, et al., 2019). In addition, multiple DCNN models that are extremely accurate in detecting weeds in bermudagrass are also discussed here.

This study uses SVM with blob analysis to test a new model for categorizing crops and weeds in digital photos, to improve automated weed management systems (S. Murawwat and S. Ahmad, et al., 2018). The accuracy achieved proves that SVM and blob analysis have an accuracy level of more than 90%. A unique technique for identifying broad-leaf weeds in the pasture is put forth by Zhang (W. Zhang et al., 2018) so that precise weed control can be accomplished with less pesticide use. To attain high detection precision and resilience in real-world contexts, both traditional machine learning algorithms and deep learning techniques have been investigated and contrasted.

A model for identifying the presence of weeds in soybean fields using deep learning has been suggested in this paper (A. dos Santos Ferreira, D. Matte Freitas, G. Gonçalves da Silva, H. Pistori, and M. Theophilo Folhes, et al., 2017). Convolution Neural Networks, which were utilized in this study, are an example of a Deep

Learning architecture that has had outstanding success with picture identification. They have used Caffe-Net architecture for the training of datasets, and for result analysis, they have compared four well-known machine learning techniques with the proposed technique. A new set of data (i.e. Carrot-Weed) for weed detection was presented by P. Lameski et al. (P. Lameski, E. Zdravevski, V. Trajkovik, and A. Kulakov, et al., 2017) under varied lighting conditions. The collection contains RGB photos of immature carrot seedlings collected in February, the author conducted a preliminary analysis of the dataset and presented preliminary results acquired using CNN designs.

Classifier-based Survey: The Survey is organized into distinct sections: disease detection in crops, as illustrated in Table 2; disease detection employing ML techniques, as shown in Table 3; and nutrient deficiency in crops, presented in Table 4.

Table 2. Literature Review

Publication Year	Existing Work Summary		
	Crop	Number of Images	Classifier
(M. Dyrmann,, et al., 2017)	Soybean	Total images= 3624	Simple linear iterative clustering
(V. Kumar, et al., 2008)	Mustard	N/A	Expert System
(A. dos Santos Ferreira,, et al., 2017)	Soybean	N/A	HIST and WDH with K means
(S. Phadikar, et al., 2013)	Rice	N/A	Rule-based methodology
(A. Johannes, et al., 2017)	Wheat	Total images=3500	Segmentation of statistics
(T. T. Dong, D. et al., 2015)	Corn	Total images=244	Classifier based on attributes
(S. Nandhini, et al., 2018)	Pomegranate, brinjal, and Tomato	Total images= 30	Compressed sensing and SVM
(A. E. Hassanien, et al., 2017)	Tomato	N/A	MFORSFS algorithm(MFO,RS,SVM)

In order to recognize and identify weeds, the author presented a novel RCNN framework in which the positive-negative problem technique was used to extract weed characteristics and PU learning was used to classify weed leaf images. The VGGNet model was utilized to identify several kinds of broadleaf weeds that are typical of the crop. The detection accuracy was 36.8% with a sample of 1,120 plants and 683 detections, suggesting that certain weeds are difficult to discern visually. As a result, for pixel-level object detection, the deep learning VGG 16 FCN algorithm was used (S. B. Jadhav, V. R. Udupi, and S. B. Patil, et al., 2020).

Additional research is required to demonstrate the effectiveness of 40 CNNs in identifying multiple weed species in targeted plants, even though they have shown promise in weed detection, verification, identification, and density estimation in various plants using pre-trained CNN techniques (S. Shorewala, A. Ashfaque, R. Sidharth, U. V et al., 2021).

Disease Detection using machine learning: Some research on the use of machine learning techniques for illness identification is covered in the part that follows. In these techniques, the data set is split into training and testing sets before being fed into the classifier.

Table 3. Based on the use of machine learning techniques for disease detection

Year of Publication	Existing Work Summary				
	Crop	Number of images		Classifier	Performance measure (Accuracy)
		Training images	Testing images		
(H. Lal, P. Meena, M. Kumar and P. Jha, et al., 2013)	Oil Palm	102	22	NN	87.75
(D. Majumdar, A. Chakraborty, D. Kole and D. Majumdar, et al., 2015)	Wheat	20	342	BPNN	84.4
(S. Biswas, h. Jagyasi, B. . P. Singh and M. Lal, et al., 2014)	Potato	Total images=27		BP Neural Network with Fuzzy C-means	100%
(K. Wahid, A. Dinh and M. Islam,, et al., 2017)	Potato	Total images = 300		SVM with RGB imaging	95%
(D. Samanta and A. Ghosh, et al., 2012)	Corn	210 210 210	43 48 47	Histogram feature-based	83.5 95.2 96.7
(L. Wang, Y. J. Shang and S. W. Zhang, et al., 2015)	Corn	90	10	PCA and K-NN Bayesian Neural Network	86.95 90 89.25
(V. K. Shah, . H. B. Prajapati and J. P. Shah, et al., 2016)	Rice	90	30	K-NN SVM	93.33 91.10
(Ministry of Electronics and Information Technology, et al., 2019)	Cucumber	250	30	ANN	Visual examination
(M. K. Asefpour, V. and J., et al., 2013)	Cucumber	150	150	PNN	90 92 92

continued on following page

Table 3. Continued

Year of Publication	Existing Work Summary				
	Crop	Number of images		Classifier	Performance measure (Accuracy)
		Training images	Testing images		
(S. Sladojevic, M. Arsenovic, A. Anderla, D. Culibrk and D. Stefanovic, et al., 2016)	Tomato	78 104 104 104 104 104	20 26 26 26 26 26	Decision Tree	70 80.7 69.2 69.2 92.3 84.6
(S. Mohanty, D. Hughes and M. Salathé, et al., 2016)	Tomato	58 32 14 22 22 32	58 32 14 22 22 32	BPNN	96.5 93.3 64.7 54.5 100 87.5
(H. Lal, P. Meena, M. Kumar, and P. Jha, et al., 2013)	Grapes	16 17	2 2	Back Propagation Neural network	100%
(M. K. Asefpour, V. and J., et al., 2013)	Citrus	20 20 20 20	20 20 20 20	Naïve Bayes K-means Discriminant analysis Random forest Tree	95 77.5 98.75 97.5
(S. Jeyalakshmi and R. Radha, et al., 2017)	Blueberry	400	400	i) HOG ii) LBP	Result: 84% accuracy index using Deep Learning
(M. G. Forero, S. Herrera-Rivera, J. Ávila-Navarro, C. A. Franco, J. Rasmussen, and J. Nielsen, et al., 2019)	Review with multiple plants	200	200 leaf	ML Techniques	Accuracies of all algorithms stated.
(S. T., S. T., S. G. G.S., S. S., and R. Kumaraswamy, et al., 2019)	Turmeric	200	200	GLCM texture analysis (K means)	91% accuracy rate

The analysis that followed covered the remaining 34 publications in this literature review. The initial assessment was centered on dates and the quantity of papers. Despite the recent publication, there has been a noticeable rise in research into the idea of deep learning for weed detection and categorization, as seen in Figure 8 a

Figure 8a. Year-by-year breakdown of articles published (Year; Number of publications).

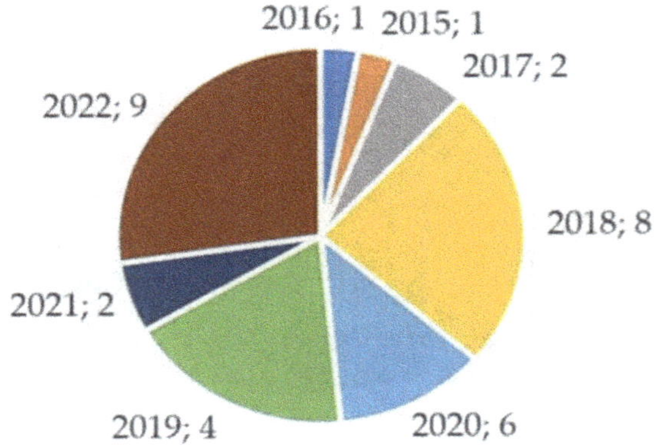

Verifying the genre of late-topic articles being published and circulated was the second study goal. Of the thirty-three publications, four discussed plant illnesses, four were fruit detection, fourteen were weed detection, and eleven were weed classification. In Figure 8 b, this is shown.

Figure 8b. Papers published by application area.

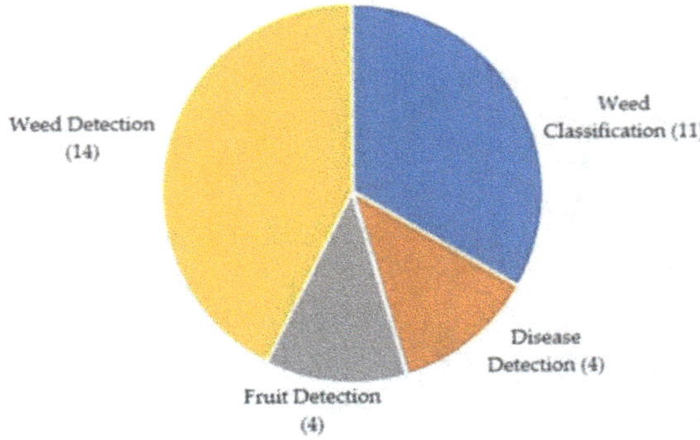

Conferences and journal publications were found here. The primary source of material for this review was published in six highly regarded journals. Only items that have been published more than twice are displayed in Figure 9 a. These were the top journals for precise farming, remote sensing, and UAV-based weed detection.

Figure 9a. Articles published for precise farming, remote sensing, and UAV-based weed detection.

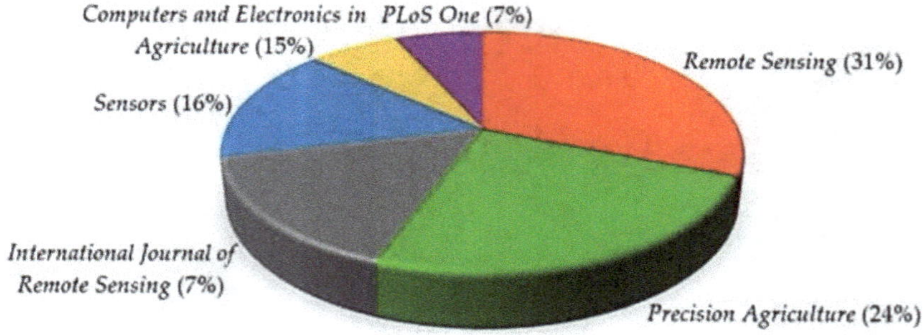

Researchers from 14 nations produced the majority of the scientific articles, according to a geographic analysis: China, Italy, the United States, Germany, India, Australia, Spain, Italy, Brazil, the Netherlands, France, Japan, Denmark, the United Kingdom, and Switzerland (Figure 9 b).

Figure 9b. Scientific publications (by June 18, 2023) written by scholars from various nations.

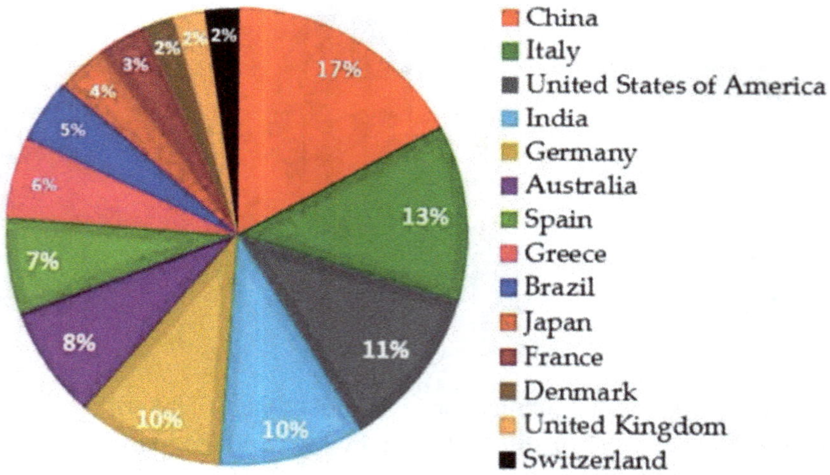

Literature survey for identifying nutrient deficiency using machine learning techniques: The section addresses the research conducted on detecting crop nutrient deficiencies through ML techniques applications.

Table 4. Based on deficiency of nutrients in crops

Year of Publication	Existing Work Summary				
	Crop	**Nutrient deficiency**	**Number of images**	**Classifier**	**Performance measure(accuracy)**
(M. A. Hairuddin, N. M. Tahir and N. S. Rizam, et al., 2011)	Oil Palm	Nitrogen, Boron, Zinc, Manganese, Potassium, phosphorous and magnesium	N/A	Fuzzy logic	Visual examination
(H. . M. Asraf, M. S. Rizam and M. T. Nooritawati, et al., 2012)	Oil Palm	Nutrient deficiency	Total images= 420	Support vector machine	95%
(K. Ż. Gołaszewska, F. G. Fordoński, and M. Wiwart, et al., 2009)	Bean, Pea, and yellow lupine	Nitrogen, magnesium, potassium, and phosphorous	N/A	Color based analysis	Visual examination
(S. Shidnal and M. V. Latte, et al., 2016)	Soyabean	Healthy, rust tan, mildew, and rust RB	Total images=1200	SVM	98%
(S. Jeyalakshmi and R. Radha, et al., 2017)	Strawberry	Iron deficiency, fungal	Iron-60 Fungal -50	Fuzzy logic classification algorithm (FLCA)	95%
(S. Laraba, S. et al., 2018)	Rice	Deficiency of Nitrogen, Potassium, Phosphorous	400		90%
(S. K. Parui, et al., 2007)	Rice	Iron, Boron, Nitrogen, Potassium, manganese, and magnesium	N/A	BPNN	88.56
(R. B M, M. Fernandez, J. Murillo and N. Arreguın, et al., 2015)	Grapevine	Potassium	25	KNN	Better than Histogram

PROBLEM DESCRIPTIONS

Weed is the undesirable plant in a particular situation or "a plant in the wrong place". In the entire world out of 2,50,000 plant species, 250 weed family species are prominent in the agricultural and non-agricultural systems. Owing to different conditions and factors around the world about 30,000 species are grouped as weeds and categorized into different families (R. Kamath, M. Balachandra, and S. Prabhu, et al., 2020). Weed as well as the crop requires nutrients such as nitrogen, phosphorus, and potassium for their growth. As there is an increase in weed growth, it shows a negative impact on (leaves, stem, root growth, etc.) crops yield (N. Son, C. Chen, C. Chen, P. T.J. of R., and undefined 2021,, et al., 2021).

The farmers are facing a problem with weed identification and recognition due to the similar homomorphic properties of weed and crop leaves. Two important factors affect crop yield, first is weed growth and another is weed density. The weed/crop plants change the morphological, visual texture, and spectral properties of leaves so it is difficult for the classification model to predict accurately with frequency changes in data (S. Haug and J. Ostermann, et al., 2015). A lot of work has been conducted on a particular crop for weed characterization but practical implementation on the mixed crop has not been done. In some cases, when the leaves overlap with each other, the morphologic feature is not properly identified (U. P. Singh, S. S. Chouhan, S. Jain, and S. Jain, et al., 2019). Also, the researchers had not paid much attention to the impact of weeds on crops due to nutrient deficiency. Another challenge in this field is the varying effect of light on the crops. The segmentation and classification method fails as a result of variations in the illumination, which also affect the colors, level of noise, 41 shadowy areas, refraction, contrary, saturation, and brilliance of the same scene. The challenge of enhancing the algorithm's effectiveness in several color space models under varied illumination circumstances is laborious (X. Ma et al., 2019).

So, to overcome this problem we have focused on weed growth and density estimation in soybean crop plants to detect and classify the weed species

Justification for Research:

a. **Motivation:** Nowadays many researchers are using Artificial Intelligence (AI) techniques in agriculture that have proven to be profitable farming. This will reduce crop costs and increase farmers' income. According to agricultural scientists, artificial intelligence will help protect crops from plague damage. Recent research and innovation in agriculture predict crop yield which is influenced by various factors. The weed is an important factor that can reduce crop yield.

Therefore, the identification and recognition of weeds is an important task in the early stage of plants. Through this technology, weeds can be predicted accurately which farmers will gain from farming. Farmers can obtain knowledge on seeds, mint, fertilizers, pesticides, and irrigation. India is almost at an estimated loss due to weeds is about 37% as shown in figure 30. For the agriculture application, a deep learning algorithm is the best algorithm for finding a better solution by using a particular problem dataset. The crop yield loss due to weeds, diseases, insects, and pests is described in Figure 10 (J. Kinyua and L. Awuah, et al., 2021).

Figure 10. Crop loss due to weeds, diseases, Insects and other pests

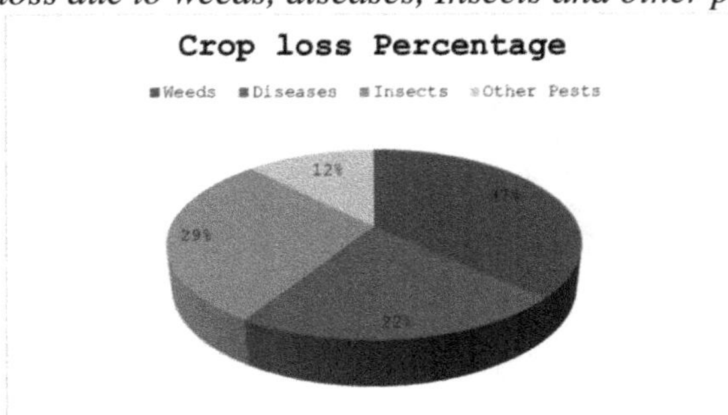

b. Research Gaps
   i. A review of the existing literature reveals several gaps and future research opportunities. Specifically, there is a notable absence of studies on disease detection in mustard crops using image analysis, unlike the more extensive research conducted on crops such as wheat, rice, and corn.
   ii. A major issue is the lack of available datasets necessary for studying disease detection in crops. Currently, there are no datasets for healthy and infected mustard leaves, which hinders progress in this area of research.
   iii. The literature indicates that Indian researchers have identified a promising opportunity to apply computer vision techniques to address significant challenges in agriculture, a focus that has been less prevalent in other countries.

MATERIALS AND METHODS

A. Site Description and Field Experiment

We used 589 photos to identify weeds in sesame crops using image processing and artificial intelligence (AI). The experiment was conducted using the weed and crop dataset that we gathered from Kaggle, which Alessandro dos Santos Ferreira founded. Figure 11 displays representative photos of the several sesame crop variants photographed for the current investigation. To evaluate the model's performance following different modifications, such data samples are used. We aimed to evaluate the efficacy and precision of multiple iterations for weed detection among the crops.

Figure 11. Crop and weed Image

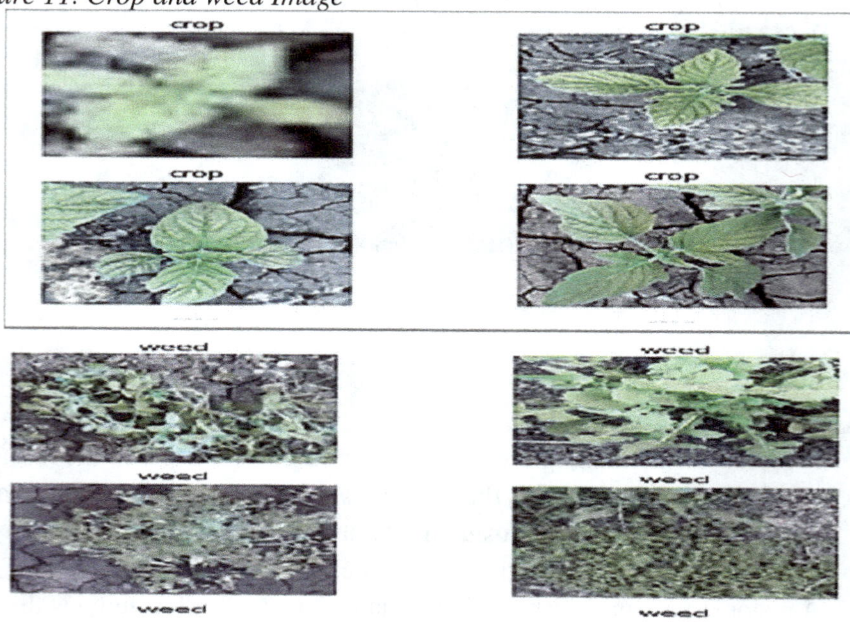

B. Capturing Images

An aggregate of 1300 RGB ordinal pictures of weeds and sesame plants, with a resolution of 4000X3000 colors, were reserved at Kaggel. We have taken 589 images to detect weeds in sesame crops through image analysis and artificial intelligence (AI). Figure 11 shows a diagram showing the steps of weed image collection.

C. Data Preprocessing

The second step towards data sample processing after image acquisition is data cleaning, which is crucial since the detection model will perform worse if any faulty photos are left in the dataset. We now have 546 photos after cleanup. The next step towards processing data samples is to convert all of our photographs to 512 X 5 12 X3 sizes because the original photo is 4000 X3000 colors in size, which is enormous and will take a very long time to train. However, training requires more than 546 images, so we must turn these 546 images into 1300 images. To increase the dataset, we used data augmentation techniques. Labeling picture data manually is an extremely tedious procedure! Drawing bounding boxes on photos—whether they are weeded or cropped—is required in this phase (figure 12).

D. Classification of sample data

For training, we have taken 1000 images to train models and 300 images to test. The object annotation can be seen in Table 5. We have taken 1637 features to train and 435 features to test models. Complete steps for Yolo Label conversion are shown below.

Figure 12. Drawing bounding boxes on photos and label visualization

Table 5. Converting data frame to Pascal format

	filename	width	height	class	xmin	ymin	xmax	ymax
0	agri_0_9354.jpeg	512	512	weed	63	120	425	442
1	agri_0_9354.jpeg	512	512	weed	0	1	180	148
2	agri_0_7574.jpeg	512	512	crop	95	167	453	469
3	agri_0_8960.jpeg	512	512	weed	52	76	422	353
4	agri_0_417.jpeg	512	512	weed	7	75	511	411
...	...	...	...	...	...	...	...	...
2067	agri_0_2825.jpeg	512	512	weed	16	144	202	303
2068	agri_0_2825.jpeg	512	512	weed	291	94	471	304
2069	agri_0_9252.jpeg	512	512	weed	247	194	384	331
2070	agri_0_9252.jpeg	512	512	weed	37	104	179	246
2071	agri_0_8141.jpeg	512	512	crop	27	51	460	498

EXPERIMENTS AND RESULTS

The presence of weeds in agriculture is undesirable. Weed consumes resources such as minerals, soil, water, and a lot more that could have been used by crops, which lowers the yield of the necessary crop. Although using pesticides to get rid of weeds is common and effective, some chemicals can stick to crops and cause issues for people. Our technology will only spray pesticides on weeds, not crops, which will lessen the issue of pesticides combining with crops and reduce pesticide waste. 1300 classifications of images of different types of weeds and sesame crops are included in this dataset. Every image has a resolution of 512 x 512 pixels. For image labeling, the YOLO format is employed.

The evaluation metrics of SVM and CNN with 1300 weed and crop datasets are calculated and analyzed here. With an accuracy of 96.85%, Table 6 shows how well the SVM algorithm performed, with a precision of .968, a TPR of .96, and obtained F1 score of 0.86 for the given dataset. It is clear from the table that out of 1300 datasets, 1259 instances is there which are precisely classified, and approx. 41 instances of a dataset are wrongly classified.

Similarly, Table 7 shows the CNN algorithm's performance, which achieved 97.5% accuracy, a precision of .97, a TPR of .97, and obtained F1 score of 0.97 for the given dataset. It is clear from the table that out of 1300 datasets, 1267 instances are there which are precisely classified and approx. 33 instances of the dataset are wrongly classified.

Table 6. SVM Evaluation metric

Overall Instances		**1300**	
Instances that are classified precisely		1,259 approx. (96.85%)	
Instances that are classified wrongly		41 approx. (3.15%)	
Accuracy	**TPR**	**Precision**	**F1 Score**
96.85%	0.96	0.968	0.86
SVM CM (Confusion Matrix)			
A		B	
1240		6	
35		19	

Table 7. Evaluation metrics of CNN Model

Overall Instances		**1300**	
Instances that are classified precisely		1,267 approx. (97.5%)	
Instances that are classified wrongly		33 approx. (2.5%)	
Accuracy	**TPR**	**Precision**	**F1 Score**
97.5%	0.97	0.97	0.97
CNN CM			
A		b	
1260		**3**	
30		07	

Table 8. Evaluation metrics of our model

Overall Instances		**1300**	
Instances that are classified precisely		1,296 approx. (99.725%)	
Instances that are classified wrongly		4 approx. (0.275%)	
Accuracy	**TPR**	**Precision**	**F1 Score**
99.725%	0.997	0.996	0.997
Proposed model CM			
A		b	
1290		**1**	
3		06	

We observed that CNN alone is not a very good approach for this kind of problem. Therefore, we have used the ensemble model which will outperform the CNN model. Ensemble models combine the predictions of multiple individual machine-learning models to make more accurate and reliable predictions. In the ensemble model with random forest, we used a weighted ensemble model in which the weight of CNN was 60% and the weight of Random Forest was 40%. In our customized ensemble model, we have taken 5 layers of CNN where, the number of input Layers = 187, 1st layer = 32, 2nd layer = 64, 3rd layer = 128, a Flatten (), 1st dense layer = 512, 2nd dense layer = 1024 and 2 output layers. Table 8 depicts the performance of our model which attained an accuracy of 99.725%, a precision of .997, a TPR of .996, and obtained F1 score of 0.997 for the same dataset. It is clear from Table 8 that out of 1300 datasets, 1296 instances is there which are precisely classified and only 4 instances of datasets are wrongly classified.

Figure 13a shows that the suggested approach offers noticeably higher accuracy outcomes than the prior best results because the work has employed an ensemble model which is a combination of two well-known machine learning models i.e. CNN + Random Forest.

Figure 13a. Accuracy comparison of various classifiers

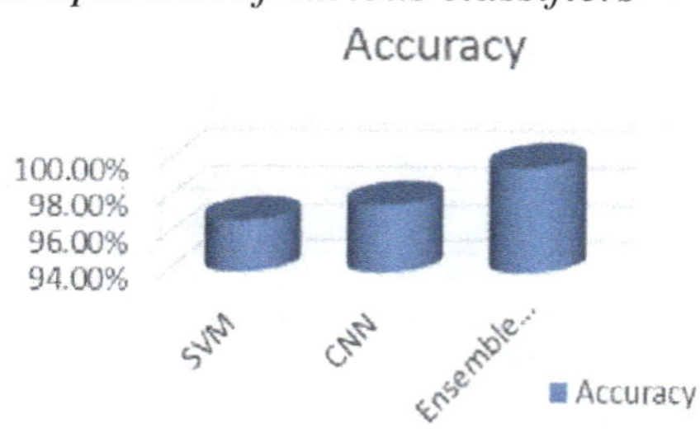

The proposed model's real-time accuracy is depicted in Figure 13 b.

Figure 13b. Real-time detection of crop and weed

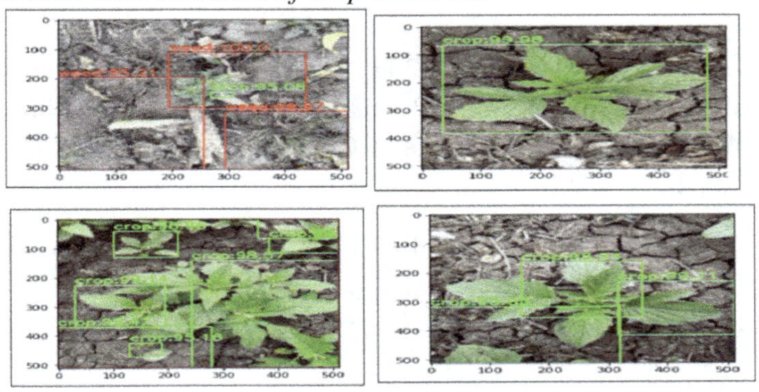

CONCLUSION

A hybrid approach that combines region-adaptive techniques with deep convolutional neural networks (CNNs) to tackle the challenges of image de-hazing and de-noising has been introduced in this paper. This method is essential for improving image quality under poor atmospheric conditions. By integrating adaptive region selection with advanced deep learning models, our approach enables more accurate and context-sensitive image processing, resulting in notable enhancements in visibility and clarity. Our experimental results show that this new method surpasses current leading techniques in both subjective and objective assessments, especially in cases with significant haze and noise. The approach's ability to dynamically adjust processing based on regional characteristics helps reduce image detail loss and artifacts, leading to more precise and natural restoration.

This research not only advances image processing methodologies but also paves the way for future studies into more sophisticated and adaptable models. Future research could investigate the incorporation of additional atmospheric correction methods and explore real-time video processing applications, which would be valuable in areas such as remote sensing, autonomous driving, and surveillance.

REFERENCES

Abouzahir, S., Sadik, M., & Sabir, E. (2022). Lightweight computer vision system for automated weed mapping. In 2022 IEEE 12th Annual Computing and Communication Workshop and Conference (CCWC) (pp. 372-376). DOI: 10.1109/CCWC54503.2022.9720800

Aitkenhead, M. J., Dalgetty, I. A., Mullins, C. E., McDonald, A. J. S., & Strachan, N. J. C. (2003). Weed and crop discrimination using image analysis and artificial intelligence methods. *Computers and Electronics in Agriculture*, 39(3), 157–171. DOI: 10.1016/S0168-1699(03)00076-0

Aji, A. F., Munajat, Q., Pratama, A. P., Kalamullah, H., Aprinaldi, J. S., & Arymurthy, A. M. (2013). Detection of palm oil leaf disease with image processing and neural network classification on mobile device. *International Journal of Computer Theory and Engineering*, 5(3), 1–5. DOI: 10.7763/IJCTE.2013.V5.743

Alam, M., Alam, M. S., Roman, M., Tufail, M., Khan, M. U., & Khan, M. T. (2020). Real-time machine-learning-based crop/weed detection and classification for variable-rate spraying in precision agriculture. In 2020 7th International Conference on Electrical and Electronics Engineering (ICEEE) (pp. 273–280). DOI: 10.1109/ICEEE49618.2020.9102505

Amara, J., Bouaziz, B., & Algergawy, A. (2017). A deep learning-based approach for banana leaf diseases classification. In Gesellschaft für Informatik Bonn (pp. 79–88).

Armstrong, D. R., Götz, M., & Nardelli, V. C., & de O. Gomes, V. E. (2021). Machine learning for identification and classification of crops and weeds. In 2021 XI Brazilian Symposium on Computing Systems Engineering (SBESC) (pp. 1-7). DOI: 10.1109/SBESC53686.2021.9628296

Arnal Barbedo, J. G. (2019). Plant disease identification from individual lesions and spots using deep learning. *Biosystems Engineering*, 180, 96–107. DOI: 10.1016/j.biosystemseng.2019.02.002

Asefpour Vakilian, M., & Massah, J. (2013). An artificial neural network approach to identify fungal diseases of cucumber (Cucumis sativus L.) plants using digital image processing. *Archiv für Phytopathologie und Pflanzenschutz*, 45(36), 1580–1588. DOI: 10.1080/03235408.2013.772321

Asraf, H. M., Rizam, M. S., & Nooritawati, M. T. (2012). Comparative study in kernel-based support vector machine of oil palm leaves nutrient disease. International Symposium on Robotics and Intelligent Sensors 2012, 41, 1353-1359. DOI: 10.1016/j.proeng.2012.07.321

Badhan, S., Desai, K., Dsilva, M., Sonkusare, R., & Weakey, S. (2021). Real-time weed detection using machine learning and stereo-vision. In 2021 6th International Conference for Convergence in Technology (I2CT) (pp. 1-5). DOI: 10.1109/I2CT51068.2021.9417989

Bah, M. D., Hafiane, A., Canals, R., & Emile, B. (2019). Deep features and one-class classification with unsupervised data for weed detection in UAV images. In *2019 Ninth International Conference on Image Processing Theory, Tools and Applications (IPTA)* (pp. 1–5). DOI: 10.1109/IPTA.2019.8936091

Bakyarani, S., Srimathi, H., & Bagavandas, M. (2019). A survey of machine learning algorithms in health care. *International Journal of Scientific & Technology Research*, 8(11), 2288–2292.

Beeharry, Y., & Bassoo, V. (2020). Performance of ANN and AlexNet for weed detection using UAV-based images. In 2020 3rd International Conference on Emerging Trends in Electrical, Electronic and Communications Engineering (ELECOM) (pp. 163–167). DOI: 10.1109/ELECOM49001.2020.9296994

Biswas, S., Jagyasi, H., Singh, B. P., & Lal, M. (2014). Severity identification of potato late blight disease from crop images captured under uncontrolled environment. In IEEE. DOI: 10.1109/IHTC.2014.7147519

Boser, B. E., Guyon, I., & Vapnik, V. (1992). A training algorithm for optimal margin classifiers. In *Proceedings of the Fifth Annual Workshop on Computational Learning Theory* (pp. 144-152). ACM Press. DOI: 10.1145/130385.130401

Boyina, L., Sandhya, G., Vasavi, S., Koneru, L., & Koushik, V. (2021). Weed detection in broad leaves using invariant U-Net model. In *2021 International Conference on Communication, Control and Information Sciences (ICCISc)* (vol. 1, pp. 1-4). DOI: 10.1109/ICCISc52257.2021.9485001

Brahimi, M., Arsenovic, M., Laraba, S., Sladojevic, S., Boukhalfa, K., & Moussaoui, A. (2018). Deep learning for plant diseases: Detection and saliency map visualization. In Springer (pp. 93–117).

Brahimi, M., Arsenovic, M., Laraba, S., Sladojevic, S., Boukhalfa, K., & Moussaoui, A. (2018). Deep learning for plant diseases: Detection and saliency map visualisation. In Springer. https://doi.org/DOI: 10.1007/978-3-319-90402-3_5

Celebi, M. E., Kingravi, H. A., & Vela, P. A. (2013). A comparative study of efficient initialization methods for the k-means clustering algorithm. *Expert Systems with Applications*, 40(1), 200–210. DOI: 10.1016/j.eswa.2012.07.021

Celebi, M. E., Kingravi, H. A., & Vela, P. A. (2013). A comparative study of efficient initialization methods for the k-means clustering algorithm. *Expert Systems with Applications*, 40(1), 200–210. DOI: 10.1016/j.eswa.2012.07.021

Cruz-Ramírez, N., Acosta-Mesa, H. G., Carrillo-Calvet, H., Nava-Fernández, L. A., & Barrientos-Martínez, R. E. (2007). Diagnosis of breast cancer using Bayesian networks: A case study. *Computers in Biology and Medicine*, 37(11), 1553–1564. DOI: 10.1016/j.compbiomed.2007.02.003 PMID: 17434159

Dong, T. T., Song, D., Xiu, C., Geng, S., & Luo, J. (2015). A curvelet-SC recognition method for maize disease. *Journal of Electrical and Computer Engineering*, 2015, 1–8. DOI: 10.1155/2015/782851

Du, Y., Zhang, G., Tsang, D., & Jawed, M. K. (2022). Deep-CNN-based robotic multi-class under-canopy weed control in precision farming. In *2022 International Conference on Robotics and Automation (ICRA)* (pp. 2273-2279). DOI: 10.1109/ICRA46639.2022.9812240

Dyrmann, M., Jørgensen, R. N., & Midtiby, H. S. (2017). Roboweedsupport- detection of weed locations in leaf occluded cereal crops using a fully convolutional neural network. *Advances in Animal Biosciences*, 8(2), 842–847. DOI: 10.1017/S2040470017000206

Ferentinos, K. P. (2018). Deep learning models for plant disease detection and diagnosis. *Computers and Electronics in Agriculture*, 145, 311–318. DOI: 10.1016/j.compag.2018.01.009

Forero, M. G., Herrera-Rivera, S., Ávila-Navarro, J., Franco, C. A., Rasmussen, J., & Nielsen, J. (2019). Color classification methods for perennial weed detection in cereal crops. In Progress in Pattern Recognition, Image Analysis, Computer Vision, and Applications (pp. 117–123).

Fuentes, A., Yoon, S., Kim, S., & Park, D. (2017). A robust deep-learning-based detector for real-time tomato plant diseases and pests recognition. *Sensors (Basel)*, 17(9), 17. DOI: 10.3390/s17092022 PMID: 28869539

Goel, S., Pandey, S., & Kaur, S. (2019). Plants disease identification and classification through leaf images. *Archives of Computational Methods in Engineering*, 26(2), 507–530. DOI: 10.1007/s11831-018-9255-6

Gołaszewska, K. Ż., Fordoński, F. G., & Wiwart, M. (2009). Early diagnostics of macronutrient deficiencies in three legume species by color image analysis. *Computers and Electronics in Agriculture*, 65(1), 125–132. DOI: 10.1016/j.compag.2008.08.003

Gui, J., Hao, L., Zhang, Q., & Bao, X. (2015). A new method for soybean leaf disease detection based on modified salient regions. *International Journal of Multimedia and Ubiquitous Engineering*, 10(6), 45–52. DOI: 10.14257/ijmue.2015.10.6.06

Gunasinghe, N., Shashi, K., Banga, S. S., You, M. P., Banga, S. K., & Barbetti, J. M. (2017). Outstanding host resistance will resolve the threat from white leaf spot disease (Pseudocercosporella capsellae) to oilseed and vegetable Brassica spp. crops. *Australasian Plant Pathology*, 46(2), 137–146. DOI: 10.1007/s13313-017-0470-7

Gunasinghe, N., You, M. P., Li, X. X., Banga, S. S., Banga, S. K., & Barbetti, M. J. (2016). New host resistances to Pseudocercosporella capsellae and implications for white leaf spot management in Brassicaceae crops. *Crop Protection (Guildford, Surrey)*, 86, 69–76. DOI: 10.1016/j.cropro.2016.04.014

Hairuddin, M. A., Tahir, N. M., & Rizam, N. S. (2011). Overview of image processing approach for nutrient deficiencies detection in Elaeis Guineensis. In IEEE.

Hameed, S., & Amin, I. (2019). Detection of weed and wheat using image processing. In *Proceedings of the 2018 IEEE 5th International Conference on Engineering Technologies and Applied Sciences (ICETAS)* (pp. 1-5). IEEE.

Hassanien, A. E., Gaber, T., Mokhtar, U., & Hefny, H. (2017). An improved moth flame optimization algorithm based on rough sets for tomato diseases detection. *Computers and Electronics in Agriculture*, 136, 86–96. DOI: 10.1016/j.compag.2017.02.026

Hassanien, A. E., Gaber, T., Mokhtar, U., & Hefny, H. (2017). An improved moth flame optimization algorithm based on rough sets for tomato diseases detection. *Computers and Electronics in Agriculture*, 136, 86–96. DOI: 10.1016/j.compag.2017.02.026

Haug, S., & Ostermann, J. (2015). A crop/weed field image dataset for the evaluation of computer vision based precision agriculture tasks. Lecture Notes in Computer Science (including Subseries Lecture Notes in Artificial Intelligence and Lecture Notes in Bioinformatics), 8928, 105–116. DOI: 10.1007/978-3-319-16220-1_8

Ibrahim, A., Anayi, F., Packianather, M., & Al-Omari, O. (2021). Novel hybrid invasive weed optimization and machine learning approach for fault detection. In 2021 56th International Universities Power Engineering Conference (UPEC) (pp. 1-6). DOI: 10.1109/UPEC50034.2021.9548171

Islam, N., Rashid, M., Wibowo, S., Xu, C., & Morshed, A. (2021). Early weed detection using image processing and machine learning techniques in an Australian chilli farm.

Jadhav, S. B., Udupi, V. R., & Patil, S. B. (2020). Identification of plant diseases using convolutional neural networks. *International Journal of Information Technology : an Official Journal of Bharati Vidyapeeth's Institute of Computer Applications and Management*. Advance online publication. DOI: 10.1007/s41870-020-00437-5

Jeyalakshmi, S., & Radha, R. (2017). A review on diagnosis of nutrient deficiency symptoms in plant. *ICTACT Journal on Image and Video Processing*, 7(4), 1515–1524. DOI: 10.21917/ijivp.2017.0216

Jeyalakshmi, S., & Radha, R. (2017). A review on diagnosis of nutrient deficiency symptoms in the plant. *ICTACT Journal on Image and Video Processing*, 7(4), 1515–1524. DOI: 10.21917/ijivp.2017.0216

Jin, X., Che, J., & Chen, Y. (2021). Weed identification using deep learning and image processing in the vegetable plantation. *IEEE Access : Practical Innovations, Open Solutions*, 9, 10940–10950. DOI: 10.1109/ACCESS.2021.3050296

Johannes, A., Picon, A., Gila, A. A., Echazarra, J., Vaamonde, S. R., & Navajas, A. D. (2017). Automatic plant disease diagnosis using mobile capture devices, applied on a wheat use case. *Computers and Electronics in Agriculture*, 138, 200–209. DOI: 10.1016/j.compag.2017.04.013

Junior, L. C. M., & Ulson, J. A. C. (2021). Real-time weed detection using computer vision and deep learning. In 2021 14th IEEE International Conference on Industry Applications (INDUSCON) (pp. 1131–1137). DOI: 10.1109/INDUSCON51756.2021.9529761

Kaarthik, K., & Anitha, M. (2016). Analysis of nutrient requirement of crops using its leaf. *Journal of Chemical and Pharmaceutical Sciences*, 8, 1–5.

Kamath, R., Balachandra, M., & Prabhu, S. (2020). Paddy crop and weed discrimination: A multiple classifier system approach. *International Journal of Agronomy*, 2020, 1–14. Advance online publication. DOI: 10.1155/2020/6474536

Kamath, R., Balachandra, M., & Prabhu, S. (2021). Classification of crop and weed from digital images: A review. In 2021 IEEE International Conference on Distributed Computing, VLSI, Electrical Circuits and Robotics (DISCOVER) (pp. 12–17). DOI: 10.1109/DISCOVER52564.2021.9663729

Kinyua, J., & Awuah, L. (2021). AI/ML in security orchestration, automation and response: Future research directions. *Intelligent Automation & Soft Computing*, 28(2), 527–545. DOI: 10.32604/iasc.2021.016240

Kumar, A., Chattopadhyay, C., Awasthi, R. P., & Meena, P. D. (2010). Alternaria blight: A chronic disease in rapeseed-mustard. *Journal of Oilseed Brassica*, 1(1), 1–11.

Kumar, G., & Kalra, R. (2016). A survey on machine learning techniques in health care industry. *International Journal of Recent Research Aspects*, 3(2), 128–132.

Kumar, V., Lehri, S., Sharma, A. K., Meena, P. D., & Kumar, A. (2008). Image based rapeseed-mustard disease expert system: An effective extension tool. *Indian Research Journal of Extension Education*, 8(2&3), 10–13.

Lal, H., Meena, P., Kumar, M., & Jha, P. (2013). Dynamics and management of Alternaria blight disease of Indian mustard (Brassica juncea) in relation to weather parameters. *Journal of Oilseed Brassica*, 4(2), 66–74.

Lameski, P., Zdravevski, E., Trajkovik, V., & Kulakov, A. (2017). Weed detection dataset with RGB images taken under variable light conditions. In ICT Innovations 2017 (pp. 112–119). DOI: 10.1007/978-3-319-67597-8_11

Lavania, S., & Matey, P. S. (2015). Novel method for weed classification in maize field using otsu and PCA implementation. In *2015 IEEE International Conference on Computational Intelligence & Communication Technology* (pp. 534-537). IEEE. DOI: 10.1109/CICT.2015.71

Li, X., Zeng, R., & Liao, H. (2016). Improving crop nutrient efficiency through root architecture modifications. *Journal of Integrative Plant Biology*, 58(3), 193–202. DOI: 10.1111/jipb.12434 PMID: 26460087

Li, Y., Al-Sarayreh, M., Irie, K., Hackell, D., Bourdot, G., Reis, M. M., & Ghamkhar, K. (2021). Identification of weeds based on hyperspectral imaging and machine learning. *Frontiers in Plant Science*, 11(January), 1–13. DOI: 10.3389/fpls.2020.611622 PMID: 33569069

Liu, B., Zhang, Y., He, D., & Li, Y. (2018). Identification of apple leaf diseases based on deep convolutional neural networks. *Symmetry*, 11(10).

Ma, X., Chen, C., Chen, P., Tain, C., Lin, Z., Zhuang, L., & Zou, X. (2019). Fully convolutional network for rice seedling and weed image segmentation at the seedling stage in paddy fields. *PLoS One*, 14(4), e0215676. Advance online publication. DOI: 10.1371/journal.pone.0215676 PMID: 30998770

Majumdar, D., Chakraborty, A., Kole, D., & Majumdar, D. (2015). An integrated digital image analysis system for detection, recognition and diagnosis of disease in wheat leaves. In *Symposium on Women in Computing and Informatics*. New York. DOI: 10.1145/2791405.2791474

Meena, P. D., Awasthi, R. P., Godika, S., Gupta, J. C., Sandhu, P. S., & Kumar, A. (2011). Eco-friendly approaches managing major diseases of Indian mustard. *World Applied Sciences Journal*, 12(8), 1192–1195.

Michalski, R. S., Carbonell, J. G., & Mitchell, T. M. (Eds.). (2013). *Machine learning: An artificial intelligence approach.* Springer Science & Business Media.

Ministry of Electronics and Information Technology. (2019, January 27). Mustard: Diseases and symptoms. Retrieved February 10, 2019, from https://vikaspedia.in/agriculture/crop-production/integrated-pest-managment/ipm-for-oilseeds/ipm-strategies-for-mustard-rapeseed/diseases-and-symptoms

Moazzam, S. I., Khan, U. S., Qureshi, W. S., Tiwana, M. I., Rashid, N., Alasmary, W. S., Iqbal, J., & Hamza, A. (2021). A patch-image based classification approach for detection of weeds in sugar beet crop. *IEEE Access : Practical Innovations, Open Solutions*, 9, 121698–121715. DOI: 10.1109/ACCESS.2021.3109015

Mohanty, S. P., Hughes, D. P., & Salathé, M. (2016). Using deep learning for image-based plant disease detection. *Frontiers in Plant Science*, 7, 1419. DOI: 10.3389/fpls.2016.01419 PMID: 27713752

Murawwat, S., & Ahmad, S. (2018). Weed detection using SVMs. *International Journal of Engineering and Advanced Technology*, 8(1), 2412–2416.

Nandhini, S., Hemalatha, R., Radha, S., & Indumathi, K. (2018). Web enabled plant disease detection system for agricultural applications using WMSN. *Wireless Personal Communications*, 102(2), 725–740. DOI: 10.1007/s11277-017-5092-4

Phadikar, S., Sil, J., & Kumar, A. D. (2013). Rice diseases classification using feature selection and rule generation techniques. *Computers and Electronics in Agriculture*, 90, 76–85. DOI: 10.1016/j.compag.2012.11.001

Rangel, B. M. S., Fernández, M. A. A., Murillo, J. C., Ortega, J. C. P., & Arreguín, J. M. R. (2016, February). KNN-based image segmentation for grapevine potassium deficiency diagnosis. In 2016 International conference on electronics, communications and computers (CONIELECOMP) (pp. 48-53). IEEE.

Samanta, D., & Ghosh, A. (2012). Histogram approach for detection of maize leaf damage. *International Journal of Computer Science and Telecommunications*, 3(12), 26–28.

Santos Ferreira, A., Matte Freitas, D., Gonçalves da Silva, G., Pistori, H., & Theophilo Folhes, M. (2017). Weed detection in soybean crops using ConvNets. *Computers and Electronics in Agriculture*, 143, 314–324. DOI: 10.1016/j.compag.2017.10.027

Sanyal, P., Bhattacharya, U., Parui, S. K., Bandyopadhyay, S. K., & Patel, S. (2007). Color texture analysis of rice leaves to diagnose deficiency in the balance of mineral levels towards improvement of crop productivity. In IEEE.

Sanyal, P., Bhattacharya, U., Parui, S. K., Bandyopadhyay, S. K., & Patel, S. (2007). Color texture analysis of rice leaves to diagnose deficiency in the balance of mineral levels towards improvement of crop productivity. In IEEE.

Sarvini, T., & Sneha, T. GS, S. G., Sushmitha, S., & Kumaraswamy, R. (2019, April). Performance comparison of weed detection algorithms. In 2019 International Conference on Communication and Signal Processing (ICCSP) (pp. 0843-0847). IEEE.

Saskatchewan Mustard Development Commission. (2017, June 1). Mustard production manual. Retrieved January 15, 2019, from https://saskmustard.com/production-manual/plant-description/types-of-mustards-and-their-uses/index.html

Selvi, C. T., Sankara Subramanian, R. S., & Ramachandran, R. (2021). Weed detection in agricultural fields using deep learning process. In 2021 7th International Conference on Advanced Computing and Communication Systems (ICACCS) (vol. 1, pp. 1470–1473). DOI: 10.1109/ICACCS51430.2021.9441683

Shah, V. K., Prajapati, H. B., & Shah, J. P. (2016). A survey on detection and classification of rice plant diseases. In IEEE. DOI: 10.1109/ICCTAC.2016.7567333

Shidnal, S., & Latte, M. V. (2016). Multiple nutrient deficiency detection in paddy leaf images using color and pattern analysis. In IEEE, India.

Shorewala, S., Ashfaque, A., Sidharth, R., & Vijay, U. (2021). Weed density and distribution estimation for precision agriculture using semi-supervised learning. *International Journal of Information Technology : an Official Journal of Bharati Vidyapeeth's Institute of Computer Applications and Management*. Advance online publication. DOI: 10.1004/s41873-020-004374

Siddiqui, S. A., Fatima, N., & Ahmad, A. (2021). Neural network-based smart weed detection system. In *2021 International Conference on Communication, Control and Information Sciences (ICCISc)* (vol. 1, pp. 1-5). https://doi.org/DOI: 10.1109/ICCISc52257.2021.9484925

Singh, U. P., Chouhan, S. S., Jain, S., & Jain, S. (2019). Multilayer convolution neural network for the classification of mango leaves infected by anthracnose disease. *IEEE Access : Practical Innovations, Open Solutions*, 7, 43721–43729. DOI: 10.1109/ACCESS.2019.2907383

Sladojevic, S., Arsenovic, M., Anderla, A., Culibrk, D., & Stefanovic, D. (2016). Deep neural networks based recognition of plant diseases by leaf image classification. *Computational Intelligence and Neuroscience*, 2016, 1–11. DOI: 10.1155/2016/3289801 PMID: 27418923

Son, N., Chen, C., & Chen, P. (2021). *A phenological object-based approach for rice crop classification using time-series Sentinel-1 synthetic aperture radar (SAR) data in Taiwan. Journal of Remote Sensing*. Retrieved from Taylor & Francis. DOI: 10.1080/01431161.2020.1862440

Tettamanzi, A. G., & Tomassini, M. (2013). *Soft computing: integrating evolutionary, neural, and fuzzy systems*. Springer Science & Business Media.

Toğaçar, M. (2022). Using DarkNet models and metaheuristic optimization methods together to detect weeds growing along with seedlings. *Ecological Informatics*, 68, 101519. DOI: 10.1016/j.ecoinf.2021.101519

Vapnik, V. (1995). *The nature of statistical learning theory*. Springer Verlag. DOI: 10.1007/978-1-4757-2440-0

Wahid, K., Dinh, A., & Islam, M. (2017). Detection of potato diseases using image segmentation and multiclass support vector machine. In IEEE.

Wang, A., Xu, Y., Wei, X., & Cui, B. (2020). Semantic segmentation of crop and weed using an encoder-decoder network and image enhancement method under uncontrolled outdoor illumination. *IEEE Access : Practical Innovations, Open Solutions*, 8, 81724–81734. Advance online publication. DOI: 10.1109/ACCESS.2020.2991354

Wang, G., Sun, Y., & Wang, J. (2017). Automatic image-based plant disease severity estimation using deep learning. *Computational Intelligence and Neuroscience*, 2017, 1–8. DOI: 10.1155/2017/2917536 PMID: 28757863

Wang, L., Shang, Y. J., & Zhang, S. W. (2015). Plant disease recognition based on plant leaf image. *The Journal of Animal & Plant Sciences*, 25, 42–45.

Yadav, S. P., Singh, D., & Kumar, S. (2015). Studies on powdery mildew of rapeseed-mustard (Brassica juncea L.) caused by Erysiphe cruciferarum and its management. *Journal of Pure & Applied Microbiology*, 9(2).

Ying, B., Xu, Y., Zhang, S., Shi, Y., & Liu, L. (2021). Weed detection in images of carrot fields based on improved YOLO v4. *TS. Traitement du Signal*, 38(2), 341–348. DOI: 10.18280/ts.380211

Yu, J., Sharpe, S. M., Schumann, A. W., & Boyd, N. S. (2019). Deep learning for image-based weed detection in turfgrass. *European Journal of Agronomy*, 104, 78–84. DOI: 10.1016/j.eja.2019.01.004

Yun, S., Xianfeng, W., Shanwen, Z., & Chuanlei, Z. (2016). PNN based crop disease recognition with leaf image features and meteorological data. *International Journal of Agricultural and Biological Engineering*, 8(4), 61–68.

Zhang, W., ... (2018). Broad-leaf weed detection in pasture. In 2018 IEEE 3rd International Conference on Image, Vision and Computing (ICIVC) (pp. 101–105). https://doi.org/DOI: 10.1109/ICIVC.2018.8492831

Chapter 10
Satellite Imagery Used for Analysis of Soil Properties

Karim Ennouri
https://orcid.org/0000-0001-6469-6455
University of Sfax, Tunisia

Monia Ennouri
https://orcid.org/0000-0001-7477-6803
Olive Tree Institute, Tunisia

Mohamed Ali Triki
Olive Tree Institute, Tunisia

ABSTRACT

The integration of remote sensing technology in soil analysis represents a monumental leap in agricultural sciences. By harnessing the power of satellite imagery, scientists can now monitor and evaluate soil conditions across vast areas with unprecedented accuracy. This non-invasive technique allows for the continuous observation of soil health, aiding in the detection of changes over time due to natural or anthropogenic factors. The data collected from these observations are crucial for developing sustainable farming practices that not only increase crop yields but also protect the environment. Furthermore, the ability to analyze soil properties on such a scale is instrumental in managing land resources effectively, ensuring that the delicate balance between food production and ecological stewardship is maintained. As the global population continues to grow, the role of remote sensing in soil conservation and agricultural productivity becomes ever more essential, offering a beacon of hope for future generations.

DOI: 10.4018/979-8-3693-6452-9.ch010

1. INTRODUCTION

The most valuable natural resource is soil, but it is limited and non-renewable. New knowledge and innovative technology are needed to protect environment while producing enough food and biomass to feed people. Knowledge of the morphological, chemical and physical properties, categorization, extent, geographical distribution and potential for various uses of soils is essential for achieving ecological crop production and preserving soil composition. Unless soil behaviour is understood, implementing scientific agriculture is extremely difficult (Ford & Quiring, 2019; Babaeian et al., 2019). Discipline and art of remote sensing consists of analysing data collected by a device that is not in close proximity to the object, place, or event being studied in order to learn more about it. The platform, the sensor and the data are the three basic components of remote sensing systems. Platforms are simply the boards on which the sensor is installed, and they can be classified as either airborne platforms (e.g., balloons, drones, and airplanes) or space-based platforms (spacecraft and satellites). The central component of a remote sensing structure, the sensor, can be passive or active and is responsible for data collection. Whereas passive sensors measure energy returned or released from ground (solar radiation), active sensors release energy and store data about reflected radiation (Pepe et al., 2018; Zhu et al., 2018). The smallest object that can be recognized serves as a portion of the spatial resolution, as example the spatial details of an image. Because the initial resolution of a picture can be converted to diverse pixel dimensions, this is frequently referred to as pixel dimension, although these words are not necessarily equivalent. The width and number of "spectral bands," also called "spectral resolution," denotes to a portion of electromagnetic spectrum that a sensor is capable to capture (Segarra et al., 2020). Besides, temporal resolution denotes to period between image captures and includes the time series of photos that are accessible for exploration (temporal handling) (Khanall et al., 2017). Aerial imaging systems typically have three main advantages: they can be positioned quickly to specific positions; they can capture high-resolution images; and can avoid unwanted atmospheric factors such as clouds. Aerial imagery is a useful substitute for ground data when determining the precision of other remote sensing data products. For purposes requiring broad geographic coverage and high spectral/temporal resolution, aerial systems are less suitable due to the significant compromises associated with these advantages. Processing, analyzing and storing data for aerial images also incurs significant financial and labour costs, therefore the images are typically costly (Shi et al., 2021).

Satellite remote sensing systems are separated in different types: coarse (for example multispectral through a spatial resolution superior than thirty metres), multispectral, hyperspatial and hyperspectral. The usefulness of each of these categories for remote sensing attributes depends on the resolution, availability and processing requirements of the system in question (Pour et al., 2021). For local mapping and accuracy assessment, hyperspatial images such as those from QuickBird, Ikonos and WorldView are a very useful alternative to aerial images due to their improved temporal and spatial resolution (all have a spatial resolution of 4 meters, with submeter preferences for WorldView or QuickBird) (Snehmani et al., 2017). The following table demonstrates the satellite remote sensing characteristics.

Table 1. spatial resolution-related terms in satellite

term of Resolution	Resolution (m)	detector example
Very low/coarse	250-1000	MODIS
Low	30-250	Landsat TM, MESS, ERS
Medium	10-30	Landsat ETM+, SPOT-4,IRS-3
High	2-10	SPOT-5, IKONS, RapidEye, Formosat-2
Very High	<1-2	Quikbird, TerraSAR-X, GeoEye, WorldView-2, Radarsat-2

2. BASIC PRINCIPLES

The process of learning about an object without touching it is called remote sensing. This involves using instruments that measure the interaction of electromagnetic energy from a source such as the sun with a real object. It reflects, transmits and/or absorbs this energy. For remote sensing, the reflectivity and absorption of electromagnetic energy at different wavelengths of an object is crucial and varies depending on the spectral range of an object being examined. For almost all types of materials, spectral reflectance can be tested in laboratory conditions or outdoors. Theoretically, objects can be recognized from remote sensing pictures established on their different spectral properties because they have characteristic spectral features (Pahlevan et al., 2017). The quantity of light leaving a target divided by the quantity of light hitting the target is called the reflectance ratio. The reflection can be diffuse or specular or both (scattered over the entire hemisphere). Remote sensing can be presented as a technique for collecting spectral data about a given object, usually a portion of the surface of Earth, using a detector situated far from the aim, usually on board a satellite, an aircraft, or newly an unmanned aerial vehicle (UAV) (Alvarez-Vanhard et al., 2021). Regarding precision agriculture, monitoring plant growth, calculating various biochemical and biophysical properties of soil and plants that may trigger or indicate plant stress, and using site-specific management recommendations for plant fertilization are just some of the uses for which remote sensing techniques can be useful.

3. HYPERSPECTRAL REMOTE SENSING

Spectral resolution is characterized by means of its ability to distinguish different wavelength ranges. A very large band of wavelengths, a few broad bands, or numerous little bands of wavelengths can form an "image" that a sensor system produces. These three image types are commonly referred to as panchromatic, multispectral, and hyperspectral, respectively. The data source for hyperspectral images typically contains hundreds of data bands, perhaps even ten or more. The bandwidth of the data typically varies between 1 and 15 nanometers. These bands afford important comprehensions about the chemical and physical characteristics of various objects in the scene that sensors were able to detect (Lu et al., 2020). The spectral remote sensing method remains constructed on the theory that the electromagnetic radiation (EMR) from sun is also transferred, or redirected and absorbed through surface being examined and that configuration of redirected portion, as saved via optical detector, depends precisely on soil characteristics as .the specific surface (vegetation, soil, etc.), as well as its physical (such as the structure of the plant and the canopy) and chemical (such as the amount of nitrogen, chlorophyll). The wavelengths in the range between 350 and 2500 nm, which includes near infrared (NIR, 750-1400 nm), visible (VIS, 380-750 nm), and fragment of the short-wave infrared, are most relevant for the investigation of vegetation (SWIR, 1400-3000 nm) (Roy et al., 2017). Nevertheless, the latter are not applicable to sensors operating outdoors, as ambient water vapour significantly interferes with solar EMR in the 1400 nm, 1900 nm and 2500 nm spectral ranges (i.e. via satellite) or airborne sensors. Vegetation surface temperature is also described using the thermal infrared spectral spectrum (TIR, 3000–15000 nm). In addition to spectral remote sensing, RADAR remote sensing can also be helpful, which uses self-emitted energy and considers wavelengths in the interval from 1 cm to 1 m to describe vegetation surfaces. Both multispectral and hyperspectral remote sensors are possible, but satellite coverage of the latter is currently quite limited (Morin et al., 2017). While hyperspectral sensors provide a large amount (between 60 and 500) of attached thin spectral bands (10–20 nm wavelength measurement), multispectral sensors provide only a small number (3 to 10) wide (20–250 nm wavelength width) spectral bands. The spectral signature of a given land cover, or the variation in reflectance (the proportion of redirected EMR to incoming EMR) with respect to wavelengths, distinguishes it from other land cover types and/or states. Healthy green vegetation emits a characteristic spectral signature that has weak EMR absorption for photosynthesis and low reflectivity in the visible spectral region, commonly known as photosynthetically active radiation (Ding et al., 2022).

The green light (from 500 to 600 nm) is captivated reasonably little, resulting in a reflection maximum in the green, delimited by reflection lacks in the red and blue areas, which is the reason that vegetation seems "green". Leaf pigments in particular are accountable for portion of this penetration, with chlorophyll a and b absorbing electromagnetic radiation in the red (600-700 nm) and blue (400-500 nm) range, while carotenoids absorb electromagnetic radiation in the blue (400-500 nm) and red range (600-700 nm) absorb range. (680-750 nm) is the range in which the reflectivity increases significantly in the range between VIS and NIR (Zahir et al., 2022). The NIR region, commonly referred to as "NIR plateau", is characterized by relatively high reflectance caused by moderate EMR absorption and increased EMR scattering, principally caused by physical properties of vegetation cells, foliage and tree crowns are regulated. The SWIR range has a medium reflectivity with three spectral ranges (approx. 1450 nm, 1950 nm and 2500 nm), which are significantly influenced by the water absorption associated with the leaf water volume. A complex interplay of many variables, some of which are interrelated and relate to the category and growth of crop, management, ecological conditions and measurement circumstances, influences the spectral signature of the crop (Mohamed et al., 2018). The physiological development of the crop (green to old age), the fraction of non-photosynthetic part such as ground and junk visible on a remote sensing picture, and the closure of the canopy (ground cover) are all directly determined by the growth phases of crop significant influence on the reflectivity of the crop. Plant emergence signals the appearance of a green reflection peak; The development of biomass contributes to an increase in the overall reflectivity in the NIR in addition to SWIR sections; and crop development, coupled with the senescence stage, leads to an increase in total reflectance throughout the 350–2500 spectral range, with a significant increase in the red part of the visible spectrum (Falcioni et al., 2022). LAI, or sided green leaf zone per unit soil surface, is often used to determine plant stand density. A higher LAI results in a substantial rise in reflectance in the NIR and to a decreased level in the SWIR1 region (1400-1900 nm), but shows a very small and often irregular change in the SWIR2 region (1900-2500 nm). While an LAI from 0 to 2 causes the red reflectivity to decrease and the green peak to appear in the visible section, an LAI > 2 is often stated to no longer have any influence on the visible region (saturation). Another structural feature of the canopy that has an important effect on the spectral response is leaf inclination, with planophilic leaves exhibiting higher reflectivity than electrophilic ones through the entire spectral range between 350 and 2500 nm (Chen et al., 2018). The next Table depicts actual satellites.

Table 2. Examples of satellite imagery

Satellite sensor	Launch	Number of bands	Spatial resolution [m]	Revisit time (days)	Biophysical factors
WorldView-2	2009	8 (B,G,R,coastal, yellow,NIR,Red-Edge, NIR2)	0.46 (pan) 1.8 (ms)	1.1	Reflection, NDVI, LAI, leaf chlorophyll (and nitrogen) concentration Classification
WorldView-3	2014	8 (B,G,R,coastal, yellow,NIR, RedEdge,NIR2) 8 SWIR 12 CAVIS	0.31 (pan) 1.24 (ms) 3.7 (short wave IR)	<1	Reflection, NDVI, LAI, leaf chlorophyll (and nitrogen) concentration Classification
QuickBird	2001	4 (B,G,R,NIR)	0.65 (pan) 2.6 (ms)	1-3.5	Reflection, LAI, NDVI Classification
GeoEye-1	2008	4 (B,G,R,NIR)	0.4 (pan)	3	Reflection, NDVI, LAI
GeoEye-2 (WorldView-4)	2016		0.3 (pan) 1.2 (ms)	<3	Reflection, LAI, NDVI, leaf chlorophyll (and nitrogen) concentration Classification
Ikonos	1999	4 (B,G,R,NIR)	1 (pan) 4 (ms)		Reflection, LAI, NDVI Classification
RapidEye (5 satellite constellation)	2008	5 (B,G,R,NIR, RedEdge)	5 (ms)	1	Reflection, LAI, NDVI, leaf chlorophyll (and nitrogen) concentration Classification
Pleiades-1A & B (2 satellite constellation)	2011/ 2012	4 (B,G,R,NIR)	0.5 (pan) 2 (ms)	1	Reflection, LAI, NDVI Classification
SkySat-1 & 2	2013/ 2014	4 (B,G,R,NIR)	0.9 (pan) 2 (ms)		
SPOT-6 & 7 constellation	2012/ 2014	4 (B,G,R,NIR)	1.5 (pan) 8 (ms)	1	
Landsat-8	2013	11 (VNIR,SWIR,TIR)	15 (pan) 30m (ms) 100m (TIR)	16	Reflection, NDVI, LAI, temperature Classification

continued on following page

Table 2. Continued

Satellite sensor	Launch	Number of bands	Spatial resolution [m]	Revisit time (days)	Biophysical factors
Aster	1999	3, 6, 5 (VNIR,SWIR,TIR)	15 (VNIR) 30 (SWIR) 90 (TIR)		
Sentinel-2A & B (2 satellite constellation)	2015/ 2016	13 (VNIR, NIR, SWIR)	10, 20, 60	<5	Reflection, NDVI, LAI, leaf chlorophyll (and nitrogen) concentration Classification

Numerous plant experiments related to nutrients, water or pathogens can also have a significant impact on plant reflectivity. A stressed crop often has a general increased reflectance in the VIS caused by a decrease in chlorophyll concentration and other photosynthetic pigments in the leaf. Greater stress leads to a comparatively greater increase in reflectance in the yellow and red areas, resultant in the typical yellow coloration of stressed plants. Whereas vigorous vegetation typically exhibits a minor green reflectance top equivalent to an obscurer green, reasonably stressed foliage habitually exhibits an amplified green reflectance top equivalent to a brighter green (He et al., 2018; Liu et al., 2017). A nitrogen-stressed crop has the unique property of showing significantly lower reflectivity in the NIR, which is primarily due to slower biomass development. In contrast, higher reflectivity in SWIR is only seen in some severe nitrogen lack cases, with SWIR2 presenting a more significant rise than SWIR1 in some mentions. According to several studies, phosphorus stress can cause some crops to develop purple leaves, leaf sheaths, or stems. This is mainly caused by increased production of anthocyanins, which strongly absorb green light, while they absorb blue and red light very slightly or not at all (Yuet al., 2019). Particularly severe water stress (senescent vegetation) can eventually lead to a decrease in reflectivity in the first portion of the NIR and expose absorption features associated with other leaf biochemicals as proteins in the SWIR, while moderate water stress normally results in a global rise in reflectivity in the entire spectral range from 350 to 2500 nm associated to well-watered foliage as well as reduction of water absorption characteristics. Since the reflectivity of crops has a strongly anisotropic behaviour, the measurement conditions, in particular the zenith angle of the sun and the sensor as well as their relative azimuth position, as well as the site topography and the associated tilt location of field can have a significant influence (Seong, 2015). Depending on type and growth of plant canopy, the relative location of plant rows is another geometric component that can have an influence on the reflectivity of the plants (distance between successive rows, height of the canopy termination and plant height). Although "atmospheric correction techniques" are often used to reduce their effects, atmospheric conditions are still a cause of reflection variations (Vanhellemont & Ruddick, 2018).

4. SPECTRAL SIGNATURE OF SOIL AND VEGETATION SPECTRA

Similar to other earth surface covers, soil is made up of a variety of different components, resulting in differences. Information is included in these versions. Both the mineralogical characteristics and the moisture content of soil influence the soil spectrum. The latter varies in space and time. Soil is traditionally analysed using chemical and physical analysis techniques. However, it is imaginable to employ hyperspectral remote sensing for ground assessment in both laboratory and field conditions. Reflection can be used to extract several indirect and direct soil characteristics in addition to soil pollution features. Nevertheless, the reflectance only represents the surface and cannot provide any information about the soil profile (Escribano et al., 2017; Nizeyimana & Petersen, 2020). The idea of using plants as markers of contaminated soil has been proposed for many years. This is due to three key factors: the plant's root system removes pollutants and releases them to above-ground parts of the plant; 1) Inorganic substances generally do not exhibit typical absorption characteristics in the VNIR-SWIR wavelength band; 2) the ground is exceptionally basic; it is often covered with vegetation; and 3) the ground is often covered.

The interaction of energy in plants is more complex than the reflection on smooth object surfaces. The green cover has a distinct colour spectrum. Reflectivity is generally small in the visible range, but green light has a higher reflectivity than red or blue light, creating the characteristic "green peak" in reflectance that gives human vision the appearance of green vegetation (Pocas et al., 2020; Xue & Su, 2017). Metabolic reactions of photosynthesis are closely related to photosynthetic pigments. In addition, they control how light moves through the sheet, which affects its optical properties such as reflectance and transmittance. This provides an approach to measure pigments without destroying them by measuring reflected radiation. Different vegetation pigment levels can be estimated to infer vegetation descriptors such as plant stress, which can be a valuable predictor of burial sites. These functions are determined from the reflectance spectra and are known as spectral vegetation indices (SVIs) (Okhrimenko et al., 2019). Numerous agricultural traits, which can be divided in two main groups, namely plant biochemical traits associated with plant chemical element content and plant biophysical traits associated with plant biomass and canopy structure, have all been assessed using remote sensing. Carotenoid, Chlorophyll, ratio (carotenoid/chlorophyll) as well as anthocyanin concentration are the most important biochemical plant properties and their relative concentrations directly affect the colour of the crop. Other biochemical plant properties include nitrogen and also crude protein concentration, the concentration of plant organisational constituents such as lignin and cellulose, the H_2O composition in addition to the starch amount (Ac et al., 2015).

Ground cover (also known as canopy, green cover, vegetation fraction or fractional cover (FCOVER)), specific leaf area, plant height, grain yield, canopy volume, flowering intensity, fraction of absorbed photosynthetically active radiation (fAPAR), and plant growing phase are examples of plant biophysical characteristics (Hu et al., 2020). The ability to extract plant properties from reflectance is limited by two main factors: reflectance tends to occur at a decreased or improved level of a particular plant trait (e.g. increased chlorophyll concentration) to go into saturation, and the reflection is the consequence of a complex relationship of many elements related to the type and development of the crop, organization, ecological conditions and measurement circumstances, which can have an identical influence on reflectivity (Chen et al., 2017). To estimate these plant properties from spectral remote sensing records, dual different methods can be employed: the extensively employed empirical statistical modeling, which is built on regression concerning spectral bands or "Spectral Vegetation Indices" (SVI) in addition to the plant properties of interest, or another popular method, physical modelling involves the application of Radiation Transfer Models (RTM) based on physical rules. To exploit the reactivity to the biochemical or else biophysical plant trait of interest whereas reducing external disparity elements, vegetation indices (VI) associate reflectance of several wavelengths (Miraglio et al., 2019). These indices can be expressed as a variance, simple ratio, normalized difference, linear combination, derivative, mixture of different indices, or other combinations. The broader "green vegetation" and its "condition" were identified using early remotely sensed vegetation indicators. They often combined a red spectral band associated with the high EMR absorption via chlorophyll through a NIR spectral band defined by an improved reflectance scattering to exploit the plant's extremely distinct spectral signature. Later developed and existing VI were employed to evaluate more precise biochemical and biophysical properties of vegetation (Sivanpillai & Whitman, 2019).

Many VIs are dedicated to assessing canopy chlorophyll amount, which is considered as a factor of foliage nitrogen nutritional condition and concentration, ever since leaf chlorophyll concentration is primarily calculated via nitrogen accessibility. It is considered a barometer of the vigour, vitality and photosynthetic action of vegetation. While it has been found that VI by means of the red and NIR spectral bands such as B. NDVI, saturates at extremely low chlorophyll concentration, VI that used the green or red edge or added a red edge to the red and NIR spectral bands presented stronger reactivity to increased chlorophyll concentration (Gao & Zhang, 2021). Figure 1 illustrates different devices in precision farming.

Figure 1. devices in precision farming

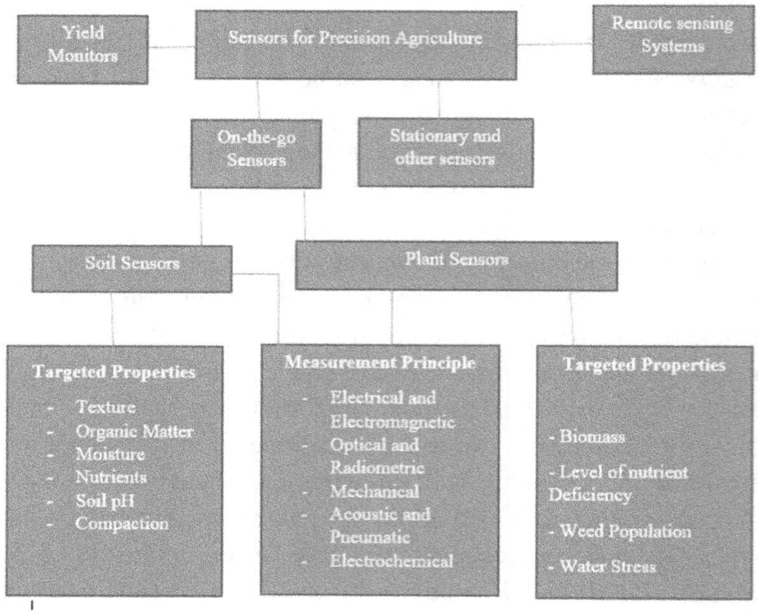

The moderately indirect changes in canopy reflectivity that accompany variations in chlorophyll present in leaf are frequently confused with significant fluctuations in plant evolution and expansion. This presents another defy for chlorophyll VI, namely to reduce its sensitivity to LAI. Some VI have been developed to evaluate additional plant pigments, particularly the ratio (carotenoids/chlorophyll), which is a good indicator of photosynthesis effectiveness and chlorophyll deterioration caused by aging or stress in plants. Some indices associate a spectral band in the 400–500 nm range, where reflectivity is low due to retention of carotenoids, with a spectral band in the 550–740 nm range, where reflectivity increases with chlorophyll deterioration (Chu et al., 2021). The NIR and red spectral bands are combined to provide the vegetation indices for LAI assessment, the latter remaining delicate even at reasonable to increased foliage densities (LAI of 2 to 6) in crops. However, at moderate to high aboveground biomass, a elementary LAI index such as the NDVI asymptotically approaches saturation, often starting at LAI values of 2 or 3. Various advanced LAI VI often include an additional green spectral band to reduce interference in sensitivity to chlorophyll concentration whereas maximizing awareness to increased LAI (Fang, 2020). Some vegetation indices such as LAI,

APAR or chlorophyll content, which are also used to evaluate green vegetation, have the express purpose of reducing the confounding effects of soil background and/or atmospheric conditions on the reflectivity of vegetation. Those that minimize the ground background effect (effects on the reflectivity of the background from non-photosynthetic resources such as ground and leaf junk) are based on the inclusion of a "ground brightness rectification factor" and are proposed to be employed in situations where there is little vegetation density and incomplete ground cover. The idea behind atmospheric adjustment is to adjust the atmosphere influence on the red spectral band using the blue spectral band. In conclusion, a sequence of indices specially designed to measure H_2O concentration in the canopy were created (Rhyma et al., 2020).

These indices usually associate a wavelength from the 800-900 nm region of the NIR spectrum, which is moderately indifferent to canopy water concentration, with a wavelength from the SWIR or NIR nearby 970 nm or 1200 nm, which corresponds to H_2O absorption and is therefore sensitive to the water concentration in the canopy. Hyperspectral devices are often found to be more efficient than multispectral sensors due to their slight and various spectral bands, although similar effectiveness is occasionally reported when comparing the relative effectiveness of different sensor types for agricultural trait assessment. The disadvantage of thermal infrared (TIR) sensors is that they are sensitive to changes in temperature and wind speed. Compared to VIS or VIS-NIR reflection, TIR reflection is intended to enable earlier or later detection of plant diseases depending on the situation (Neinavaz et al., 2021; Delogu et al., 2021).

5. MAPPING CONSIDERATIONS, METHODS, AND PRODUCTS

To produce precise and accurate products, it is necessary to develop unique large-scale representing techniques that can efficiently assimilate data sources and investigation techniques. Most map-related image analysis methods are developed to address the blended pixel effect. Conventional pixel-based classifiers can regularly produce acceptable outcomes as this effect is significantly decreased in hyperspatial images (Zou & Greenberg, 2019). The three most essential things to think about when mapping are: - The spatial extent - The imagery/image exploration methods (how will the coverage be represented), and - The end users/uses. These include techniques for supervised and unsupervised categorization on the basis of spectral characteristics of picture. In supervised classification, the specialist chooses a subset of pixels that correspond to a particular type or classes to serve as the algorithm's "training data." The algorithm categorizes the picture through comparing other pixels with the spectral properties of the training set (Richards,

2022). In contrast, unsupervised classification requires the analyst to assemble or split discrete "clusters" of pixels based on spectral properties to represent the required modules (Zhanget al., 2019). However, these methods overlook the particular shape and texture qualities present in a hyperspatial image and are susceptible to inaccuracies arising from spectral overlap among modules and other vegetation categories. (For example, compared to other vegetation, tree canopies often have a spherical shape and a distinct texture.) This was addressed by the development of an influential advanced classification method named object-based image analysis (OBIA), which influences all data enclosed in a hyperspatial picture amid the classification procedure (Hossain & Chen, 2019). Because of segmenting the picture into categories of "objects" whose attributes are determined by the operator based on form, texture, and spectral features, OBIA approaches overcome pixel-based limitations. Categorization is then performed using a hierarchical "ruleset" created by the expert that uses thresholds for the same properties of the objects (additional data such as digital elevation models can also be included). OBIA has been used successfully in numerous studies to provide accurate, reasonably thorough type-level classifications, and comparisons with pixel-based classifiers generally show that OBIA is higher. These methods have only recently been applied to large-scale mapping because they are most effective on hyperspatial images. For images with lower spatial resolution, such as multispectral photography, which is required to cover large areas, the blended pixel effect is obviously more severe. To categorize coverage in these situations, image analysis approaches rely primarily on temporal and spectral properties of the picture as well as auxiliary information (Kucharczyk et al., 2020). Figure 2 resumes remote crop observing.

Figure 2. remote crop observing

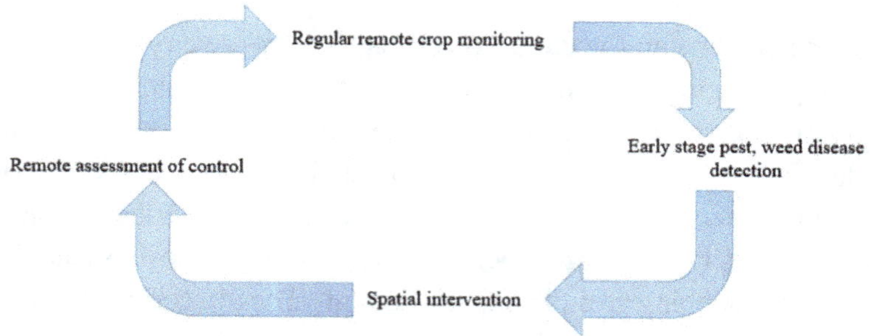

The most popular method is probably hierarchical mapping, which classifies the image into different land cover types based on thresholds of particular spectral bands or band ratios. The remaining layered pixels are then subjected to categorization.

According to other studies, multiple imaging data (multitemporal photography) can significantly improve this approach to capture phonologically significant variation induced by leafing out and senescence (Zeng et al., 2020; Mechravar et al., 2023; Neyns & Canters, 2022). Spectral Mixture Analyses (SMA), which divide mixed pixels into linear or nonlinear arrangements of "end members" (i.e., spectra of an identified coverage type, such as a "pure" pixel of a specific tree species), are more sophisticated mapping techniques using images with low coverage spatial and high spectral resolution (Duan et al., 2019). These techniques use machine learning to measure the proportion of tree species or types (Dai et al., 2020). Historically, hyperspectral imaging was the only suitable format for these methods and species-level plotting generally.

6. ALGORITHMS AND SATELLITE IMAGERIES FOR ANALYSIS OF SOIL PROPERTIES.

The integration of image analysis algorithms with satellite imagery has modernised the science of soil, offering a non-invasive method to assess and monitor soil properties over large areas. Recently, the usage of Geospatial Artificial Intelligence (GeoAI) combined with satellite imagery has shown promising results in predicting soil physical properties crucial for agricultural management and ecological protection. For instance, a study conducted in Iran utilized 317 soil samples to calibrate and validate the prediction models based on data from Landsat-8 satellite pictures, topographic factors from the Shuttle Radar Topography Mission (SRTM), and meteorological factors (Hosseini et al., 2023). The researchers employed a hybrid model combining Convolutional Neural Networks (CNN) with Random Forest (RF) algorithms, which outperformed traditional methods in predicting soil texture components such as clay, silt, and sand. Convolutional Neural Networks (CNN) and Random Forest (RF) algorithms are powerful tools in the domain of machine learning, offering significant advancements in soil analysis. CNNs, known for their prowess in image recognition tasks, can be utilized to analyse soil images, identifying patterns and characteristics that are not easily discernible by human eyes. On the other hand, RF algorithms, which operate through the creation of multiple decision trees to perform classification tasks, have been successfully applied to classify soils based on various parameters such as grain size distribution. These techniques not only augment the precision of soil classification but also contribute to the development of predictive models that can forecast soil fertility and suitability for diverse crop categories (Boston et al., 2022). The integration of these methods can lead to more informed decisions in agriculture, environmental management, and geotechnical engineering, ultimately contributing to sustainable land use and increased agricultural productivity. Figure 3 demonstrates the integration of data system.

Figure 3. Integration of system for rotating data into information

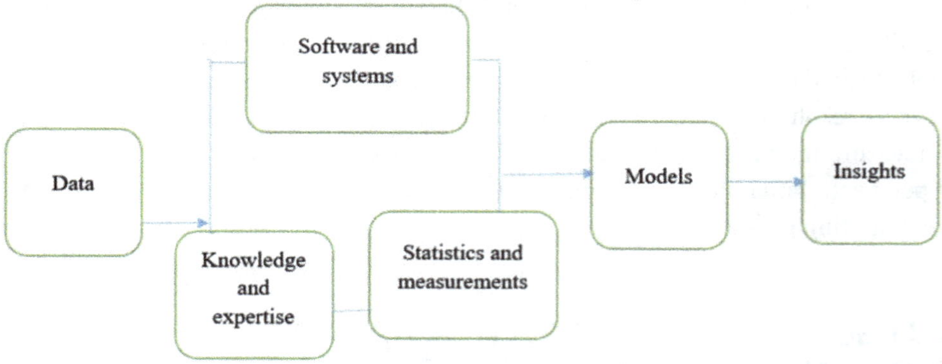

Another innovative approach involves the Enhancement and Analysis of Hyper-spectral Satellite Images for Soil Study and Behaviour (EAHSB), which comprises pre-processing, segmentation, feature extraction, and classification steps. This method enhances the spatial information of soil properties by addressing the limitations posed by the presence of objects other than soil in the imagery. The EAHSB model employs improved anisotropic filtering for image pre-processing, followed by segmentation using an enhanced spatial constraint kernel Fuzzy c-means (KFCM) (Malik et al., 2024). The classification of soil behaviour is then achieved through a hybrid model that combines Deep Residual Network (DRN) in addition to developed Recurrent Neural Network (RNN), providing detailed insights into soil moisture, vegetation indices, and pH levels. Deep Residual Networks (DRNs) and Recurrent Neural Networks (RNNs) are proving to be highly effective in the field of soil analysis (Singh & Kasana, 2022). DRNs, with their deep learning capabilities, are being used to classify soil types and predict soil properties, enhancing the performance of IoT-based big data environments in agriculture. On the other hand, RNNs, principally Long Short-Term Memory (LSTM) networks, are more appropriate for temporal records and are being utilized for soil moisture prediction, offering insights into soil moisture dynamics crucial for agricultural planning (Filipovic et al., 2022). These advanced neural networks are transforming the way soil data is analysed, leading to more informed decisions in land management and farming practices. The application of multispectral and hyperspectral imaging from satellites like Sentinel-2 has also been instrumental in digital soil mapping. These technologies enable the detailed analysis of soil properties by capturing data across various spectral bands, which can be associated with soil parameters such as moisture concentration, organic carbon levels, and nutrient accessibility. For example, a study on the coast of Egypt utilized Vis-NIR spectroscopy and Landsat-8 OLI records to draw ground modules under dry conditions, demonstrating the potential of satellite imagery in enhancing soil agricultural maps (Mohamed et al., 2020).

The developments in satellite picture analysis algorithms have not only improved the precision of soil property predictions but also facilitated the creation of more detailed and dynamic soil maps. These maps are invaluable for precision agriculture, allowing for targeted interventions and better resource management. As satellite technology continues to evolve, the potential applications of these algorithms in soil science are bound to expand, offering new opportunities for ecological land use and environmental stewardship (Xie et al., 2020). The use of multisource data-based soil property analysis is also gaining traction. A complete extraction of ecological variables, comprising complex wavebands and vegetation indices, was conducted using the Google Earth platform. This method allowed for a detailed assessment of soil properties, highlighting the importance of selecting appropriate features for accurate soil analysis (Yan et al., 2023).

These case studies underscore the importance of satellite image-based soil property analysis in enhancing our understanding of soil behaviour and its implications for environmental and agricultural decision-making. By leveraging advanced imaging techniques and analytical models, researchers can obtain detailed and accurate information about soil conditions, leading to more knowledgeable and ecological land supervision practices. The integration of satellite data with traditional soil analysis methods holds great promise for the future of soil science and the preservation of our natural resources.

In conclusion, the case study of image analysis algorithms applied to satellite images for soil properties analysis exemplifies the transformative impact of remote sensing technology in soil science. By joining the influence of advanced algorithms and satellite data, researchers and practitioners can gain a deeper understanding of soil dynamics, leading to more informed decisions in agriculture and land management. The future of soil analysis is poised for further innovation as these technologies become more sophisticated and accessible to the global community.

7. CONCLUSION AND FUTURE TRENDS

The first problem is that farmers are not aware of the imaging options available in the evolving industry. They also have no idea what kind of imagery to choose or how to order imagery from archived records for a specific use. In order for consumers and farmers to choose and purchase suitable images, image suppliers and various vendors must create simple instructions. One of the problems is the timely collection and delivery of the photos. In agriculture, significant differences in soil moisture, vegetal nutrition, crop development and yields can occur because of environmental changes. Due to the rapid changes in plant populations, monitoring of plants must be continuous. Satellite and remote imaging have been successfully used in crop forecasting. The agricultural sector will see an 80-90% market increase due to the recent development of small unmanned aircraft systems (UAS), making it the largest commercial market for their use (Association of Unmanned Vehicle Systems International, 2013). UAS is a great substitute for crop observing because of its low cost, high spatial resolution imaging, and easy image capture. Applications of UAS in agriculture include spraying of fungicides, herbicides and pesticides, aerial dispersal of seeds, measurement of plant and environmental parameters, monitoring of plant physiological properties, leaf area index (LAI), diseases and plant stress, observing plant development and harvest, eliminating rainwater (temperature, humidity, etc.).

REFERENCES

Ač, A., Malenovský, Z., Olejníčková, J., Gallé, A., Rascher, U., & Mohammed, G. (2015). Meta-analysis assessing potential of steady-state chlorophyll fluorescence for remote sensing detection of plant water, temperature and nitrogen stress. *Remote Sensing of Environment*, 168, 420–436. DOI: 10.1016/j.rse.2015.07.022

Alvarez-Vanhard, E., Corpetti, T., & Houet, T. (2021). UAV satellite synergies for optical remote sensing applications: A literature review, *Science of remote sensing*, 3, 100019.

Babaeian, E., Sadeghi, M., Jones, S. B., Montzka, C., Vereecken, H., & Tuller, M. (2019). Ground, proximal, and satellite remote sensing of soil moisture. *Reviews of Geophysics*, 57(2), 530–616. DOI: 10.1029/2018RG000618

Boston, T., Van Dijk, A., Larraondo, P. R., & Thackway, R. (2022). Comparing CNNs and random forests for Landsat image segmentation trained on a large proxy land cover dataset. *Remote Sensing (Basel)*, 14(14), 3396. DOI: 10.3390/rs14143396

Chen, H., Huang, W., Li, W., Niu, Z., Zhang, L., & Xing, S. (2018). Estimation of LAI in winter wheat from multi-angular hyperspectral VNIR data: Effects of view angles and plant architecture. *Remote Sensing (Basel)*, 10(10), 1630. DOI: 10.3390/rs10101630

Chen, J., Zhu, W., Tian, Y. Q., Yu, Q., Zheng, Y., & Huang, L. (2017). Remote estimation of colored dissolved organic matter and chlorophyll-a in Lake Huron using Sentinel-2 measurements. *Journal of Applied Remote Sensing*, 11(3), 036007. DOI: 10.1117/1.JRS.11.036007

Chu, H. J., He, Y. C., Chusnah, W. N. U., Jaelani, L. M., & Chang, C. H. (2021). Multi-reservoir water quality mapping from remote sensing using spatial regression. *Sustainability (Basel)*, 13(11), 6416. DOI: 10.3390/su13116416

Dai, J., Roberts, D. A., Stow, D. A., An, L., Hall, S. J., Yabiku, S. T., & Kyriakidis, P. C. (2020). Mapping understory invasive plant species with field and remotely sensed data in Chitwan, Nepal. *Remote Sensing of Environment*, 250, 112037. DOI: 10.1016/j.rse.2020.112037

Das, B., Mahajan, G. R., & Singh, R. (2018). Hyperspectral Remote Sensing: Use in Detecting Abiotic Stresses in Agriculture. In Bal, S., Mukherjee, J., Choudhury, B., & Dhawan, A. (Eds.), *Advances in Crop Environment Interaction*. Springer. DOI: 10.1007/978-981-13-1861-0_12

Delogu, E., Olioso, A., Alliès, A., Demarty, J., & Boulet, G. (2021). Evaluation of multiple methods for the production of continuous evapotranspiration estimates from TIR remote sensing. *Remote Sensing (Basel)*, 13(6), 1086. DOI: 10.3390/rs13061086

Ding, X., He, X., Bai, Y., Li, J., Xu, Y., Wang, X., & Zhu, Q. (2022). Requirement on the diurnal observation frequency for satellite remote sensing of photosynthetically available radiation. *Optics Express*, 30(6), 9021–9034. DOI: 10.1364/OE.451207 PMID: 35299341

Duan, B., Fang, S., Zhu, R., Wu, X., Wang, S., Gong, Y., & Peng, Y. (2019). Remote estimation of rice yield with unmanned aerial vehicle (UAV) data and spectral mixture analysis. *Frontiers in Plant Science*, 10, 204. DOI: 10.3389/fpls.2019.00204 PMID: 30873194

Escribano, P., Schmid, T., Chabrillat, S., Rodríguez-Caballero, E., & García, M. (2017). Optical remote sensing for soil mapping and monitoring. In *Soil mapping and process modeling for sustainable land use management* (pp. 87–125). Elsevier. DOI: 10.1016/B978-0-12-805200-6.00004-9

Falcioni, R., Gonçalves, J. V. F., Oliveira, K. M. D., Antunes, W. C., & Nanni, M. R. (2022). VIS-NIR-SWIR Hyperspectroscopy Combined with Data Mining and Machine Learning for Classification of Predicted Chemometrics of Green Lettuce. *Remote Sensing (Basel)*, 14(24), 6330. DOI: 10.3390/rs14246330

Fang, H. (2020). Development and validation of satellite leaf area index (LAI) products in China. *Remote Sensing Technology and Application*, 35(5), 990–1003.

Filipović, N., Brdar, S., Mimić, G., Marko, O., & Crnojević, V. (2022). Regional soil moisture prediction system based on Long Short-Term Memory network. *Biosystems Engineering*, 213, 30–38. DOI: 10.1016/j.biosystemseng.2021.11.019

Ford, T. W., & Quiring, S. M. (2019). Comparison of contemporary *in situ*, model, and satellite remote sensing soil moisture with a focus on drought monitoring. *Water Resources Research*, 55(2), 1565–1582. DOI: 10.1029/2018WR024039

Gao, F., & Zhang, X. (2021). Mapping crop phenology in near real-time using satellite remote sensing: Challenges and opportunities. *Yaogan Xuebao*, •••, 2021.

Gerhards, M., Schlerf, M., Mallick, K., & Udelhoven, T. (2019). Challenges and Future Perspectives of Multi-/Hyperspectral Thermal Infrared Remote Sensing for Crop Water-Stress Detection: A Review. *Remote Sensing (Basel)*, 11(10), 1240. DOI: 10.3390/rs11101240

He, L., Coburn, C. A., Wang, Z. J., Feng, W., & Guo, T. C. (2018). Reduced prediction saturation and view effects for estimating the leaf area index of winter wheat. *IEEE Transactions on Geoscience and Remote Sensing*, 57(3), 1637–1652. DOI: 10.1109/TGRS.2018.2868138

Hossain, M. D., & Chen, D. (2019). Segmentation for Object-Based Image Analysis (OBIA): A review of algorithms and challenges from remote sensing perspective. *ISPRS Journal of Photogrammetry and Remote Sensing*, 150, 115–134. DOI: 10.1016/j.isprsjprs.2019.02.009

Hosseini, F. S., Seo, M. B., Razavi-Termeh, S. V., Sadeghi-Niaraki, A., Jamshidi, M., & Choi, S. M. (2023). Geospatial Artificial Intelligence (GeoAI) and Satellite Imagery Fusion for Soil Physical Property Predicting. *Sustainability (Basel)*, 15(19), 14125. DOI: 10.3390/su151914125

Hu, Q., Yang, J., Xu, B., Huang, J., Memon, M. S., Yin, G., Zeng, Y., Zhao, J., & Liu, K. (2020). Evaluation of global decametric-resolution LAI, FAPAR and FVC estimates derived from Sentinel-2 imagery. *Remote Sensing (Basel)*, 12(6), 912. DOI: 10.3390/rs12060912

Khanal, S., Fulton, J., & Shearer, S. (2017). An overview of current and potential applications of thermal remote sensing in precision agriculture. *Computers and Electronics in Agriculture*, 139, 22–32. DOI: 10.1016/j.compag.2017.05.001

Kucharczyk, M., Hay, G. J., Ghaffarian, S., & Hugenholtz, C. H. (2020). Geographic object-based image analysis: A primer and future directions. *Remote Sensing (Basel)*, 12(12), 2012. DOI: 10.3390/rs12122012

Liu, N., Budkewitsch, P., & Treitz, P. (2017). Examining spectral reflectance features related to Arctic percent vegetation cover: Implications for hyperspectral remote sensing of Arctic tundra. *Remote Sensing of Environment*, 192, 58–72. DOI: 10.1016/j.rse.2017.02.002

Lu, B., Dao, P. D., Liu, J., He, Y., & Shang, J. (2020). Recent advances of hyperspectral imaging technology and applications in agriculture. *Remote Sensing (Basel)*, 12(16), 2659. DOI: 10.3390/rs12162659

Malik, V., Mittal, R., Kaur, A., Singla, G., Mittal, A., & Singh, M. (2024). Enhancement and analysis of hyperspectral satellite images for Soil Study and Behavior. *Multimedia Tools and Applications*, 83(11), 33879–33902. DOI: 10.1007/s11042-023-16729-4

Mehravar, S., Razavi-Termeh, S. V., Moghimi, A., Ranjgar, B., Foroughnia, F., & Amani, M. (2023). Flood susceptibility mapping using multi-temporal SAR imagery and novel integration of nature-inspired algorithms into support vector regression. *Journal of Hydrology (Amsterdam)*, 617, 129100. DOI: 10.1016/j.jhydrol.2023.129100

Miraglio, T., Adeline, K., Huesca, M., Ustin, S., & Briottet, X. (2019). Monitoring LAI, chlorophylls, and carotenoids content of a woodland savanna using hyperspectral imagery and 3D radiative transfer modelling. *Remote Sensing (Basel)*, 12(1), 28. DOI: 10.3390/rs12010028

Mohamed, E. S., Baroudy, A. A. E., El-Beshbeshy, T., Emam, M., Belal, A. A., Elfadaly, A., & Lasaponara, R. (2020). Vis-nir spectroscopy and satellite landsat-8 oli data to map soil nutrients in arid conditions: A case study of the northwest coast of egypt. *Remote Sensing (Basel)*, 12(22), 3716. DOI: 10.3390/rs12223716

Mohamed, E. S., Saleh, A. M., Belal, A. B., & Gad, A. (2018). Application of near-infrared reflectance for quantitative assessment of soil properties. *The Egyptian Journal of Remote Sensing and Space Sciences*, 21(1), 1–14. DOI: 10.1016/j.ejrs.2017.02.001

Morin, M., Lawrence, R., Repasky, K., Sterling, T., McCann, C., & Powell, S. (2017). Agreement analysis and spatial sensitivity of multispectral and hyperspectral sensors in detecting vegetation stress at management scales. *Journal of Applied Remote Sensing*, 11(4), 046025. DOI: 10.1117/1.JRS.11.046025

Neinavaz, E., Schlerf, M., Darvishzadeh, R., Gerhards, M., & Skidmore, A. K. (2021). Thermal infrared remote sensing of vegetation: Current status and perspectives. *International Journal of Applied Earth Observation and Geoinformation*, 102, 102415. DOI: 10.1016/j.jag.2021.102415

Neyns, R., & Canters, F. (2022). Mapping of urban vegetation with high-resolution remote sensing: A review. *Remote Sensing (Basel)*, 14(4), 1031. DOI: 10.3390/rs14041031

Nizeyimana, E., & Petersen, G. W. (2020). Remote sensing applications to soil degradation assessments. In *Methods for assessment of soil degradation* (pp. 393–405). CRC Press. DOI: 10.1201/9781003068716-20

Okhrimenko, M., Coburn, C., & Hopkinson, C. (2019). Multi-spectral lidar: Radiometric calibration, canopy spectral reflectance, and vegetation vertical SVI profiles. *Remote Sensing (Basel)*, 11(13), 1556. DOI: 10.3390/rs11131556

Pahlevan, N., Smith, B., Binding, C., & O'Donnell, D. M. (2017). Spectral band adjustments for remote sensing reflectance spectra in coastal/inland waters. *Optics Express*, 25(23), 28650–28667. DOI: 10.1364/OE.25.028650

Pepe, M., Fregonese, L., & Scaioni, M. (2018). Planning airborne photogrammetry and remote-sensing missions with modern platforms and sensors. *European Journal of Remote Sensing*, 51(1), 412–436. DOI: 10.1080/22797254.2018.1444945

Pôças, I., Calera, A., Campos, I., & Cunha, M. (2020). Remote sensing for estimating and mapping single and basal crop coefficients: A review on spectral vegetation indices approaches. *Agricultural Water Management*, 233, 106081. DOI: 10.1016/j.agwat.2020.106081

Pour, A. B., Zoheir, B., Pradhan, B., & Hashim, M. (2021). Editorial for the special issue: Multispectral and hyperspectral remote sensing data for mineral exploration and environmental monitoring of mined areas. *Remote Sensing (Basel)*, 13(3), 519. DOI: 10.3390/rs13030519

Rhyma, P. P., Norizah, K., Hamdan, O., Faridah-Hanum, I., & Zulfa, A. W. (2020). Integration of normalised different vegetation index and Soil-Adjusted Vegetation Index for mangrove vegetation delineation. *Remote Sensing Applications: Society and Environment*, 17, 100280. DOI: 10.1016/j.rsase.2019.100280

Richards, J. A. (2022). Remote sensing digital image analysis, springer, pp. 263-367. DOI: 10.1007/978-3-030-82327-6

Roy, P. S., Behera, M. D., & Srivastav, S. K. (2017). Satellite remote sensing: Sensors, applications and techniques. *Proceedings of the National Academy of Sciences. India. Section A, Physical Sciences*, 87(4), 465–472. DOI: 10.1007/s40010-017-0428-8

Segarra, J., Buchaillot, M. L., Araus, J. L., & Kefauver, S. C. (2020). Remote sensing for precision agriculture: Sentinel-2 improved features and applications. *Agronomy (Basel)*, 10(5), 641. DOI: 10.3390/agronomy10050641

Seong, J. C. (2015). Sun position calculator (SPC) for Landsat imagery with geodetic latitudes. *Computers & Geosciences*, 85, 68–74. DOI: 10.1016/j.cageo.2015.09.011

Sheirich, P. (2009). An Engineering Trade Space Analysis for a Space-Based Hyperspectral Chromotomographic Scanner.

Shi, Q., Liu, M., Li, S., Liu, X., Wang, F., & Zhang, L. (2021). A deeply supervised attention metric-based network and an open aerial image dataset for remote sensing change detection. *IEEE Transactions on Geoscience and Remote Sensing*, 60, 1–16.

Singh, S., & Kasana, S. S. (2022). Quantitative estimation of soil properties using hybrid features and RNN variants. *Chemosphere*, 287, 131889. DOI: 10.1016/j.chemosphere.2021.131889 PMID: 34461337

Sivanpillai, R., & Whitman, R. D. (2019). Relating leaf spectral reflectance to its color: An inquiry-based activity to enhance understanding of electromagnetic radiation. *Science Activities*, 56(1), 19–26. DOI: 10.1080/00368121.2019.1638746

Snehmani, G. A., Ganju, A., Kumar, S., Srivastava, P. K., & Hari Ram, R. P. (2017). A comparative analysis of pansharpening techniques on QuickBird and WorldView-3 images. *Geocarto International*, 32(11), 1268–1284. DOI: 10.1080/10106049.2016.1206627

Vanhellemont, Q., & Ruddick, K. (2018). Atmospheric correction of metre-scale optical satellite data for inland and coastal water applications. *Remote Sensing of Environment*, 216, 586–597. DOI: 10.1016/j.rse.2018.07.015

Xie, H., Zhang, Y., Zeng, X., & He, Y. (2020). Sustainable land use and management research: A scientometric review. *Landscape Ecology*, 35(11), 2381–2411. DOI: 10.1007/s10980-020-01002-y

Xue, J., & Su, B. (2017). Significant remote sensing vegetation indices: A review of developments and applications. *Journal of Sensors*, 2017, 2017. DOI: 10.1155/2017/1353691

Yan, Y., Xin, Z., Bai, X., Zhan, H., Xi, J., Xie, J., & Cheng, Y. (2023). Analysis of growing season normalized difference vegetation index variation and its influencing factors on the Mongolian Plateau based on Google Earth Engine. *Plants*, 12(13), 2550. DOI: 10.3390/plants12132550 PMID: 37447111

Yue, J., Tian, J., Tian, Q., Xu, K., & Xu, N. (2019). Development of soil moisture indices from differences in water absorption between shortwave-infrared bands. *ISPRS Journal of Photogrammetry and Remote Sensing*, 154, 216–230. DOI: 10.1016/j.isprsjprs.2019.06.012

Zahir, S. A. D. M., Omar, A. F., Jamlos, M. F., Azmi, M. A. M., & Muncan, J. (2022). A review of visible and near-infrared (Vis-NIR) spectroscopy application in plant stress detection. *Sensors and Actuators. A, Physical*, 338, 113468. DOI: 10.1016/j.sna.2022.113468

Zeng, Y., Ning, Z., Liu, P., Luo, P., Zhang, Y., & He, G. (2020). A mosaic method for multi-temporal data registration by using convolutional neural networks for forestry remote sensing applications. *Computing*, 102(3), 795–811. DOI: 10.1007/s00607-019-00716-5

Zhang, L., Du, B., You, J., & Tao, D. (2019). Hyperspectral image unsupervised classification by robust manifold matrix factorization. *Information Sciences*, 485, 154–169. DOI: 10.1016/j.ins.2019.02.008

Zhu, L., Suomalainen, J., Liu, J., Hyyppä, J., Kaartinen, H., & Haggren, H. (2018). A review: Remote sensing sensors. Multi-purposeful application of geospatial data, 19.

Zou, Y., & Greenberg, J. A. (2019). A spatialized classification approach for land cover mapping using hyperspatial imagery. *Remote Sensing of Environment*, 232, 111248. DOI: 10.1016/j.rse.2019.111248

Compilation of References

3DFLOW. (2024). 3DF ZEPHYR. Retrieved March 27, 2024, from https://www.3dflow.net/

Abd Manaf, M. N. H., Juraimi, A. S., Che'Ya, N. N., Su, A. S. M., Roslim, M. H. M., Ahmad, A., & Noor, N. M. (2020). Predictive Modelling for Rice Weeds in Climate Change: A Review. Advances in Agricultural and Food Research Journal.

AbdelRahman, M. A. (2023). An overview of land degradation, desertification and sustainable land management using GIS and remote sensing applications. *Rendiconti Lincei. Scienze Fisiche e Naturali*, 34(3), 767–808. DOI: 10.1007/s12210-023-01155-3

Abouzahir, S., Sadik, M., & Sabir, E. (2022). Lightweight computer vision system for automated weed mapping. In 2022 IEEE 12th Annual Computing and Communication Workshop and Conference (CCWC) (pp. 372-376). DOI: 10.1109/CCWC54503.2022.9720800

Ač, A., Malenovský, Z., Olejníčková, J., Gallé, A., Rascher, U., & Mohammed, G. (2015). Meta-analysis assessing potential of steady-state chlorophyll fluorescence for remote sensing detection of plant water, temperature and nitrogen stress. *Remote Sensing of Environment*, 168, 420–436. DOI: 10.1016/j.rse.2015.07.022

Adami, A., Cerato, I., D'Annibale, E., Demetrescu, E., & Ferdani, D. (2014). Different photogrammetric approaches to 3D survey of the Mausoleum of Romulus in Rome. *2014Eurographics Workshop on Graphics and Cultural Heritage, GCH 2014*, 10. DOI: 10.2312/gch.20141300

Adediji, A., & Adepoju, K. A. (2010). *Assessment of Revised Universal Soil Loss Equation (RUSLE) in Kastina Area, Kastina State of Nigeria using Remote Sensing (RS) and Geographic Information System (GIS) Environmental impact of metal recycling industry view project*. https://www.researchgate.net/publication/49593936

Agarwal, S., Furukawa, Y., Sna, N., Curless, B., Seitz, S. M., & Szeliski, R. (2010). Reconstructing Rome. *Computer*, 43(6), 40–47. DOI: 10.1109/MC.2010.175

Agrawal, S., & Gupta, R. D. (2017). Web GIS and its architecture: A review. *Arabian Journal of Geosciences*, 10(23), 518. DOI: 10.1007/s12517-017-3296-2

Aguilera, D. G., & Lahoz, J. G. (2006). Laser scanning or image-based modeling? A comparative through the modelization of San Nicolas Church. *International Archives of Photogrammetry and Remote Sensing*, 36(B5).

Ahamrouni, J. (1996). *Erosion hydrique dans le bassin versant de la Moulouya (Maroc Oriental) : Recherche des zones sources d'envasement de la retenue du barrage Mohamed V*. Université Cheikh Anta Diop, Faculté des Sciences et Techniques.

Ahmad Fuad, N., Yusoff, A. R., Ismail, Z., & Majid, Z. (2018). Comparing the performance of point cloud registration methods for landslide monitoring using mobile laser scanning data. *The International Archives of the Photogrammetry. Remote Sensing and Spatial Information Sciences, XLII*, 4(W9), 11–21. DOI: 10.5194/isprs-archives-XLII-4-W9-11-2018

Ahmad, H. R., Mehmood, K., Bibi, S., Bashir, M. H., & Siddique, A. (2024). Revealing air quality dynamics: Leveraging remote sensing and GIS for active monitoring and risk evaluation. In *Geographical Information Science* (pp. 417–428). Elsevier., DOI: 10.1016/B978-0-443-13605-4.00021-7

Ait Brahim, L., Chotin, P., Hinaj, S., Abdelouafi, A., El Adraoui, A., Nakcha, C., Dhont, D., Charroud, M., Soussey Alaoui, F., Amrhar, M., Bouaza, A., Tabyaoui, H., & Chaouni, A. (2002). Paleostress evolution in the Moroccan African margin from Triassic to Present. *Tectonophysics*, 357(1-4), 187–205. DOI: 10.1016/S0040-1951(02)00368-2

Ait-Fora, A. (1995). *Modélisation spatiale de l'érosion hydrique dans un bassin versant du Rif marocain : validation de l'approche géomatique par la sédimentologie, les traceurs radioactifs et la susceptibilité magnétique des sédiments*. Université de Sherbrooke.

Aitkenhead, M. J., Dalgetty, I. A., Mullins, C. E., McDonald, A. J. S., & Strachan, N. J. C. (2003). Weed and crop discrimination using image analysis and artificial intelligence methods. *Computers and Electronics in Agriculture*, 39(3), 157–171. DOI: 10.1016/S0168-1699(03)00076-0

Aji, A. F., Munajat, Q., Pratama, A. P., Kalamullah, H., Aprinaldi, J. S., & Arymurthy, A. M. (2013). Detection of palm oil leaf disease with image processing and neural network classification on mobile device. *International Journal of Computer Theory and Engineering*, 5(3), 1–5. DOI: 10.7763/IJCTE.2013.V5.743

Al Karkouri J (2016). *Les milieux montagneux marocains à l'épreuve du changement climatique (cas de la montagne rifaine)*. Hespéris-Tamuda LII (1) (2017): 237-267.

Al Karkouri, J. (2003). *Dégradation du milieu naturel dans le bassin de Beni Boufrah (Rif central-Maroc) : analyse des facteurs et des processus, essai de quantification et modélisation spatiale*. Université Mohamed V, Faculté des Lettres.

Alam, M., Alam, M. S., Roman, M., Tufail, M., Khan, M. U., & Khan, M. T. (2020). Real-time machine-learning-based crop/weed detection and classification for variable-rate spraying in precision agriculture. In 2020 7th International Conference on Electrical and Electronics Engineering (ICEEE) (pp. 273–280). DOI: 10.1109/ICEEE49618.2020.9102505

Aliaâ, H., Kassou, N., & Kacimi, I. (2017) Using the SLEUTH Model to Evaluated and Analyze the Impact of Urban Sprawl on Agricultural Land in the cities of Kenitra and Mehdia (Morocco). In: Proceedings of the 2nd International Conference on Computing and Wireless Communication Systems. ACM, Larache Morocco, pp 1–7 https://doi.org/DOI: 10.1145/3167486.31675

Ali, B. (2015). *Territorial Policies of the Spanish Colonial Authorities in Northern Morocco 1912-1956*. Al-Hidaya Press.

Alidoost, F., & Arefi, H. (2017). Comparison of uas-based photogrammetry software for 3d point cloud generation: a survey over a historical site. *ISPRS Annals of the Photogrammetry. Remote Sensing and Spatial Information Sciences, IV*, 4(W4), 55–61. DOI: 10.5194/isprs-annals-IV-4-W4-55-2017

Allahyari, A., & Farjad, B. (2022). Investigation of sand drift potential and sand dune mobility in Khartouran Erg, Sabzevar, NE Iran. *Aeolian Research*, 57, 100807. DOI: 10.1016/j.aeolia.2022.100807

Alvarez-Vanhard, E., Corpetti, T., & Houet, T. (2021). UAV satellite synergies for optical remote sensing applications: A literature review, *Science of remote sensing*, 3, 100019.

Amani, M., Ghorbanian, A., Ahmadi, S. A., Kakooei, M., Moghimi, A., Mirmazloumi, S. M., Moghaddam, S. H. A., Mahdavi, S., Ghahremanloo, M., Parsian, S., Wu, Q., & Brisco, B. (2020). Google Earth Engine Cloud Computing Platform for Remote Sensing Big Data Applications: A Comprehensive Review. *IEEE Journal of Selected Topics in Applied Earth Observations and Remote Sensing*, 13, 5326–5350. DOI: 10.1109/JSTARS.2020.3021052

Amara, J., Bouaziz, B., & Algergawy, A. (2017). A deep learning-based approach for banana leaf diseases classification. In Gesellschaft für Informatik Bonn (pp. 79–88).

Amiri F, Shariff AR, Tabatabaie T (2012). Monitoring land suitability for mixed livestock grazing using Geographic Information System (GIS). *Application of Geographic Information Systems*, 241–266.

Anderson, J. R., Hardy, E. E., Roach, J. T., & Witmer, R. E. (1961). *A land use and land cover classification system for use with remote sensor data.*

Andrienko, G. Andrienko, N., Dykes, J., Mountain, D., Noy, P., Gahegan, M., … Theus, M. (2005). Creating Instruments for Ideation. In *Exploring Geovisualization* (pp. 101–125). Elsevier. DOI: 10.1016/B978-008044531-1/50423-1

Andrienko, G., Andrienko, N., Jankowski, P., Keim, D., Kraak, M.-J., MacEachren, A., & Wrobel, S. (2007). Geovisual analytics for spatial decision support: Setting the research agenda. *International Journal of Geographical Information Science*, 21(8), 839–857. DOI: 10.1080/13658810701349011

Antova, G. (2019). Application of Areal Change Detection Methods Using Point Clouds Data. *IOP Conference Series. Earth and Environmental Science*, 221, 012082. DOI: 10.1088/1755-1315/221/1/012082

Anyimah, F. O.Jnr, Osei, E. M.Jnr, & Nyamekye, C. (2021). Detection of stress areas in cocoa farms using GIS and remote sensing: A case study of Offinso Municipal & Offinso North district, Ghana. *Environmental Challenges*, 4, 100087. DOI: 10.1016/j.envc.2021.100087

Armstrong, D. R., Götz, M., & Nardelli, V. C., & de O. Gomes, V. E. (2021). Machine learning for identification and classification of crops and weeds. In 2021 XI Brazilian Symposium on Computing Systems Engineering (SBESC) (pp. 1-7). DOI: 10.1109/SBESC53686.2021.9628296

Arnal Barbedo, J. G. (2019). Plant disease identification from individual lesions and spots using deep learning. *Biosystems Engineering*, 180, 96–107. DOI: 10.1016/j.biosystemseng.2019.02.002

Arumugam, T., Kinattinkara, S., Velusamy, S., Shanmugamoorthy, M., & Murugan, S. (2023). GIS based landslide susceptibility mapping and assessment using weighted overlay method in Wayanad: A part of Western Ghats, Kerala. *Urban Climate*, 49, 101508. DOI: 10.1016/j.uclim.2023.101508

Asefpour Vakilian, M., & Massah, J. (2013). An artificial neural network approach to identify fungal diseases of cucumber (Cucumis sativus L.) plants using digital image processing. *Archiv für Phytopathologie und Pflanzenschutz*, 45(36), 1580–1588. DOI: 10.1080/03235408.2013.772321

Asraf, H. M., Rizam, M. S., & Nooritawati, M. T. (2012). Comparative study in kernel-based support vector machine of oil palm leaves nutrient disease. International Symposium on Robotics and Intelligent Sensors 2012, 41, 1353-1359. DOI: 10.1016/j.proeng.2012.07.321

Atsmon, G., Nehurai, O., Kizel, F., Eizenberg, H., & Nisim Lati, R. (2022). Hyperspectral imaging facilitates early detection of Orobanche cumana below-ground parasitism on sunflower under field conditions. *Computers and Electronics in Agriculture*, 196, 106881. DOI: 10.1016/j.compag.2022.106881

Azaiez, N. A. (2024). Insights into soil loss in the Jouah basin using EPM model and Caesium 137 as an isotopic tracer. *Egyptian Journal of Soil Science*, 64(1), 0. DOI: 10.21608/ejss.2023.245181.1681

Baas, A. C. W. (2008). Challenges in wind and water erosion research. *Geomorphology*, 100(3-4), 246–257. DOI: 10.1016/j.geomorph.2008.01.018

Babaeian, E., Sadeghi, M., Jones, S. B., Montzka, C., Vereecken, H., & Tuller, M. (2019). Ground, proximal, and satellite remote sensing of soil moisture. *Reviews of Geophysics*, 57(2), 530–616. DOI: 10.1029/2018RG000618

Báčová, M., Krása, J., Devátý, J., & Kavka, P. (2019). A GIS method for volumetric assessments of erosion rills from digital surface models. *European Journal of Remote Sensing, 52*(sup1), 96–107. DOI: 10.1080/22797254.2018.1543556

Badhan, S., Desai, K., Dsilva, M., Sonkusare, R., & Weakey, S. (2021). Real-time weed detection using machine learning and stereo-vision. In 2021 6th International Conference for Convergence in Technology (I2CT) (pp. 1-5). DOI: 10.1109/I2CT51068.2021.9417989

Bagnold, R. A. (1941). *The physics of blown sand and desert dunes*. Methuen.

Bah, M. D., Hafiane, A., Canals, R., & Emile, B. (2019). Deep features and one-class classification with unsupervised data for weed detection in UAV images. In *2019 Ninth International Conference on Image Processing Theory, Tools and Applications (IPTA)* (pp. 1–5). DOI: 10.1109/IPTA.2019.8936091

Bakyarani, S., Srimathi, H., & Bagavandas, M. (2019). A survey of machine learning algorithms in health care. *International Journal of Scientific & Technology Research*, 8(11), 2288–2292.

Baltsavias, E. P. (1999). A comparison between photogrammetry and laser scanning. *ISPRS Journal of Photogrammetry and Remote Sensing*, 54(2–3), 83–94. DOI: 10.1016/S0924-2716(99)00014-3

Banos, A., & Thévenin, T. (Eds.). (2011). *Geographical information and urban transport systems*. ISTE.

Barathon, J. J. (1987). *Bassins et littoraux du rif oriental (maroc) : evolution morphoclimatique et tectonique depuis le neogene superieur*. Université de Poitiers.

Barbosa, A., Trevisan, R., Hovakimyan, N., & Martin, N. F. (2020). Modeling yield response to crop management using convolutional neural networks. *Computers and Electronics in Agriculture*, 170, 105197. DOI: 10.1016/j.compag.2019.105197

Barchyn, T. E., & Hugenholtz, C. H. (2012). Predicting dune field state change with time-series vegetation data. *Earth Surface Processes and Landforms*, 37(8), 749–758. DOI: 10.1002/esp.3199

Basha, S. M., Rajput, D. S., J, J., Somula, R. S., & Ram, S. (2020). Principles and Practices of Making Agriculture Sustainable: Crop Yield prediction using Random Forest. *Scalable Computing: Practice and Experience*, 21(4), 591–599. DOI: 10.12694/scpe.v21i4.1714

Bayram, B., Nemli, G., Özkan, T., Oflaz, O. E., Kankotan, B., & Çetin, İ. (2015). Comparison Of Laser Scanning And Photogrammetry And Their Use For Digital Recording Of Cultural Monument Case Study: Byzantine Land Walls-Istanbul. *ISPRS Annals of the Photogrammetry. ISPRS Annals of the Photogrammetry, Remote Sensing and Spatial Information Sciences*, 5(W3), 17–24. DOI: 10.5194/isprsannals-II-5-W3-17-2015

Beeharry, Y., & Bassoo, V. (2020). Performance of ANN and AlexNet for weed detection using UAV-based images. In 2020 3rd International Conference on Emerging Trends in Electrical, Electronic and Communications Engineering (ELECOM) (pp. 163–167). DOI: 10.1109/ELECOM49001.2020.9296994

Bemporad, G. A., Alterach, J., Amighetti, F. F., Peviani, M., & Saccardo, I. (1997). A distributed approach for sediment yield evaluation in Alpine regions. ... *Journal of Hydrology (Amsterdam)*, 197(1-4), 370–392. DOI: 10.1016/0022-1694(95)02978-8

Benalla, M., Alem, E. M., Rognon, P., Desjardins, R., Hilali, A., & Khardi, A. (2003). Les dunes du Tafilalet (Maroc): Dynamique éolienne et ensablement des palmeraies. *Sécheresse, 14*(2), 73–83. ISSN 1147-7806.

Benchelha, N., Bezza, M., Belbounaguia, N., Benchelha, S., & Benchelha, M. (2022). Modeling Dynamic Urban Growth Using Cellular Automata and Geospatial Technique: Case of Casablanca in Morocco. *International Journal of Geoinformatics*. Advance online publication. DOI: 10.52939/ijg.v18i5.2369

Ben-Dor, G., Ogulenko, A., Klein, I., & Benenson, I. (2022). Modeling the Effect of Congestion Charge and Parking Pricing on Urban Traffic: Example of Jerusalem. *AGILE: GIScience Series*, 3, 1–4. DOI: 10.5194/agile-giss-3-26-2022

Bennani, M., Jawab, F., Amegouz, D., Hani, Y., & Elmhamedi, A. (2020). The impact of congestion on the cost of transport in urban areas: Application for the location of an urban distribution center. *2020 IEEE 13th International Colloquium of Logistics and Supply Chain Management (LOGISTIQUA)*, 1–6. DOI: 10.1109/LOGISTIQUA49782.2020.9353949

Benzakour, S. (1978). *Essai sur la politique urbaine au Maroc, 1912-1975: sur le rôle de l'État*. Éditions maghrébines.

Berk, R. (2004). *Regression Analysis: A Constructive Critique*. SAGE Publications, Inc., DOI: 10.4135/9781483348834

Bertrand, J. M. (1977). *Pratique de la ville* (Masson, C. G., Ed.).

Bhambri, D., Dhanoa, D. I. S., Sinha, D. V. K., & Kaur, E. J. (2020). Paddy Crop Production Analysis Based on SVM and KNN Classifier. [IJRTE]. *International Journal of Recent Technology and Engineering*, 8(5), 2790–2793. DOI: 10.35940/ijrte.D8650.018520

Bibhuti Bhusan, S., & Giri, R. (2023). Application of Remote Sensing & Geographic Information Systems for Agriculture-A Review. In *Advances in Agricultural Technology*.

Biswas, S., Jagyasi, H., Singh, B. P., & Lal, M. (2014). Severity identification of potato late blight disease from crop images captured under uncontrolled environment. In IEEE. DOI: 10.1109/IHTC.2014.7147519

Bonczak, B., & Kontokosta, C. E. (2019). Large-scale parameterization of 3D building morphology in complex urban landscapes using aerial LiDAR and city administrative data. *Computers, Environment and Urban Systems*, 73, 126–142. DOI: 10.1016/j.compenvurbsys.2018.09.004

Boser, B. E., Guyon, I., & Vapnik, V. (1992). A training algorithm for optimal margin classifiers. In *Proceedings of the Fifth Annual Workshop on Computational Learning Theory* (pp. 144-152). ACM Press. DOI: 10.1145/130385.130401

Boston, T., Van Dijk, A., Larraondo, P. R., & Thackway, R. (2022). Comparing CNNs and random forests for Landsat image segmentation trained on a large proxy land cover dataset. *Remote Sensing (Basel)*, 14(14), 3396. DOI: 10.3390/rs14143396

Bourke, M. C., Fenton, L. K., & Balme, M. R. (2022). Dune Activity on Mars: Insights from Terrestrial Analog Studies. *Journal of Geophysical Research: Planets, 127*(4), e2021JE007108. https://doi.org/DOI: 10.1029/2021JE007108

Boyina, L., Sandhya, G., Vasavi, S., Koneru, L., & Koushik, V. (2021). Weed detection in broad leaves using invariant U-Net model. In *2021 International Conference on Communication, Control and Information Sciences (ICCISc)* (vol. 1, pp. 1-4). DOI: 10.1109/ICCISc52257.2021.9485001

Brahimi, M., Arsenovic, M., Laraba, S., Sladojevic, S., Boukhalfa, K., & Moussaoui, A. (2018). Deep learning for plant diseases: Detection and saliency map visualisation. In Springer. https://doi.org/DOI: 10.1007/978-3-319-90402-3_5

Brahimi, M., Arsenovic, M., Laraba, S., Sladojevic, S., Boukhalfa, K., & Moussaoui, A. (2018). Deep learning for plant diseases: Detection and saliency map visualization. In Springer (pp. 93–117).

Breiman, L. (2001). Random Forests. *Machine Learning*, 45(1), 5–32. https://doi.org/http://dx.doi.org/10.1023/A:1010933404324. DOI: 10.1023/A:1010933404324

Breiman, L., Friedman, J. H., Olshen, R. A., & Stone, C. J. (2017). *Classification and Regression Trees*. Routledge., DOI: 10.1201/9781315139470

Brown, C. F., Brumby, S. P., Guzder-Williams, B., Birch, T., Hyde, S. B., Mazzariello, J., Czerwinski, W., Pasquarella, V. J., Haertel, R., Ilyushchenko, S., Schwehr, K., Weisse, M., Stolle, F., Hanson, C., Guinan, O., Moore, R., & Tait, A. M. (2022). Dynamic World, Near real-time global 10 m land use land cover mapping. *Scientific Data*, 9(1), 1–17. DOI: 10.1038/s41597-022-01307-4 PMID: 35013360

Brown, M., & Lowe, D. G. (2005). Unsupervised 3D Object Recognition and Reconstruction in Unordered Datasets. *Fifth International Conference on 3-D Digital Imaging and Modeling (3DIM'05)*, 56–63. IEEE. DOI: 10.1109/3DIM.2005.81

Bulgakov, A., Sayfeddine, D., Bock, T., & Fares, A. (2020). Generation of Orthomosaic Model for Construction Site using Unmanned Aerial Vehicle. *Proceedings of the 37th International Symposium on Automation and Robotics in Construction, ISARC 2020: From Demonstration to Practical Use - To New Stage of Construction Robot*. DOI: 10.22260/ISARC2020/0124

Bullard, J. E. (1997). A note on the use of the "Fryberger method" for evaluating potential sand transport by wind. *Journal of Sedimentary Research*, 67(3), 499–501. DOI: 10.1306/D42685A9-2B26-11D7-8648000102C1865D

Burns, J. H. R., & Delparte, D. (2017). Comparison of commercial structure-from-motion photogrammety software used for underwater three-dimensional modeling of coral reef environments. *The International Archives of the Photogrammetry. Remote Sensing and Spatial Information Sciences, XLII*, 2(W3), 127–131. DOI: 10.5194/isprs-archives-XLII-2-W3-127-2017

Callot, Y. (1987). *Geomorphology and paleoenvironments of the Saharan Atlas in the Western Grand Erg: Wind dynamics and Holocene paleolakes* (Doctoral thesis). Pierre and Marie Curie University.

Callot, Y., & Oulehri, T. (2015). Géodynamique des sables éoliens dans le Nord-Ouest saharien: Relations entre aérologie et géomorphologie. *Geodinamica Acta*, 9(1), 1–12. DOI: 10.1080/09853111.1996.11417259

Campbell, J. B., Wynne, R. H., & Thomas, V. A. (2023). *Introduction to Remote Sensing* (6th ed.). The Guilford Press.

Caras, T., Lati, R. N., Holland, D., Dubinin, V. M., Hatib, K., Shulner, I., Keiesar, O., Liddor, G., & Paz-Kagan, T. (2024). Monitoring the effects of weed management strategies on tree canopy structure and growth using UAV-LiDAR in a young almond orchard. *Computers and Electronics in Agriculture*, 216, 108467. DOI: 10.1016/j.compag.2023.108467

Carrasco, L., O'Neil, A. W., Morton, R. D., & Rowland, C. S. (2019). Evaluating Combinations of Temporally Aggregated Sentinel-1, Sentinel-2 and Landsat 8 for Land Cover Mapping with Google Earth Engine. *Remote Sensing (Basel)*, 11(3), 288. DOI: 10.3390/rs11030288

Celebi, M. E., Kingravi, H. A., & Vela, P. A. (2013). A comparative study of efficient initialization methods for the k-means clustering algorithm. *Expert Systems with Applications*, 40(1), 200–210. DOI: 10.1016/j.eswa.2012.07.021

Chan, J. C.-W., & Paelinckx, D. (2008). Evaluation of Random Forest and Adaboost tree-based ensemble classification and spectral band selection for ecotope mapping using airborne hyperspectral imagery. *Remote Sensing of Environment*, 112(6), 2999–3011. DOI: 10.1016/j.rse.2008.02.011

Chan, J. J., & Chye, S. W. (2023). Impact of Smart City Initiatives on Urban Planning Strategies in Singapore: An In-Depth Analysis of Technology-Driven Solutions and Their Influence on Sustainable Development and Quality of Life. *Journal of Strategy and Management*, 7(7), 11–21. DOI: 10.53819/81018102t4200

Chen, H., Huang, W., Li, W., Niu, Z., Zhang, L., & Xing, S. (2018). Estimation of LAI in winter wheat from multi-angular hyperspectral VNIR data: Effects of view angles and plant architecture. *Remote Sensing (Basel)*, 10(10), 1630. DOI: 10.3390/rs10101630

Chen, J., Zhu, W., Tian, Y. Q., Yu, Q., Zheng, Y., & Huang, L. (2017). Remote estimation of colored dissolved organic matter and chlorophyll-a in Lake Huron using Sentinel-2 measurements. *Journal of Applied Remote Sensing*, 11(3), 036007. DOI: 10.1117/1.JRS.11.036007

Chen, X., Zhang, K., Chuai, G., Gao, W., Si, Z., Hou, Y., & Liu, X. (2023). Urban Area Characterization and Structure Analysis: A Combined Data-Driven Approach by Remote Sensing Information and Spatial–Temporal Wireless Data. *Remote Sensing (Basel)*, 15(4), 1041. DOI: 10.3390/rs15041041

Chen, Z., Jiang, Y., & Sun, D. (2020). Discrimination and Prediction of Traffic Congestion States of Urban Road Network Based on Spatio-Temporal Correlation. *IEEE Access : Practical Innovations, Open Solutions*, 8, 3330–3342. DOI: 10.1109/ACCESS.2019.2959125

Chen, Z., Wang, J., & Jin, J. (2023). Fully automated proximal hyperspectral imaging system for high-resolution and high-quality in vivo soybean phenotyping. *Precision Agriculture*, 24(6), 2395–2415. DOI: 10.1007/s11119-023-10045-5

Cherlet, M., Hutchinson, C., Reynolds, J., Hill, J., Sommer, S., & von Maltitz, G. (2018). *World atlas of desertification*. Publication Office of the European Union., DOI: 10.2760/06292

Chouki, M. (2011). *La Ville Marocaine, Essai de lecture synthétique*. Edition Dar Attaouhidi.

Chu, H. J., He, Y. C., Chusnah, W. N. U., Jaelani, L. M., & Chang, C. H. (2021). Multi-reservoir water quality mapping from remote sensing using spatial regression. *Sustainability (Basel)*, 13(11), 6416. DOI: 10.3390/su13116416

CloudCompare. (2024). CloudCompare. Retrieved March 27, 2024, from https://www.danielgm.net/cc/

Cramer, M., Stallmann, D., & Haala, N. (2000). Direct georeferencing using gps/inertial exterior orientations for photogrammetric applications. *International Archives of Photogrammetry and Remote Sensing, 33*(B3/1; PART 3), 198–205.

Crivelari-Costa, P. M., Lima, M., La Scala Jr, N., Rossi, F. S., Della-Silva, J. L., Dalagnol, R., Teodoro, P. E., Teodoro, L. P. R., de Oliveira, G., de Oliveira Junior, J. F., & da Silva Junior, C. A. (2023). Changes in carbon dioxide balance associated with land use and land cover in Brazilian Legal Amazon based on remotely sensed imagery. *Remote Sensing (Basel)*, 15(11), 2780. DOI: 10.3390/rs15112780

Cruz-Ramírez, N., Acosta-Mesa, H. G., Carrillo-Calvet, H., Nava-Fernández, L. A., & Barrientos-Martínez, R. E. (2007). Diagnosis of breast cancer using Bayesian networks: A case study. *Computers in Biology and Medicine*, 37(11), 1553–1564. DOI: 10.1016/j.compbiomed.2007.02.003 PMID: 17434159

da Silva, R. M., Lopes, A. G., & Santos, C. A. G. (2023). Deforestation and fires in the Brazilian Amazon from 2001 to 2020: Impacts on rainfall variability and land surface temperature. *Journal of Environmental Management*, 326, 116664. DOI: 10.1016/j.jenvman.2022.116664 PMID: 36370609

Daghmoumi, R. (1994). Contribution of geology and GIS to the study of the watersheds of Aoudour (Central Rif) (Doctoral dissertation, Thesis 3rd cycle, Univ. Mohammed V, Fac. Sci. Rabat,(162p)).

Dai, J., Roberts, D. A., Stow, D. A., An, L., Hall, S. J., Yabiku, S. T., & Kyriakidis, P. C. (2020). Mapping understory invasive plant species with field and remotely sensed data in Chitwan, Nepal. *Remote Sensing of Environment*, 250, 112037. DOI: 10.1016/j.rse.2020.112037

Dalal, Y. M., Naveen, S., & Um, A. K. (2023). Unveiling Traffic Patterns for Effective Traffic Management System. *2023 International Conference on Network, Multimedia and Information Technology (NMITCON)*, 1–6. DOI: 10.1109/NMITCON58196.2023.10276327

Damoah-Afari, P., Osei, J. D., Yevugah, L. L., Boakye, L., & Amaning, K. (2023). Integrating GIS and Satellite Remote Sensing to Assess the Geospatial Dynamics of Urban Footprints in the Dormaa Central Municipality in Ghana. *Journal of Energy Engineering and Thermodynamics(JEET) ISSN 2815-0945, 3*(02), Article 02. DOI: 10.55529/jeet.32.22.39

Das, B., Mahajan, G. R., & Singh, R. (2018). Hyperspectral Remote Sensing: Use in Detecting Abiotic Stresses in Agriculture. In Bal, S., Mukherjee, J., Choudhury, B., & Dhawan, A. (Eds.), *Advances in Crop Environment Interaction*. Springer. DOI: 10.1007/978-981-13-1861-0_12

Dayal, K. R., Raghavendra, S., Pande, H., Tiwari, P. S., & Chauhan, I. (2017). Comparative analysis of 3d point clouds generated from a freeware and terrestrial laser scanner. *The International Archives of the Photogrammetry. Remote Sensing and Spatial Information Sciences, XLII*, 4(W2), 67–71. DOI: 10.5194/isprs-archives-XLII-4-W2-67-2017

de Paula Sousa Júnior, V., Sparacino, J., de Espindola, G. M., & Sousa de Assis, R. J. (2022). Land-Use and Land-Cover Dynamics in the Brazilian Caatinga Dry Tropical Forest. *Conservation*, 2(4), 739–752. DOI: 10.3390/conservation2040048

De Ploey, J., Imeson, A., & Oldeman, L.-R. (1991). Soil erosion, soil degradation and climatic change. In Brouwer, F.-M., Thomas, A.-J., & Chadwick, M.-J. (Eds.), *Land use changes in Europe* (pp. 275–292). Kluwer Academic Publishers. DOI: 10.1007/978-94-011-3290-9_12

De Vente, J., Poesen, J., Arabkhedri, M., & Verstraeten, G. (2007). The sediment delivery problem revisited. *Progress in Physical Geography*, 31(2), 155–178. DOI: 10.1177/0309133307076485

Debesa, G., Gebre, S. L., Melese, A., Regassa, A., & Teka, S. (2020). GIS and remote sensing-based physical land suitability analysis for major cereal crops in Dabo Hana district, South-West Ethiopia. *Cogent Food & Agriculture*, 6(1), 1780100. DOI: 10.1080/23311932.2020.1780100

Delarue, J. M. (2008). La Politique Urbaine. Migrations Société, Regards croisés sur les politiques françaises d'immigration. Vol. 20. n. 117-118.

Delogu, E., Olioso, A., Alliès, A., Demarty, J., & Boulet, G. (2021). Evaluation of multiple methods for the production of continuous evapotranspiration estimates from TIR remote sensing. *Remote Sensing (Basel)*, 13(6), 1086. DOI: 10.3390/rs13061086

Department of Natural Resources and Environment of Ba Ria - Vung Tau Province. (2019). *Project "Action Plan for Biodiversity in Ba Ria - Vung Tau Province to 2020 and Vision to 2030."*

Department of Natural Resources and Environment of Ba Ria-Vung Tau Province. (1995). *Land Use Status Map of 1995.*

Department of Natural Resources and Environment of Ba Ria-Vung Tau Province. (2005). *Land Use Status Map of 2005.*

Department of Natural Resources and Environment of Ba Ria-Vung Tau Province. (2015). *Land Use Status Map of 2015.*

Department of Natural Resources and Environment of Ba Ria-Vung Tau Province. (2020). *Land Use Status Map of 2020.*

Department of Natural Resources and Environment of Ba Ria-Vung Tau Province. (2022). *Electronic Atlas of Ba Ria-Vung Tau Province.* http://atlas.baria-vungtau.gov.vn

Dethier, J. (1970). *Soixante ans d'urbanisme au Maroc*. Princeton University.

Dharani, M. K., Thamilselvan, R., Natesan, P., Kalaivaani, P., & Santhoshkumar, S. (2021). Review on Crop Prediction Using Deep Learning Techniques. *Journal of Physics: Conference Series*, 1767(1), 012026. DOI: 10.1088/1742-6596/1767/1/012026

Di Gregorio, A., & Jansen, L. J. M. (1998). *Land Cover Classification System (LCCS)*. Classification Concepts and User Manual.

Ding, X., He, X., Bai, Y., Li, J., Xu, Y., Wang, X., & Zhu, Q. (2022). Requirement on the diurnal observation frequency for satellite remote sensing of photosynthetically available radiation. *Optics Express*, 30(6), 9021–9034. DOI: 10.1364/OE.451207 PMID: 35299341

Dong, T. T., Song, D., Xiu, C., Geng, S., & Luo, J. (2015). A curvelet-SC recognition method for maize disease. *Journal of Electrical and Computer Engineering*, 2015, 1–8. DOI: 10.1155/2015/782851

Downs, A. (2004). *Still stuck in traffic: Coping with peak-hour traffic congestion.* Brookings Institution Press.

Dragičević, N., Barbara, K., & Nevenka, O. (2016). A Review of the Gavrilović Method (Erosion Potential Method). *Application*, 68, 715–725. DOI: 10.14256/JCE.1602.2016

Dragičević, N., Karleuša, B., & Ožanić, N. (2018). Modification of erosion potential method using climate and land cover parameters. *Geomatics, Natural Hazards & Risk*, 9(1), 1085–1105. DOI: 10.1080/19475705.2018.1496483

Dragičević, N., Whyatt, D., Davies, G., Karleuša, B., & Ožanić, N. (2014). Erosion model sensitivity to land cover inputs: case study of the Dubračina catchment, Croatia. *In:Proc. 22nd Annual Conf. GISRUK 2014,* Glasgow, 340–348.

Droj, G., Droj, L., Badea, A.-C., & Dragomir, P. I. (2023). GIS-Based Urban Traffic Assessment in a Historical European City under the Influence of Infrastructure Works and COVID-19. *Applied Sciences (Basel, Switzerland)*, 13(3), 1355. DOI: 10.3390/app13031355

Duan, B., Fang, S., Zhu, R., Wu, X., Wang, S., Gong, Y., & Peng, Y. (2019). Remote estimation of rice yield with unmanned aerial vehicle (UAV) data and spectral mixture analysis. *Frontiers in Plant Science*, 10, 204. DOI: 10.3389/fpls.2019.00204 PMID: 30873194

Du, Y., Zhang, G., Tsang, D., & Jawed, M. K. (2022). Deep-CNN-based robotic multi-class under-canopy weed control in precision farming. In *2022 International Conference on Robotics and Automation (ICRA)* (pp. 2273-2279). DOI: 10.1109/ICRA46639.2022.9812240

Dyrmann, M., Jørgensen, R. N., & Midtiby, H. S. (2017). Roboweedsupport- detection of weed locations in leaf occluded cereal crops using a fully convolutional neural network. *Advances in Animal Biosciences*, 8(2), 842–847. DOI: 10.1017/S2040470017000206

EarthExplorer (usgs.gov). 20/08/2023.

Edsall, R. M. (2005). Challenging Conventions for Geovisualization Interface Design: Implications for Exploratory Data Analysis. *Ninth International Conference on Information Visualisation (IV'05)*, 642–647. IEEE. DOI: 10.1109/IV.2005.31

El Cheikhi, N. E., & El Abdlaoui, M. (1997). Social, Economic, and Spatial Transformations in Urban Centers (Martil, M'diq, and Fnideq) on the Northern Tetouan Coast and Adjacent Areas. *Faculty of Arts and Human Sciences Journal, 8.*

El Gharbaoui A (1981). *La terre et l'Homme dans la péninsule tingitane. Etude sur l'Homme et le milieu dans le Rif occidental.* Travaux de l'Institut Scientifique, n° 15.

Ellenberger, K. (2017). Virtual and Augmented Reality in Public Archaeology Teaching. *Advances in Archaeological Practice*, 5(3), 305–309. DOI: 10.1017/aap.2017.20

Eltner, A., & Schneider, D. (2015). Analysis of Different Methods for 3D Reconstruction of Natural Surfaces from Parallel-Axes. *The Photogrammetric Record*, 30(151), 279–299. DOI: 10.1111/phor.12115

Escribano, P., Schmid, T., Chabrillat, S., Rodríguez-Caballero, E., & García, M. (2017). Optical remote sensing for soil mapping and monitoring. In *Soil mapping and process modeling for sustainable land use management* (pp. 87–125). Elsevier. DOI: 10.1016/B978-0-12-805200-6.00004-9

Eskicioglu, A. M., & Fisher, P. S. (1995). Image quality measures and their performance. *IEEE Transactions on Communications*, 43(12), 2959–2965. DOI: 10.1109/26.477498

European Commission. (2023). *GHSL Data Package 2023*. Publications Office of the European Union.

Falcioni, R., Gonçalves, J. V. F., Oliveira, K. M. D., Antunes, W. C., & Nanni, M. R. (2022). VIS-NIR-SWIR Hyperspectroscopy Combined with Data Mining and Machine Learning for Classification of Predicted Chemometrics of Green Lettuce. *Remote Sensing (Basel)*, 14(24), 6330. DOI: 10.3390/rs14246330

Fanetti, D., & Vezzoli, L. (2007). Sediment input and evolution of lacustrine deltas: The Breggia and Greggio rivers case study (lake Como, Italy). *Quaternary International*, 173-174, 113–124. DOI: 10.1016/j.quaint.2007.02.008

Fang, H. (2020). Development and validation of satellite leaf area index (LAI) products in China. *Remote Sensing Technology and Application*, 35(5), 990–1003.

Fan, P., Chen, J., Fung, C., Naing, Z., Ouyang, Z., Nyunt, K. M., Myint, Z. N., Qi, J., Messina, J. P., Myint, S. W., & Peter, B. G. (2022). Urbanization, economic development, and environmental changes in transitional economies in the global south: A case of Yangon. *Ecological Processes*, 11(1), 65. DOI: 10.1186/s13717-022-00409-6 PMID: 36397837

Fan, P., Ouyang, Z., Nguyen, D. D., Nguyen, T. T. H., Park, H., & Chen, J. (2019). Urbanization, economic development, environmental and social changes in transitional economies: Vietnam after Doimoi. *Landscape and Urban Planning*, 187, 145–155. DOI: 10.1016/j.landurbplan.2018.10.014

Faruque, M. J., Vekerdy, Z., Hasan, M. Y., Islam, K. Z., Young, B., Ahmed, M. T., Monir, M. U., Shovon, S. M., Kakon, J. F., & Kundu, P. (2022). Monitoring of land use and land cover changes by using remote sensing and GIS techniques at human-induced mangrove forests areas in Bangladesh. *Remote Sensing Applications: Society and Environment*, 25, 100699. DOI: 10.1016/j.rsase.2022.100699

Fassi, F., Fregonese, L., Ackermann, S., & De Troia, V. (2013). Comparison between laser scanning and automated 3d modelling techniques to reconstruct complex and extensive cultural heritage areas. *The International Archives of the Photogrammetry. Remote Sensing and Spatial Information Sciences, XL*, 5(W1), 73–80. DOI: 10.5194/isprsarchives-XL-5-W1-73-2013

Ferentinos, K. P. (2018). Deep learning models for plant disease detection and diagnosis. *Computers and Electronics in Agriculture*, 145, 311–318. DOI: 10.1016/j.compag.2018.01.009

Filipović, N., Brdar, S., Mimić, G., Marko, O., & Crnojević, V. (2022). Regional soil moisture prediction system based on Long Short-Term Memory network. *Biosystems Engineering*, 213, 30–38. DOI: 10.1016/j.biosystemseng.2021.11.019

Floreano, I. X., & de Moraes, L. A. F. (2021). Land use/land cover (LULC) analysis (2009–2019) with Google Earth Engine and 2030 prediction using Markov-CA in the Rondônia State, Brazil. *Environmental Monitoring and Assessment*, 193(4), 239. DOI: 10.1007/s10661-021-09016-y PMID: 33783626

Fonstad, M. A., Dietrich, J. T., Courville, B. C., Jensen, J. L., & Carbonneau, P. E. (2013). Topographic structure from motion: A new development in photogrammetric measurement. *Earth Surface Processes and Landforms*, 38(4), 421–430. DOI: 10.1002/esp.3366

Foody, G. M. (2004). Thematic Map Comparison. *Photogrammetric Engineering and Remote Sensing*, 70(5), 627–633. DOI: 10.14358/PERS.70.5.627

Ford, T. W., & Quiring, S. M. (2019). Comparison of contemporary *in situ*, model, and satellite remote sensing soil moisture with a focus on drought monitoring. *Water Resources Research*, 55(2), 1565–1582. DOI: 10.1029/2018WR024039

Forero, M. G., Herrera-Rivera, S., Ávila-Navarro, J., Franco, C. A., Rasmussen, J., & Nielsen, J. (2019). Color classification methods for perennial weed detection in cereal crops. In Progress in Pattern Recognition, Image Analysis, Computer Vision, and Applications (pp. 117–123).

Franco-Plata, R., Miranda-Vázquez, C., Solares-Hernández, H., Manzano-Solís, L. R., Bâ, K. M., & Expósito-Castillo, J. L. (2013). Implementing into GIS a Tool to Automate the Calculation of Physiographic Parameters of River Basins. *Open Journal of Modern Hydrology*, 03(02), 67–74. DOI: 10.4236/ojmh.2013.32009

Friedl, M. A., & Brodley, C. E. (1997). Decision tree classification of land cover from remotely sensed data. *Remote Sensing of Environment*, 61(3), 399–409. DOI: 10.1016/S0034-4257(97)00049-7

Fryberger, S. G. (1979). Dune form and wind regime. In McKee, E. D. (Ed.), U.S. Geological Survey Professional Paper: Vol. 1052. *A study of global sand sea* (pp. 137–169). Government Printing Office.

Fryberger, S. G. (1986). Dune form and wind regime. In McKee, M. E. (Ed.), U.S. Geological Survey Professional Paper: Vol. 1052. *Sand seas* (pp. 137–169).

Fuentes, A., Yoon, S., Kim, S., & Park, D. (2017). A robust deep-learning-based detector for real-time tomato plant diseases and pests recognition. *Sensors (Basel)*, 17(9), 17. DOI: 10.3390/s17092022 PMID: 28869539

Gabara, G., & Sawicki, P. (2023). CRBeDaSet: A Benchmark Dataset for High Accuracy Close Range 3D Object Reconstruction. *Remote Sensing (Basel)*, 15(4), 1116. DOI: 10.3390/rs15041116

Gagliolo, S., Ausonio, E., Federici, B., Ferrando, I., Passoni, D., & Sguerso, D. (2018). 3D cultural heritage documentation: a comparison between different photogrammetric software and their products. *The International Archives of the Photogrammetry. Remote Sensing and Spatial Information Sciences, XLII*, 2, 347–354. DOI: 10.5194/isprs-archives-XLII-2-347-2018

Gao, F., & Zhang, X. (2021). Mapping crop phenology in near real-time using satellite remote sensing: Challenges and opportunities. *Yaogan Xuebao*, 2021.

Gavrilović S (1972). Inzenjering o bujic nimtokovima i eroziji. Beograd. Serbia. *Izgradnja (Special Issue).*

Gavrilović, Z., Stefanovic, M., Milojevic, M., & Bacevac, J. (2006). Erosion Potential Method, An Important Support for Integrate Water Ressource Management. *Presented atXXIII Conference of the Danubian Countries on the Hydrological Forecasting and Hydrological Bases of Water Management,* Bled, Slovenia.

Geist, H. J., & Lambin, E. F. (2004). Dynamic causal patterns of desertification. *Bioscience*, 54(9), 817–829. DOI: 10.1641/0006-3568(2004)054[0817:DCPOD]2.0.CO;2

Gerhards, M., Schlerf, M., Mallick, K., & Udelhoven, T. (2019). Challenges and Future Perspectives of Multi-/Hyperspectral Thermal Infrared Remote Sensing for Crop Water-Stress Detection: A Review. *Remote Sensing (Basel)*, 11(10), 1240. DOI: 10.3390/rs11101240

Gervais-Lambony, P. (2001). « La ségrégation dans la grande ville, un essai de définition », in Gervais-Lambony Marie-Anne, Les très grandes villes dans le monde. p. 33-38.

Ghatrehsamani, S., Jha, G., Dutta, W., Molaei, F., Nazrul, F., Fortin, M., Bansal, S., Debangshi, U., & Neupane, J. (2023). Artificial Intelligence Tools and Techniques to Combat Herbicide Resistant Weeds—A Review. *Sustainability (Basel)*, 15(3), 1843. DOI: 10.3390/su15031843

Gherib, A., Ghazali, M. B., & Bouzid, S. (2023). Monitoring desertification in North Africa: Causes and solutions. *Environmental Science and Pollution Research International*, 30, 31395–31405. DOI: 10.1007/s11356-023-27883-9

Ghimire, B., Rogan, J., & Miller, J. (2010). Contextual land-cover classification: Incorporating spatial dependence in land-cover classification models using random forests and the Getis statistic. *Remote Sensing Letters*, 1(1), 45–54. DOI: 10.1080/01431160903252327

Giri, C. P. (Ed.). (2012). *Remote Sensing of Land Use and Land Cover - Principles and Applications* (1st ed.). CRC Press., DOI: 10.1201/b11964

Gitis, V., & Derendyaev, A. (2023). A Technology for Seismogenic Process Monitoring and Systematic Earthquake Forecasting. *Remote Sensing (Basel)*, 15(8), 2171. DOI: 10.3390/rs15082171

Glaeser, E. (2011). Triumph of The City, How our greatest invention makes us richer, smarter, greener, healthier, and happier. Ed. Penguin Books.

Globevnik, L., Holjevic, D., Petkovsek, G., & Rubinic, J. (2003). Applicability of the Gavrilovic method in erosion calculation using spatial data manipulation techniques, erosion prediction in Ungauged Basins: Integrating methods and techniques. *IAHS Publ*, (279), 224–233.

Goel, S., Pandey, S., & Kaur, S. (2019). Plants disease identification and classification through leaf images. *Archives of Computational Methods in Engineering*, 26(2), 507–530. DOI: 10.1007/s11831-018-9255-6

Gołaszewska, K. Ż., Fordoński, F. G., & Wiwart, M. (2009). Early diagnostics of macronutrient deficiencies in three legume species by color image analysis. *Computers and Electronics in Agriculture*, 65(1), 125–132. DOI: 10.1016/j.compag.2008.08.003

Gorelick, N., Hancher, M., Dixon, M., Ilyushchenko, S., Thau, D., & Moore, R. (2017). Google Earth Engine: Planetary-scale geospatial analysis for everyone. *Remote Sensing of Environment*, 202, 18–27. DOI: 10.1016/j.rse.2017.06.031

Grewal, K. K., Sehra, S. S., Singh, J., & Rai, H. S. (2020). An augmented reality based tool for contribution to OpenStreetMap. *International Journal of Information Technology : an Official Journal of Bharati Vidyapeeth's Institute of Computer Applications and Management*, 12(2), 345–352. DOI: 10.1007/s41870-019-00403-w

Grimm, A. (1980). Der Ursprung des Wortes PHOTOGRAMMETRIE. *International Archives of Photogrammetry, 23*(Part B10, Commission VI), 323–330.

Gui, J., Hao, L., Zhang, Q., & Bao, X. (2015). A new method for soybean leaf disease detection based on modified salient regions. *International Journal of Multimedia and Ubiquitous Engineering*, 10(6), 45–52. DOI: 10.14257/ijmue.2015.10.6.06

Gunasinghe, N., Shashi, K., Banga, S. S., You, M. P., Banga, S. K., & Barbetti, J. M. (2017). Outstanding host resistance will resolve the threat from white leaf spot disease (Pseudocercosporella capsellae) to oilseed and vegetable Brassica spp. crops. *Australasian Plant Pathology*, 46(2), 137–146. DOI: 10.1007/s13313-017-0470-7

Gunasinghe, N., You, M. P., Li, X. X., Banga, S. S., Banga, S. K., & Barbetti, M. J. (2016). New host resistances to Pseudocercosporella capsellae and implications for white leaf spot management in Brassicaceae crops. *Crop Protection (Guildford, Surrey)*, 86, 69–76. DOI: 10.1016/j.cropro.2016.04.014

Guo, L., Chehata, N., Mallet, C., & Boukir, S. (2011). Relevance of airborne lidar and multispectral image data for urban scene classification using Random Forests. *ISPRS Journal of Photogrammetry and Remote Sensing*, 66(1), 56–66. DOI: 10.1016/j.isprsjprs.2010.08.007

Haala, N., & Cavegn, S. (2016). High density aerial image matching: state-of-the-art and future prospects. *ISPRS - International Archives of the Photogrammetry. Remote Sensing and Spatial Information Sciences, XLI*, B4, 625–630. DOI: 10.5194/isprsarchives-XLI-B4-625-2016

Haghizadeh, A., Teang Shui, L., & Godarzi, E. (2009). Forecasting sediment with erosion potential method with emphasis on Land use changes at basin, electronic. *Journal of Geotechnical Engineering*, 14, 1–12.

Hairuddin, M. A., Tahir, N. M., & Rizam, N. S. (2011). Overview of image processing approach for nutrient deficiencies detection in Elaeis Guineensis. In IEEE.

Hakim, W. L., & Lee, C. W. (2020). A review on remote sensing and GIS applications to monitor natural disasters in Indonesia. *Korean Journal of Remote Sensing, 36*(6_1), 1303-1322. DOI: 10.7780/kjrs.2020.36.6.1.3

Hameed, Q. A., Hussein, H. A., Ahmed, M. A., & Omar, M. B. (2022). Development of Augmented Reality-based object recognition mobile application with Vuforia. *Journal of Algebraic Statistics*, 13(2), 2039–2046.

Hameed, S., & Amin, I. (2019). Detection of weed and wheat using image processing. In *Proceedings of the 2018 IEEE 5th International Conference on Engineering Technologies and Applied Sciences (ICETAS)* (pp. 1-5). IEEE.

Hansen, M. C., & DeFries, R. S. (2004). Detecting Long-term Global Forest Change Using Continuous Fields of Tree-Cover Maps from 8-km Advanced Very High Resolution Radiometer (AVHRR) Data for the Years 1982?99. *Ecosystems (New York, N.Y.)*, 7(7), 695–716. DOI: 10.1007/s10021-004-0243-3

Hansman, R. J., & Ring, U. (2019). Workflow: From photo-based 3-D reconstruction of remotely piloted aircraft images to a 3-D geological model. *Geosphere*, 15(4), 1393–1408. DOI: 10.1130/GES02031.1

Hassanien, A. E., Gaber, T., Mokhtar, U., & Hefny, H. (2017). An improved moth flame optimization algorithm based on rough sets for tomato diseases detection. *Computers and Electronics in Agriculture*, 136, 86–96. DOI: 10.1016/j.compag.2017.02.026

Hassani, F. (2015). Documentation of cultural heritage; techniques, potentials, and constraints. *The International Archives of the Photogrammetry. The International Archives of the Photogrammetry, Remote Sensing and Spatial Information Sciences*, 5(W7), 207–214. DOI: 10.5194/isprsarchives-XL-5-W7-207-2015

Haug, S., & Ostermann, J. (2015). A crop/weed field image dataset for the evaluation of computer vision based precision agriculture tasks. Lecture Notes in Computer Science (including Subseries Lecture Notes in Artificial Intelligence and Lecture Notes in Bioinformatics), 8928, 105–116. DOI: 10.1007/978-3-319-16220-1_8

He, L., Coburn, C. A., Wang, Z. J., Feng, W., & Guo, T. C. (2018). Reduced prediction saturation and view effects for estimating the leaf area index of winter wheat. *IEEE Transactions on Geoscience and Remote Sensing*, 57(3), 1637–1652. DOI: 10.1109/TGRS.2018.2868138

Hensesns, J. (1971). La villa comme habitat urbain. *Bulletin Economique et Social du Maroc*, 122.

Heusch B (1970). *L'érosion du pré-rif : une étude quantitative de l'érosion hydraulique dans les collines marneuses du pré-rif occidental.*

High Commission for Water and Forests and the Fight against Desertification — ERA-LEARN.

Hossain, M. D., & Chen, D. (2019). Segmentation for Object-Based Image Analysis (OBIA): A review of algorithms and challenges from remote sensing perspective. *ISPRS Journal of Photogrammetry and Remote Sensing*, 150, 115–134. DOI: 10.1016/j.isprsjprs.2019.02.009

Hosseini, F. S., Seo, M. B., Razavi-Termeh, S. V., Sadeghi-Niaraki, A., Jamshidi, M., & Choi, S. M. (2023). Geospatial Artificial Intelligence (GeoAI) and Satellite Imagery Fusion for Soil Physical Property Predicting. *Sustainability (Basel)*, 15(19), 14125. DOI: 10.3390/su151914125

Hu, Q., Yang, J., Xu, B., Huang, J., Memon, M. S., Yin, G., Zeng, Y., Zhao, J., & Liu, K. (2020). Evaluation of global decametric-resolution LAI, FAPAR and FVC estimates derived from Sentinel-2 imagery. *Remote Sensing (Basel)*, 12(6), 912. DOI: 10.3390/rs12060912

Ibrahim, A., Anayi, F., Packianather, M., & Al-Omari, O. (2021). Novel hybrid invasive weed optimization and machine learning approach for fault detection. In 2021 56th International Universities Power Engineering Conference (UPEC) (pp. 1-6). DOI: 10.1109/UPEC50034.2021.9548171

Islam, N., Rashid, M., Wibowo, S., Xu, C., & Morshed, A. (2021). Early weed detection using image processing and machine learning techniques in an Australian chilli farm.

Islam, M. R., Khan, M. N. I., Khan, M. Z., & Roy, B. (2021). A three decade assessment of forest cover changes in Nijhum dwip national park using remote sensing and GIS. *Environmental Challenges*, 4, 100162. DOI: 10.1016/j.envc.2021.100162

Issanova, G., Tussupbekova, G., & Rau, S. (2023). Exploring the complexities of sand dune transformation: The role of anthropogenic degradation and climatic conditions. *Frontiers in Environmental Science*, 11, 1089321. DOI: 10.3389/fenvs.2023.1089321

Jadhav, S. B., Udupi, V. R., & Patil, S. B. (2020). Identification of plant diseases using convolutional neural networks. *International Journal of Information Technology : an Official Journal of Bharati Vidyapeeth's Institute of Computer Applications and Management*. Advance online publication. DOI: 10.1007/s41870-020-00437-5

Jensen, J. R. (1996). *Introductory Digital Image Processing: A Remote Sensing Perspective* (2nd ed.). Prentice Hall.

Jeyalakshmi, S., & Radha, R. (2017). A review on diagnosis of nutrient deficiency symptoms in plant. *ICTACT Journal on Image and Video Processing*, 7(4), 1515–1524. DOI: 10.21917/ijivp.2017.0216

Jiachen, L., Mengqing, M., Xin, X., & Wenhui, R. (2021). The Spatial Effect of Shared Mobility on Urban Traffic Congestion: Evidence from Chinese Cities. *Sustainability (Basel)*, 13(24), 14065. DOI: 10.3390/su132414065

Jin, J., Rong, D., Pang, Y., Ye, P., Ji, Q., Wang, X., Wang, G., & Wang, F.-Y. (2022). An Agent-Based Traffic Recommendation System: Revisiting and Revising Urban Traffic Management Strategies. *IEEE Transactions on Systems, Man, and Cybernetics. Systems*, 52(11), 7289–7301. DOI: 10.1109/TSMC.2022.3177027

Jin, X., Che, J., & Chen, Y. (2021). Weed identification using deep learning and image processing in the vegetable plantation. *IEEE Access : Practical Innovations, Open Solutions*, 9, 10940–10950. DOI: 10.1109/ACCESS.2021.3050296

Joftic, L., Milliman, J., & Sestini, G. (1992). *Climate change and the Mediterranean, UNEPE*. Arnoid Pub.

Johannes, A., Picon, A., Gila, A. A., Echazarra, J., Vaamonde, S. R., & Navajas, A. D. (2017). Automatic plant disease diagnosis using mobile capture devices, applied on a wheat use case. *Computers and Electronics in Agriculture*, 138, 200–209. DOI: 10.1016/j.compag.2017.04.013

Junior, L. C. M., & Ulson, J. A. C. (2021). Real-time weed detection using computer vision and deep learning. In 2021 14th IEEE International Conference on Industry Applications (INDUSCON) (pp. 1131–1137). DOI: 10.1109/INDUSCON51756.2021.9529761

Justice, C. O., & Townshend, J. R. G. (1981). Integrating ground data with remote sensing. In Townshend, J. R. G. (Ed.), *Terrain Analysis and Remote Sensing* (pp. 38–58). Allen & Unwin.

Kaarthik, K., & Anitha, M. (2016). Analysis of nutrient requirement of crops using its leaf. *Journal of Chemical and Pharmaceutical Sciences*, 8, 1–5.

Kadobayashi, R., Kochi, N., Otani, H., & Furukawa, R. (2004). Comparison and evaluation of laser scanning and photogrammetry and their combined use for digital recording of cultural heritage. *The International Archives of the Photogrammetry, Remote Sensing and Spatial Information Sciences*, 35(5), 401–406.

Kakoti, M., & Sarma, H. H. (2024). Applications of geospatial and remote sensing technologies in management of natural resources. *Indian Farmer*, 11(07), 230–235.

Kalman, R. (1967). Essai d'évaluation pour le pré-Rif du facteur couverture végétale de la formule de Wischmeier de calcul de l'érosion. *Rapport Pour l'administration de La Forêt et d'eau, Rabat, 1*(12).

Kamath, R., Balachandra, M., & Prabhu, S. (2021). Classification of crop and weed from digital images: A review. In 2021 IEEE International Conference on Distributed Computing, VLSI, Electrical Circuits and Robotics (DISCOVER) (pp. 12–17). DOI: 10.1109/DISCOVER52564.2021.9663729

Kamath, R., Balachandra, M., & Prabhu, S. (2020). Paddy crop and weed discrimination: A multiple classifier system approach. *International Journal of Agronomy*, 2020, 1–14. Advance online publication. DOI: 10.1155/2020/6474536

Kaminsky, R. S., Snavely, N., Seitz, S. M., & Szeliski, R. (2009). Alignment of 3D point clouds to overhead images. *2009 IEEE Computer Society Conference on Computer Vision and Pattern Recognition Workshops*, 63–70. IEEE. DOI: 10.1109/CVPRW.2009.5204180

Kant, C., Kumar, G., & Meena, R. S. (2023). Modeling morphometric and geomorphological parameters of mountainous river basin for water resource management using remote sensing and GIS approach. *Modeling Earth Systems and Environment*, 9(2), 2151–2163. DOI: 10.1007/s40808-022-01614-0

Kareem, I. J., Jasim, G. S., Ali, H. A., & Amin, R. M. (2024). Estimating the Extent of Water Erosion in Darbandikhan Lake Using a Model Gavrilović Method (EPM) (Erosion Potential Method). *International Journal of Religion*, 5(9), 358–369. DOI: 10.61707/neq6c157

Karimi, H., Ghadirifaraz, B., Shetab Boushehri, S. N., Hosseininasab, S.-M., & Rafiei, N. (2022). Reducing traffic congestion and increasing sustainability in special urban areas through one-way traffic reconfiguration. *Transportation*, 49(1), 37–60. DOI: 10.1007/s11116-020-10162-4

Karmaoui, A., Moumane, A., El Jaafari, S., Menouni, A., Al Karkouri, J., Yacoubi, M., & Hajji, L. (2023). Thirty Years of Change in the Land Use and Land Cover of the Ziz Oases (Pre-Sahara of Morocco) Combining Remote Sensing, GIS, and Field Observations. *Land (Basel)*, 12(12), 2127. DOI: 10.3390/land12122127

Kayimierski, L. D., Irigoyen, M., Re, M., Menendey, A. N., Spalletti, P., & Brea, J. D. (2013). Impact of climate change on sediment yield from the upper Plata basin. *International Journal of River Basin Management*, 11, 1–11.

Kazhdan, M., & Hoppe, H. (2013). Screened poisson surface reconstruction. *ACM Transactions on Graphics*, 32(3), 1–13. DOI: 10.1145/2487228.2487237

Kedron, P., Zhao, Y., & Frazier, A. E. (2019). Three dimensional (3D) spatial metrics for objects. *Landscape Ecology*, 34(9), 2123–2132. DOI: 10.1007/s10980-019-00861-4

Khanal, S., Fulton, J., & Shearer, S. (2017). An overview of current and potential applications of thermal remote sensing in precision agriculture. *Computers and Electronics in Agriculture*, 139, 22–32. DOI: 10.1016/j.compag.2017.05.001

Khan, U., Khan, M. K., Latif, M. A., Naveed, M., Alam, M. M., Khan, S. A., & Su'ud, M. M. (2024). A Systematic Literature Review of Machine Learning and Deep Learning Approaches for Spectral Image Classification in Agricultural Applications Using Aerial Photography. *Computers, Materials & Continua*, 78(3), 2967–3000. DOI: 10.32604/cmc.2024.045101

Kim, T., & Im, Y.-J. (2003). Automatic satellite image registration by combination of matching and random sample consensus. *IEEE Transactions on Geoscience and Remote Sensing*, 41(5), 1111–1117. DOI: 10.1109/TGRS.2003.811994

Kingsland, K. (2020). Comparative analysis of digital photogrammetry software for cultural heritage. *Digital Applications in Archaeology and Cultural Heritage*, 18, e00157. DOI: 10.1016/j.daach.2020.e00157

Kinyua, J., & Awuah, L. (2021). AI/ML in security orchestration, automation and response: Future research directions. *Intelligent Automation & Soft Computing*, 28(2), 527–545. DOI: 10.32604/iasc.2021.016240

Kok, Z. H., Mohamed Shariff, A. R., Alfatni, M. S. M., & Khairunniza-Bejo, S. (2021). Support Vector Machine in Precision Agriculture: A review. *Computers and Electronics in Agriculture*, 191, 106546. DOI: 10.1016/j.compag.2021.106546

Kraak, M.-J. (2003). Geovisualization illustrated. *ISPRS Journal of Photogrammetry and Remote Sensing*, 57(5–6), 390–399. DOI: 10.1016/S0924-2716(02)00167-3

Kraak, M.-J. (2009). Geovisualization. In *International Encyclopedia of Human Geography* (pp. 468–480). Elsevier., DOI: 10.1016/B978-008044910-4.00033-X

Kucharczyk, M., Hay, G. J., Ghaffarian, S., & Hugenholtz, C. H. (2020). Geographic object-based image analysis: A primer and future directions. *Remote Sensing (Basel)*, 12(12), 2012. DOI: 10.3390/rs12122012

Kulwant, M., & Patel, D. (2024). Application of Remote Sensing, GIS, and AI Techniques in the Agricultural Sector. In S. Gupta, S. K. Himanshu, & P. K. Gupta, *Agri-Tech Approaches for Nutrients and Irrigation Water Management* (1st ed., pp. 248–271). CRC Press. DOI: 10.1201/9781003441175-15

Kumar, A., Chattopadhyay, C., Awasthi, R. P., & Meena, P. D. (2010). Alternaria blight: A chronic disease in rapeseed-mustard. *Journal of Oilseed Brassica*, 1(1), 1–11.

Kumar, G., & Kalra, R. (2016). A survey on machine learning techniques in health care industry. *International Journal of Recent Research Aspects*, 3(2), 128–132.

Kumar, S. K., & Babu, S. D. B. (2016). A web GIS based decision support system for agriculture crop monitoring system-A case study from part of Medak District. *J Remote Sensing & GIS*, 5(4), 177–197.

Kumar, V., Lehri, S., Sharma, A. K., Meena, P. D., & Kumar, A. (2008). Image based rapeseed-mustard disease expert system: An effective extension tool. *Indian Research Journal of Extension Education*, 8(2&3), 10–13.

Lague, D., Brodu, N., & Leroux, J. (2013). Accurate 3D comparison of complex topography with terrestrial laser scanner: Application to the Rangitikei canyon (N-Z). *ISPRS Journal of Photogrammetry and Remote Sensing*, 82, 10–26. DOI: 10.1016/j.isprsjprs.2013.04.009

Lakra, D., Singh, S. K., Gupta, S. K., & Kanga, S. (2024). Enhancing integrated resource management through remote sensing and GIS. *Journal of Geography and Cartography*, 7(1), 4265. DOI: 10.24294/jgc.v7i1.4265

Lal, H., Meena, P., Kumar, M., & Jha, P. (2013). Dynamics and management of Alternaria blight disease of Indian mustard (Brassica juncea) in relation to weather parameters. *Journal of Oilseed Brassica*, 4(2), 66–74.

Lambin, E. F., & Meyfroidt, P. (2011). Global land use change, economic globalization, and the looming land scarcity. *Proceedings of the National Academy of Sciences of the United States of America*, 108(9), 3465–3472. DOI: 10.1073/pnas.1100480108 PMID: 21321211

Lameski, P., Zdravevski, E., Trajkovik, V., & Kulakov, A. (2017). Weed detection dataset with RGB images taken under variable light conditions. In ICT Innovations 2017 (pp. 112–119). DOI: 10.1007/978-3-319-67597-8_11

Lancaster, N. (1983). Controls of dune morphology in the Namib sand sea. In *Developments in sedimentology* (Vol. 38, pp. 261–289). Elsevier., DOI: 10.1177/030913338901300105

Lane, L. J., Hernandez, M., & Nichols, M. (1997). Processes controlling sediment yield from watersheds as functions of spatial scale. *Environmental Modelling & Software*, 12(4), 355–369. DOI: 10.1016/S1364-8152(97)00027-3

Latifi, H., Nothdurft, A., & Koch, B. (2010). Non-parametric prediction and mapping of standing timber volume and biomass in a temperate forest: Application of multiple optical/LiDAR-derived predictors. *Forestry*, 83(4), 395–407. DOI: 10.1093/forestry/cpq022

Lavania, S., & Matey, P. S. (2015). Novel method for weed classification in maize field using otsu and PCA implementation. In *2015 IEEE International Conference on Computational Intelligence & Communication Technology* (pp. 534-537). IEEE. DOI: 10.1109/CICT.2015.71

Lazarevic, R. (1985). The new method for erosion coefficient determination – Z. *Erosion Bulletin*, 13, 54–61.

Le Bissonnais, Y., Dubreuil, N., Daroussin, J., & Gorce, M. (2005). Modélisation et cartographie de l'aléa d'érosion des sols à l'échelle régionale. *Etude et Gestion Des Sols*, 11, 307–321.

Lee, J.-H., & Sull, S. (2019). Regression Tree CNN for Estimation of Ground Sampling Distance Based on Floating-Point Representation. *Remote Sensing (Basel)*, 11(19), 2276. DOI: 10.3390/rs11192276

Li, J., Chen, D., Yin, X., & Li, Z. (2024). Performance evaluation of semi-supervised learning frameworks for multi-class weed detection. *Frontiers in Plant Science*, 15, 1396568. DOI: 10.3389/fpls.2024.1396568 PMID: 39228840

Lin, F.-C., Chung, L.-K., Wang, C.-J., Ku, W.-Y., & Chou, T.-Y. (2013). Storage and processing of massive remote sensing images using a novel cloud computing platform. *GIScience & Remote Sensing*, 50(3), 322–336. DOI: 10.1080/15481603.2013.810976

Lin, F., Zhang, D., Huang, Y., Wang, X., & Chen, X. (2017). Detection of Corn and Weed Species by the Combination of Spectral, Shape and Textural Features. *Sustainability (Basel)*, 9(8), 1335. DOI: 10.3390/su9081335

Liu, B., Zhang, Y., He, D., & Li, Y. (2018). Identification of apple leaf diseases based on deep convolutional neural networks. *Symmetry*, 11(10).

Liu, J., Kang, H., Tao, W., Li, H., He, D., Ma, L., Tang, H., Wu, S., Yang, K., & Li, X. (2023). A spatial distribution – Principal component analysis (SD-PCA) model to assess pollution of heavy metals in soil. *The Science of the Total Environment*, 859, 160112. DOI: 10.1016/j.scitotenv.2022.160112 PMID: 36375553

Liu, N., Budkewitsch, P., & Treitz, P. (2017). Examining spectral reflectance features related to Arctic percent vegetation cover: Implications for hyperspectral remote sensing of Arctic tundra. *Remote Sensing of Environment*, 192, 58–72. DOI: 10.1016/j.rse.2017.02.002

Liu, Q., Liu, Z., Xu, Y., Xiong, W., Yang, J., & Wang, Q. (2020). Toward Identifying the Urban Community Structure from Population Flow and Public Services Distribution. *2020 IEEE International Conference on Big Data (Big Data)*, 3133–3142. DOI: 10.1109/BigData50022.2020.9378225

Livingstone, I. (1989). Monitoring surface change on a Namib linear dune. *Earth Surface Processes and Landforms*, 14(4), 317–332. DOI: 10.1002/esp.3290140407

Li, X., Zeng, R., & Liao, H. (2016). Improving crop nutrient efficiency through root architecture modifications. *Journal of Integrative Plant Biology*, 58(3), 193–202. DOI: 10.1111/jipb.12434 PMID: 26460087

Li, Y., Al-Sarayreh, M., Irie, K., Hackell, D., Bourdot, G., Reis, M. M., & Ghamkhar, K. (2021). Identification of weeds based on hyperspectral imaging and machine learning. *Frontiers in Plant Science*, 11(January), 1–13. DOI: 10.3389/fpls.2020.611622 PMID: 33569069

Long, N. Q., Goyal, R., Bui, L. K., Cuong, C. X., Van Canh, L., Minh, N. Q., & Bui, X. N. (2023). Optimal Choice of the Number of Ground Control Points for Developing Precise DSM Using Light-Weight UAV in Small and Medium-Sized Open-Pit Mine. *Archives of Mining Sciences*. https://doi.org/DOI: 10.24425/ams.2021.138594

Loulad, S., Nguyen, T. T., Simou, M. R., Rhinane, H., & Buerkert, A. (2023). Monitoring rural-urban transformation in the coastal region of Rabat-Sale-Kenitra, Morocco. *PLoS One*, 18(8), e0290829. DOI: 10.1371/journal.pone.0290829 PMID: 37651348

Loveland, T. R., & Belward, A. S. (1997). The International Geosphere Biosphere Programme Data and Information System global land cover data set (DISCover). *Acta Astronautica, 41*(4–10), 681–689. DOI: 10.1016/S0094-5765(98)00050-2

Lu, B., Dao, P. D., Liu, J., He, Y., & Shang, J. (2020). Recent advances of hyperspectral imaging technology and applications in agriculture. *Remote Sensing (Basel)*, 12(16), 2659. DOI: 10.3390/rs12162659

Luo, K., Pan, H., Zhang, Y., & Guan, T. (2019). Partial bundle adjustment for accurate three-dimensional reconstruction. *IET Computer Vision*, 13(7), 666–675. DOI: 10.1049/iet-cvi.2018.5564

Majumdar, D., Chakraborty, A., Kole, D., & Majumdar, D. (2015). An integrated digital image analysis system for detection, recognition and diagnosis of disease in wheat leaves. In *Symposium on Women in Computing and Informatics*. New York. DOI: 10.1145/2791405.2791474

Malik, V., Mittal, R., Kaur, A., Singla, G., Mittal, A., & Singh, M. (2024). Enhancement and analysis of hyperspectral satellite images for Soil Study and Behavior. *Multimedia Tools and Applications*, 83(11), 33879–33902. DOI: 10.1007/s11042-023-16729-4

Manaouch, M., Zouagui, A., & Fenjiro, I. (2021). A review of soil erosion modeling by R/USLE in Morocco: Achievements and limits. In *E3S Web of Conferences* (Vol. 234, p. 00067). EDP Sciences. DOI: 10.1051/e3sconf/202123400067

Marais-Sicre, C., Queguiner, S., Bustillo, V., Lesage, L., Barcet, H., Pelle, N., Breil, N., & Coudert, B. (2024). Sun/Shade Separation in Optical and Thermal UAV Images for Assessing the Impact of Agricultural Practices. *Remote Sensing (Basel)*, 16(8), 1436. DOI: 10.3390/rs16081436

Marelli, D., Morelli, L., Farella, E. M., Bianco, S., Ciocca, G., & Remondino, F. (2023). ENRICH: Multi-purposE dataset for beNchmaRking In Computer vision and pHotogrammetry. *ISPRS Journal of Photogrammetry and Remote Sensing*, 198, 84–98. DOI: 10.1016/j.isprsjprs.2023.03.002

Martino, P., & Panagiotis, P. (2023). *GHS-BUILT-S R2023A - GHS built-up surface grid, derived from Sentinel2 composite and Landsat, multitemporal (1975-2030)*. European Commission, Joint Research Centre. https://doi.org/DOI: 10.2905/9F06F36F-4B11-47EC-ABB0-4F8B7B1D7

Marzouk, M., Attia, K., & Azab, S. (2021). Assessment of coastal vulnerability to climate change impacts using GIS and remote sensing: A case study of Al-Alamein New City. *Journal of Cleaner Production*, 290, 125723. DOI: 10.1016/j.jclepro.2020.125723

Ma, S., Liu, S., Gao, Z., Wang, X., Ma, S., & Wang, S. (2024). Water Deficit Diagnosis of Winter Wheat Based on Thermal Infrared Imaging. *Plants*, 13(3), 361. DOI: 10.3390/plants13030361 PMID: 38337894

Masrour, A. (2012). *Le Jurassique moyen-supérieur du Rif externe et l'avant pays oriental: Stratigraphie, sédimentologie et modélisations paléogéographiques* (9783838181042nd ed.).

Mastrorocco, G., Salvini, R., Esposito, G., & Seddaiu, M. (2016). 3D point cloud analysis for surface roughness measurement: Application of UAV photogrammetry. *Rendiconti Online Della Società Geologica Italiana*, 41, 313–316. DOI: 10.3301/ROL.2016.156

Maurer, G. (1968). *Les montagnes du Rif central. Étude géomorphologique*. Université Paris-Diderot.

Maurya, R. K., & Kulkarni, S. T. (2021). Multidimensional spatial clustering and visualization of 3D topographic relief data. *International Journal of Information Technology : an Official Journal of Bharati Vidyapeeth's Institute of Computer Applications and Management*, 13(2), 581–592. DOI: 10.1007/s41870-020-00595-6

Ma, X., Chen, C., Chen, P., Tain, C., Lin, Z., Zhuang, L., & Zou, X. (2019). Fully convolutional network for rice seedling and weed image segmentation at the seedling stage in paddy fields. *PLoS One*, 14(4), e0215676. Advance online publication. DOI: 10.1371/journal.pone.0215676 PMID: 30998770

McElwee, P. (2009). Reforesting "Bare Hills" in Vietnam: Social and Environmental Consequences of the 5 Million Hectare Reforestation Program. *Ambio*, 38(6), 325–333. DOI: 10.1579/08-R-520.1 PMID: 19860156

Meena, P. D., Awasthi, R. P., Godika, S., Gupta, J. C., Sandhu, P. S., & Kumar, A. (2011). Eco-friendly approaches managing major diseases of Indian mustard. *World Applied Sciences Journal*, 12(8), 1192–1195.

Mehravar, S., Razavi-Termeh, S. V., Moghimi, A., Ranjgar, B., Foroughnia, F., & Amani, M. (2023). Flood susceptibility mapping using multi-temporal SAR imagery and novel integration of nature-inspired algorithms into support vector regression. *Journal of Hydrology (Amsterdam)*, 617, 129100. DOI: 10.1016/j.jhydrol.2023.129100

Merlin, P., & Choay, F. (2010). Dictionnaire de l'Urbanisme et de l'Aménagement. (Ed. Quadrige / Presses Universitaires de France (PUF), 3e édition.

Merzouk A, Bonn F, Badraoui M, Benkhaiti M, Tahri M (1992). Développement des SIG pour la mise en valeur agricole des terres arides et semi-arides : la problématique et les attentes. *Géo. Observateur. 2*, 77–84.

Mesrar, L., Akdim, M., Lakrim, M., Aroussi, O. E., Akhrif, I., & Jabrane, R. (2015). The physico-mechanical characteristics of the clays in and after doping with metal oxide Al2O3 in The region of Fez (Morocco). *Present Environment and Sustainable Development*, 9(1), 103–110. DOI: 10.1515/pesd-2015-0006

Metashape. (2024). Agisoft. Retrieved March 27, 2024, from https://www.agisoft.com/

Meydenbauer, A. (1867). Die photometrographie. *Wochenblatt Des Architektenvereins Zu Berlin*, 1(14), 125–126.

Meyfroidt, P., & Lambin, E. F. (2008). The causes of the reforestation in Vietnam. *Land Use Policy*, 25(2), 182–197. DOI: 10.1016/j.landusepol.2007.06.001

Michalski, R. S., Carbonell, J. G., & Mitchell, T. M. (Eds.). (2013). *Machine learning: An artificial intelligence approach*. Springer Science & Business Media.

MicMac. (2024). MicMac. Retrieved March 27, 2024, from https://micmac.ensg.eu/

Milenkovic, M., & Ressl, C. (2013). Surface Roughness from Point Clouds-A Multi-Scale Analysis. *EGU General Assembly*, 15, 13570.

Milevski, I. (2015). General geomorphological characteristics of the Republic of Macedonia. *Geographical Reviews*, 48, 5–25.

Ministry of Electronics and Information Technology. (2019, January 27). Mustard: Diseases and symptoms. Retrieved February 10, 2019, from https://vikaspedia.in/agriculture/crop-production/integrated-pest-managment/ipm-for-oilseeds/ipm-strategies-for-mustard-rapeseed/diseases-and-symptoms

Miraglio, T., Adeline, K., Huesca, M., Ustin, S., & Briottet, X. (2019). Monitoring LAI, chlorophylls, and carotenoids content of a woodland savanna using hyperspectral imagery and 3D radiative transfer modelling. *Remote Sensing (Basel)*, 12(1), 28. DOI: 10.3390/rs12010028

Moazzam, S. I., Khan, U. S., Qureshi, W. S., Tiwana, M. I., Rashid, N., Alasmary, W. S., Iqbal, J., & Hamza, A. (2021). A patch-image based classification approach for detection of weeds in sugar beet crop. *IEEE Access : Practical Innovations, Open Solutions*, 9, 121698–121715. DOI: 10.1109/ACCESS.2021.3109015

Mohamed, E. S., Baroudy, A. A. E., El-Beshbeshy, T., Emam, M., Belal, A. A., Elfadaly, A., & Lasaponara, R. (2020). Vis-nir spectroscopy and satellite landsat-8 oli data to map soil nutrients in arid conditions: A case study of the northwest coast of egypt. *Remote Sensing (Basel)*, 12(22), 3716. DOI: 10.3390/rs12223716

Mohamed, E. S., Saleh, A. M., Belal, A. B., & Gad, A. (2018). Application of near-infrared reflectance for quantitative assessment of soil properties. *The Egyptian Journal of Remote Sensing and Space Sciences*, 21(1), 1–14. DOI: 10.1016/j.ejrs.2017.02.001

Mohammed Almatar, K. (2023). Traffic congestion patterns in the urban road network: (Dammam metropolitan area). *Ain Shams Engineering Journal*, 14(3), 101886. DOI: 10.1016/j.asej.2022.101886

Mohanty, S. P., Hughes, D. P., & Salathé, M. (2016). Using deep learning for image-based plant disease detection. *Frontiers in Plant Science*, 7, 1419. DOI: 10.3389/fpls.2016.01419 PMID: 27713752

Morgan, R. P. C. (1986). Soil erosion and conservation. *Rev. and Enl. Ed. of: Soil Erosion*, 256–295.

Morin, M., Lawrence, R., Repasky, K., Sterling, T., McCann, C., & Powell, S. (2017). Agreement analysis and spatial sensitivity of multispectral and hyperspectral sensors in detecting vegetation stress at management scales. *Journal of Applied Remote Sensing*, 11(4), 046025. DOI: 10.1117/1.JRS.11.046025

Mostafa, E., Li, X., Sadek, M., & Dossou, J. F. (2021). Monitoring and forecasting of urban expansion using machine learning-based techniques and remotely sensed data: A case study of gharbia governorate, Egypt. *Remote Sensing (Basel)*, 13(22), 4498. DOI: 10.3390/rs13224498

Moukhchane, M., Bouhiassa, S., & Chalouan, A. (1998). Cartographic and magnetic approach for the identification of sediment sources: case of the Nakhla watershed (Rif, Morocco). Science and Planetary Change/Drought, 9 (3), 227-232.

Moumane, A., Al Karkouri, J., Benmansour, A., El Ghazali, F. E., Fico, J., Karmaoui, A., & Batchi, M. (2022). Monitoring long-term land use, land cover change, and desertification in the Ternata oasis, Middle Draa Valley, Morocco. *Remote Sensing Applications: Society and Environment*, 26, 100745. DOI: 10.1016/j.rsase.2022.100745

Moumane, A., El Ghazali, F. E., Al Karkouri, J., Delorme, J., Batchi, M., Chafiki, D., & Karmaoui, A. (2021). Monitoring spatiotemporal variation of groundwater level and salinity under land use change using integrated field measurements, GIS, geostatistical, and remote-sensing approach: Case study of the Feija aquifer, Middle Draa watershed, Moroccan Sahara. *Environmental Monitoring and Assessment*, 193(12), 769. DOI: 10.1007/s10661-021-09581-2 PMID: 34735624

Mountrakis, G., Watts, R., Luo, L., & Wang, J. (2009). Developing Collaborative Classifiers using an Expert-based Model. *Photogrammetric Engineering and Remote Sensing*, 75(7), 831–843. DOI: 10.14358/PERS.75.7.831

Müllerová, J., Brundu, G., Große-Stoltenberg, A., Kattenborn, T., & Richardson, D. M. (2023). Pattern to process, research to practice: Remote sensing of plant invasions. *Biological Invasions*, 25(12), 3651–3676. DOI: 10.1007/s10530-023-03150-z

Murawwat, S., & Ahmad, S. (2018). Weed detection using SVMs. *International Journal of Engineering and Advanced Technology*, 8(1), 2412–2416.

Murtiyoso, A., Grussenmeyer, P., Börlin, N., Vandermeerschen, J., & Freville, T. (2018). Open Source and Independent Methods for Bundle Adjustment Assessment in Close-Range UAV Photogrammetry. *Drones (Basel)*, 2(1), 3. DOI: 10.3390/drones2010003

Nandhini, S., Hemalatha, R., Radha, S., & Indumathi, K. (2018). Web enabled plant disease detection system for agricultural applications using WMSN. *Wireless Personal Communications*, 102(2), 725–740. DOI: 10.1007/s11277-017-5092-4

Narkhede, U. P., & Adhiya, K. P. (2014). Evaluation of Modified K-Means Clustering Algorithm in Crop Prediction. *International Journal of Advanced Computer Research*, 4(3), 799–807.

Nasraoui, R., & Benzidiya, K. (2024) The Impact of Industrial Activities on the Land Surface Temperature (LST) in the Urban Environment of the City of Kenitra (Morocco). In: Kahime K, El Yamani M, Pouffary S (eds) Climate Change Effects and Sustainability Needs. Springer Nature Switzerland, Cham, pp 163–180https:// DOI: 10.1007/978-3-031-59603-2_11

Nebel, S., Beege, M., Schneider, S., & Rey, G. D. (2020). A Review of Photogrammetry and Photorealistic 3D Models in Education From a Psychological Perspective. *Frontiers in Education*, 5, 144. DOI: 10.3389/feduc.2020.00144

Neinavaz, E., Schlerf, M., Darvishzadeh, R., Gerhards, M., & Skidmore, A. K. (2021). Thermal infrared remote sensing of vegetation: Current status and perspectives. *International Journal of Applied Earth Observation and Geoinformation*, 102, 102415. DOI: 10.1016/j.jag.2021.102415

Nelson, E. S. (1996). How Maps Work: Representation, Visualization, and Design. *Cartographic Perspectives*, (24), 27–30. DOI: 10.14714/CP24.757

Nex, F., Gerke, M., Remondino, F., Przybilla, H.-J., Bäumker, M., & Zurhorst, A. (2015). ISPRS BENCHMARK FOR MULTI-PLATFORM PHOTOGRAMMETRY. *ISPRS Annals of the Photogrammetry. Remote Sensing and Spatial Information Sciences, II*, 3(W4), 135–142. DOI: 10.5194/isprsannals-II-3-W4-135-2015

Nex, F., & Remondino, F. (2014). UAV for 3D mapping applications: A review. *Applied Geomatics*, 6(1), 1–15. DOI: 10.1007/s12518-013-0120-x

Neyns, R., & Canters, F. (2022). Mapping of urban vegetation with high-resolution remote sensing: A review. *Remote Sensing (Basel)*, 14(4), 1031. DOI: 10.3390/rs14041031

Niederheiser, R., Mokroš, M., Lange, J., Petschko, H., Prasicek, G., & Elberink, S. O. (2016). DERIVING 3D POINT CLOUDS FROM TERRESTRIAL PHOTOGRAPHS - COMPARISON OF DIFFERENT SENSORS AND SOFTWARE. *ISPRS - International Archives of the Photogrammetry. Remote Sensing and Spatial Information Sciences, XLI*, B5, 685–692. DOI: 10.5194/isprsarchives-XLI-B5-685-2016

Nikolov, I., & Madsen, C. (2016). Benchmarking Close-range Structure from Motion 3D Reconstruction Software Under Varying Capturing Conditions. In *Lecture Notes in Computer Science (including subseries Lecture Notes in Artificial Intelligence and Lecture Notes in Bioinformatics)* (pp. 15–26). DOI: 10.1007/978-3-319-48496-9_2

Nizeyimana, E., & Petersen, G. W. (2020). Remote sensing applications to soil degradation assessments. In *Methods for assessment of soil degradation* (pp. 393–405). CRC Press. DOI: 10.1201/9781003068716-20

Noi Phan, T., Kuch, V., & Lehnert, L. W. (2020). Land cover classification using google earth engine and random forest classifier-the role of image composition. *Remote Sensing (Basel)*, 12(15), 2411. Advance online publication. DOI: 10.3390/rs12152411

Okhrimenko, M., Coburn, C., & Hopkinson, C. (2019). Multi-spectral lidar: Radiometric calibration, canopy spectral reflectance, and vegetation vertical SVI profiles. *Remote Sensing (Basel)*, 11(13), 1556. DOI: 10.3390/rs11131556

Oliensis, J. (2000). A Critique of Structure-from-Motion Algorithms. *Computer Vision and Image Understanding*, 80(2), 172–214. DOI: 10.1006/cviu.2000.0869

Oniga, V.-E., Breaban, A.-I., & Statescu, F. (2018). Determining the Optimum Number of Ground Control Points for Obtaining High Precision Results Based on UAS Images. *International Electronic Conference on Remote Sensing*, 352. MDPI. DOI: 10.3390/ecrs-2-05165

Padró, J.-C., Muñoz, F.-J., Planas, J., & Pons, X. (2019). Comparison of four UAV georeferencing methods for environmental monitoring purposes focusing on the combined use with airborne and satellite remote sensing platforms. *International Journal of Applied Earth Observation and Geoinformation*, 75, 130–140. DOI: 10.1016/j.jag.2018.10.018

Pádua, L., Adão, T., Hruška, J., Marques, P., Sousa, A., Morais, R., & Peres, E. (2018). UAS-based photogrammetry of cultural heritage sites. *Proceedings of the International Conference on Geoinformatics and Data Analysis*, 72–76. New York, NY, USA: ACM. DOI: 10.1145/3220228.3220243

Pahlevan, N., Smith, B., Binding, C., & O'Donnell, D. M. (2017). Spectral band adjustments for remote sensing reflectance spectra in coastal/inland waters. *Optics Express*, 25(23), 28650–28667. DOI: 10.1364/OE.25.028650

Papadopoulou, E.-E., Vasilakos, C., Zouros, N., & Soulakellis, N. (2021). DEM-Based UAV Flight Planning for 3D Mapping of Geosites: The Case of Olympus Tectonic Window, Lesvos, Greece. *ISPRS International Journal of Geo-Information*, 10(8), 535. DOI: 10.3390/ijgi10080535

Parra, L. (2022). Remote sensing and GIS in environmental monitoring. *Applied Sciences (Basel, Switzerland)*, 12(16), 8045. DOI: 10.3390/app12168045

Pavlova-Traykova, E. (2024). ASSESSMENT SOIL EROSION RISK IN FOREST TERRITORIES IN SELSKA RIVER. Forest Science (0861-007X), 60.

People's Committee of Ba Ria - Vung Tau Province. (2023). *Comprehensive Report on the Provincial Planning of Ba Ria - Vung Tau for the Period 2021-2030, Vision to 2050.*

Pepe, M., Fregonese, L., & Scaioni, M. (2018). Planning airborne photogrammetry and remote-sensing missions with modern platforms and sensors. *European Journal of Remote Sensing*, 51(1), 412–436. DOI: 10.1080/22797254.2018.1444945

Pérez-Cutillas, P., Pérez-Navarro, A., Conesa-García, C., Zema, D. A., & Amado-Álvarez, J. P. (2023). What is going on within google earth engine? A systematic review and meta-analysis. *Remote Sensing Applications: Society and Environment, 29*(September 2022), 100907. DOI: 10.1016/j.rsase.2022.100907

Phadikar, S., Sil, J., & Kumar, A. D. (2013). Rice diseases classification using feature selection and rule generation techniques. *Computers and Electronics in Agriculture*, 90, 76–85. DOI: 10.1016/j.compag.2012.11.001

Phan, D. C., Trung, T. H., Truong, V. T., Sasagawa, T., Vu, T. P. T., Bui, D. T., Hayashi, M., Tadono, T., & Nasahara, K. N. (2021). First comprehensive quantification of annual land use/cover from 1990 to 2020 across mainland Vietnam. *Scientific Reports*, 11(1), 1–20. DOI: 10.1038/s41598-021-89034-5 PMID: 33976255

PIX4D. (2024). PIX4D. Retrieved March 27, 2024, from https://www.pix4d.com/

Pôças, I., Calera, A., Campos, I., & Cunha, M. (2020). Remote sensing for estimating and mapping single and basal crop coefficients: A review on spectral vegetation indices approaches. *Agricultural Water Management*, 233, 106081. DOI: 10.1016/j.agwat.2020.106081

Ponto, K., & Tredinnick, R. (2022). Opportunities for utilizing consumer grade 3D capture tools for insurance documentation. *International Journal of Information Technology : an Official Journal of Bharati Vidyapeeth's Institute of Computer Applications and Management*, 14(6), 2757–2766. DOI: 10.1007/s41870-022-01040-6

Potapov, P., Hansen, M. C., Pickens, A., Hernandez-Serna, A., Tyukavina, A., Turubanova, S., Zalles, V., Li, X., Khan, A., Stolle, F., Harris, N., Song, X.-P., Baggett, A., Kommareddy, I., & Kommareddy, A. (2022). The Global 2000-2020 Land Cover and Land Use Change Dataset Derived From the Landsat Archive: First Results. *Frontiers in Remote Sensing*, 3(April), 1–22. DOI: 10.3389/frsen.2022.856903

Pour, A. B., Zoheir, B., Pradhan, B., & Hashim, M. (2021). Editorial for the special issue: Multispectral and hyperspectral remote sensing data for mineral exploration and environmental monitoring of mined areas. *Remote Sensing (Basel)*, 13(3), 519. DOI: 10.3390/rs13030519

Prajwal, V., Abhay, R., Rohith Gowda, D., & Savitha, G. (2023). AI-Driven Urban Traffic Optimization to Assess Complex Traffic Patterns for Public Traffic Control and Mobility. *International Journal for Research in Applied Science and Engineering Technology*, 11(11), 794–798. DOI: 10.22214/ijraset.2023.56630

Prudente, V. H. R., Skakun, S., Oldoni, L. V., Xaud, H. A., Xaud, M. R., Adami, M., & Sanches, I. D. A. (2022). Multisensor approach to land use and land cover mapping in Brazilian Amazon. *ISPRS Journal of Photogrammetry and Remote Sensing*, 189, 95–109. DOI: 10.1016/j.isprsjprs.2022.04.025

Pujol, B. (2019). *Machine learning for early detection of traffic congestion using public transport traffic data*. https://api.semanticscholar.org/CorpusID:199011854

Rahman, M. N. (2019). Urban Expansion Analysis and Land Use Changes in Rangpur City Corporation Area, Bangladesh, using Remote Sensing (RS) and Geographic Information System (GIS) Techniques. *Geosfera Indonesia*, 4(3), 217. DOI: 10.19184/geosi.v4i3.13921

Rangel, B. M. S., Fernández, M. A. A., Murillo, J. C., Ortega, J. C. P., & Arreguín, J. M. R. (2016, February). KNN-based image segmentation for grapevine potassium deficiency diagnosis. In 2016 International conference on electronics, communications and computers (CONIELECOMP) (pp. 48-53). IEEE.

Rani, S. V. J., Kumar, P. S., Priyadharsini, R., Srividya, S. J., & Harshana, S. (2022). Automated weed detection system in smart farming for developing sustainable agriculture. *International Journal of Environmental Science and Technology*, 19(9), 9083–9094. DOI: 10.1007/s13762-021-03606-6

Remondino, F., Spera, M. G., Nocerino, E., Menna, F., & Nex, F. (2014). State of the art in high density image matching. *The Photogrammetric Record*, 29(146), 144–166. DOI: 10.1111/phor.12063

Reulke, R., & Döring, T. (2006). Computer Vision im Kontext von Photographie und Photogrammetrie. In Reisig, W., & Freytag, J.-C. (Eds.), *Informatik* (pp. 315–328). Springer Berlin Heidelberg., Retrieved from http://link.springer.com/10.1007/3-540-32743-6_13 DOI: 10.1007/3-540-32743-6_13

Rhyma, P. P., Norizah, K., Hamdan, O., Faridah-Hanum, I., & Zulfa, A. W. (2020). Integration of normalised different vegetation index and Soil-Adjusted Vegetation Index for mangrove vegetation delineation. *Remote Sensing Applications: Society and Environment*, 17, 100280. DOI: 10.1016/j.rsase.2019.100280

Richards, J. A. (2022). Remote sensing digital image analysis, springer, pp. 263-367. DOI: 10.1007/978-3-030-82327-6

Rieucau, J. (2004). Dualisme urbain et glissement du centre décisionnel, de la cité prétouristique à la ville touristique, dans l'Espagne méditerranéenne du début du XXIe siècle. *Cahiers de Geographie de Quebec*, 48(133), 47–70. DOI: 10.7202/009762ar

Ritsema, C. J., Albaladejo, J., Castaldi, F., & Keesstra, S. D. (2022). Climatic and land-use drivers of desertification. *Journal of Arid Environments*, 201, 104689. DOI: 10.1016/j.jaridenv.2022.104689

RJ, K. (2024). Invasive weed optimization with deep transfer learning for multispectral image classification model. Multimedia Tools and Applications, 83(15), 45519-45534.

Rogan, J., Miller, J., Stow, D., Franklin, J., Levien, L., & Fischer, C. (2003). Land-Cover Change Monitoring with Classification Trees Using Landsat TM and Ancillary Data. *Photogrammetric Engineering and Remote Sensing*, 69(7), 793–804. DOI: 10.14358/PERS.69.7.793

Roose, E., & Barthès, B. (2001). Organic matter management for soil conservation and productivity restoration in Africa: a contribution from Francophone research. In *Managing Organic Matter in Tropical Soils: Scope and Limitations* (pp. 159–170). Springer Netherlands. https://doi.org/DOI: 10.1007/978-94-017-2172-1_15

Roose, E., Sabir, M., Arabi, M., Morsli, B., & Mazour, M. (2012). Soixante années de recherches en coopération sur l'érosion hydrique et la lutte antiérosive au Maghreb. *Physio-Géo*, 6(Volume 6), 43–69. DOI: 10.4000/physio-geo.2319

Roslim, M. H. M., Juraimi, A. S., Che'Ya, N. N., Sulaiman, N., Manaf, M. N. H. A., Ramli, Z., & Motmainna, M. (1809). Che'Ya, N. N., Sulaiman, N., Manaf, M. N. H. A., Ramli, Z., & Motmainna, Mst. (2021). Using Remote Sensing and an Unmanned Aerial System for Weed Management in Agricultural Crops: A Review. *Agronomy (Basel)*, 11(9), 1809. Advance online publication. DOI: 10.3390/agronomy11091809

Roy, P. S., Behera, M. D., & Srivastav, S. K. (2017). Satellite remote sensing: Sensors, applications and techniques. *Proceedings of the National Academy of Sciences. India. Section A, Physical Sciences*, 87(4), 465–472. DOI: 10.1007/s40010-017-0428-8

Rozenfeld, S., Kalo, N., Naor, A., Dag, A., Edan, Y., & Alchanatis, V. (2024). Thermal imaging for identification of malfunctions in subsurface drip irrigation in orchards. *Precision Agriculture*, 25(2), 1038–1066. DOI: 10.1007/s11119-023-10104-x

Sabrou, F., & Mazzine, H. (2020). La politique urbaine au Maroc: entre la ségrégation ethnique et socio-spatiale ». Ouvrage collectif: La gestion des espaces urbaines au Maroc dans un contexte mouvant. Centre d'Etudes et de recherches Takamul (D. Assalam, Ed.)., PP 187-204.

Sadiki, A., Bouhlassa, S., Auajjar, J., Faleh, A., & Macaire, JJ (2004). Use of a GIS for the assessment and mapping of erosion risks by the Universal Soil Loss Equation in the Eastern Rif (Morocco): case of the Boussouab wadi watershed. *Bulletin of the Scientific Institute, Rabat, Earth Sciences section, 26* (2004), 69-79.

Salhi, A., Benabdelouahab, T., El Hasnaoui, Y., El Moussaoui, M., El Mourabit, A., & Himi, M. (2019). Soil Erosion Assessment and Farmer's Perception in South Mediterranean Basins: A Moroccan Case Study. The 2nd Euro-Mediterranean Conference for Environmental Integration (EMCEI). Springer, Sousse, Tunisia.

Salhi, A., Benabdelouahab, T., Martin-Vide, J., Okacha, A., El Hasnaoui, Y., El Mousaoui, M., El Morabit, A., Himi, M., Benabdelouahab, S., Lebrini, Y., Boudhar, A., & Casas Ponsati, A. (2020c). Bridging the gap of perception is the only way to align soil protection actions. *The Science of the Total Environment*, 718, 137421. DOI: 10.1016/j.scitotenv.2020.137421 PMID: 32105933

Salhi, A., Martin-Vide, J., Benhamrouche, A., Benabdelouahab, S., Himi, M., Benabdelouahab, T., & Casas Ponsati, A. (2019). Rainfall distribution and trends of the daily precipitation concentration index in northern Morocco: A need for an adaptive environmental policy. *SN Applied Sciences*, 1(3), 277. DOI: 10.1007/s42452-019-0290-1

Samanta, D., & Ghosh, A. (2012). Histogram approach for detection of maize leaf damage. *International Journal of Computer Science and Telecommunications*, 3(12), 26–28.

Sampurno, R. M., & Ahamed, T. (2024). Challenges in Orchard Weed Management: Perspectives on the Use of 3D Cameras and LiDAR to Develop a Low-Cost Small-Scale Robotic Weeder. In Ahamed, T. (Ed.), *IoT and AI in Agriculture* (pp. 229–247). Springer Nature Singapore., DOI: 10.1007/978-981-97-1263-2_14

Sankaranarayanan, J., Samet, H., & Varshney, A. (2007). A fast all nearest neighbor algorithm for applications involving large point-clouds. *Computers & Graphics*, 31(2), 157–174. DOI: 10.1016/j.cag.2006.11.011

Santos Ferreira, A., Matte Freitas, D., Gonçalves da Silva, G., Pistori, H., & Theophilo Folhes, M. (2017). Weed detection in soybean crops using ConvNets. *Computers and Electronics in Agriculture*, 143, 314–324. DOI: 10.1016/j.compag.2017.10.027

Sanyal, P., Bhattacharya, U., Parui, S. K., Bandyopadhyay, S. K., & Patel, S. (2007). Color texture analysis of rice leaves to diagnose deficiency in the balance of mineral levels towards improvement of crop productivity. In IEEE.

Sarvini, T., & Sneha, T. GS, S. G., Sushmitha, S., & Kumaraswamy, R. (2019, April). Performance comparison of weed detection algorithms. In 2019 International Conference on Communication and Signal Processing (ICCSP) (pp. 0843-0847). IEEE.

Saskatchewan Mustard Development Commission. (2017, June 1). Mustard production manual. Retrieved January 15, 2019, from https://saskmustard.com/production-manual/plant-description/types-of-mustards-and-their-uses/index.html

Schöning, J., & Heidemann, G. (2015). Evaluation of Multi-view 3D Reconstruction Software. In *Lecture Notes in Computer Science (including subseries Lecture Notes in Artificial Intelligence and Lecture Notes in Bioinformatics)* (pp. 450–461). DOI: 10.1007/978-3-319-23117-4_39

Sebbar A (2013). *Etude de la variabilité et de l'évolution de la pluviométrie au Maroc (1935-2005): Réactualisation de la carte des précipitations* [Université Hassan II]. https://doi.org/.DOI: 10.1684/sec.2009.0169

Segarra, J. (2024). Satellite Imagery in Precision Agriculture. In P. M. Priyadarshan, S. M. Jain, S. Penna, & J. M. Al-Khayri (Eds.), *Digital Agriculture* (pp. 325–340). Springer International Publishing. DOI: 10.1007/978-3-031-43548-5_10

Segarra, J., Buchaillot, M. L., Araus, J. L., & Kefauver, S. C. (2020). Remote sensing for precision agriculture: Sentinel-2 improved features and applications. *Agronomy (Basel)*, 10(5), 641. DOI: 10.3390/agronomy10050641

Seiche, A. T., Wittstruck, L., & Jarmer, T. (2024). Weed Detection from Unmanned Aerial Vehicle Imagery Using Deep Learning—A Comparison between High-End and Low-Cost Multispectral Sensors. *Sensors (Basel)*, 24(5), 1544. DOI: 10.3390/s24051544 PMID: 38475081

Seitz, L., Haas, C., Noack, M., & Wieprecht, S. (2018). From picture to porosity of river bed material using Structure-from-Motion with Multi-View-Stereo. *Geomorphology*, 306, 80–89. DOI: 10.1016/j.geomorph.2018.01.014

Selvi, C. T., Sankara Subramanian, R. S., & Ramachandran, R. (2021). Weed detection in agricultural fields using deep learning process. In 2021 7th International Conference on Advanced Computing and Communication Systems (ICACCS) (vol. 1, pp. 1470–1473). DOI: 10.1109/ICACCS51430.2021.9441683

Seong, J. C. (2015). Sun position calculator (SPC) for Landsat imagery with geodetic latitudes. *Computers & Geosciences*, 85, 68–74. DOI: 10.1016/j.cageo.2015.09.011

Setiyadi, S., Mukhtar, H., & Cahyadi, W. A. (2021). A Comparative Study of Affordable Photogrammetry Software for Reconstructing 3D Model of a Human Foot. *2021 IEEE 7th International Conference on Smart Instrumentation, Measurement and Applications (ICSIMA)*, 286–291. IEEE. DOI: 10.1109/ICSIMA50015.2021.9526314

Shaban, A., & Khawlie, M. (1998). Geo-environmental assessment of riparian zones under extreme climatic events: a case study of representative rivers in Lebanon. Mediterranean rivers and riparian zones-processes and management symposium, Zaragosa, Spain, 2 1 September-2 October 1998, 25 p.

Shafiei, S., Gu, Z., Grzybowska, H., & Cai, C. (2023). Impact of self-parking autonomous vehicles on urban traffic congestion. *Transportation*, 50(1), 183–203. DOI: 10.1007/s11116-021-10241-0

Shah, V. K., Prajapati, H. B., & Shah, J. P. (2016). A survey on detection and classification of rice plant diseases. In IEEE. DOI: 10.1109/ICCTAC.2016.7567333

Sheirich, P. (2009). An Engineering Trade Space Analysis for a Space-Based Hyperspectral Chromotomographic Scanner.

Shen, Y., Lindenbergh, R., & Wang, J. (2016). Change Analysis in Structural Laser Scanning Point Clouds: The Baseline Method. *Sensors (Basel)*, 17(12), 26. DOI: 10.3390/s17010026 PMID: 28029121

Shidnal, S., & Latte, M. V. (2016). Multiple nutrient deficiency detection in paddy leaf images using color and pattern analysis. In IEEE, India.

Shi, Q., Liu, M., Li, S., Liu, X., Wang, F., & Zhang, L. (2021). A deeply supervised attention metric-based network and an open aerial image dataset for remote sensing change detection. *IEEE Transactions on Geoscience and Remote Sensing*, 60, 1–16.

Shi, Z. H., Fang, N. F., Wu, F. Z., Wang, L., Yue, B. J., & Wu, G. L. (2002). Assessment of Erosion Risk with the Rusle and Gis in the Middle and Lower Reaches of Hanjiang River. *Proceeding of the 12th ISCO Conference,* Beijing, 73–78.

Shorewala, S., Ashfaque, A., Sidharth, R., & Vijay, U. (2021). Weed density and distribution estimation for precision agriculture using semi-supervised learning. *International Journal of Information Technology : an Official Journal of Bharati Vidyapeeth's Institute of Computer Applications and Management*. Advance online publication. DOI: 10.1004/s41873-020-004374

Siddiqui, S. A., Fatima, N., & Ahmad, A. (2021). Neural network-based smart weed detection system. In *2021 International Conference on Communication, Control and Information Sciences (ICCISc)* (vol. 1, pp. 1-5). https://doi.org/DOI: 10.1109/ICCISc52257.2021.9484925

Singh, D., Dahiya, M., Kumar, R., & Nanda, C. (2021). Sensors and systems for air quality assessment monitoring and management: A review. *Journal of Environmental Management*, 289, 112510. DOI: 10.1016/j.jenvman.2021.112510 PMID: 33827002

Singh, G., Slonecki, T., Wadl, P., Flessner, M., Sosnoskie, L., Hatterman-Valenti, H., Gage, K., & Cutulle, M. (2024). Implementing Digital Multispectral 3D Scanning Technology for Rapid Assessment of Hemp (Cannabis sativa L.) Weed Competitive Traits. *Remote Sensing (Basel)*, 16(13), 2375. DOI: 10.3390/rs16132375

Singh, S. P., Jain, K., & Mandla, V. R. (2014). 3D Scene Reconstruction from Video Camera for Virtual 3D City Modeling. *American Journal of Engineering Research*, 140–148.

Singh, S., & Kasana, S. S. (2022). Quantitative estimation of soil properties using hybrid features and RNN variants. *Chemosphere*, 287, 131889. DOI: 10.1016/j.chemosphere.2021.131889 PMID: 34461337

Singh, U. P., Chouhan, S. S., Jain, S., & Jain, S. (2019). Multilayer convolution neural network for the classification of mango leaves infected by anthracnose disease. *IEEE Access : Practical Innovations, Open Solutions*, 7, 43721–43729. DOI: 10.1109/ACCESS.2019.2907383

Singla, K., & Nand, P. (2023). Optimizing deep learning architectures for novel view synthesis: Investigating the impact of NeRF MLP parameters on complex scenes. *International Journal of Information Technology : an Official Journal of Bharati Vidyapeeth's Institute of Computer Applications and Management*. Advance online publication. DOI: 10.1007/s41870-023-01470-w

Sivanpillai, R., & Whitman, R. D. (2019). Relating leaf spectral reflectance to its color: An inquiry-based activity to enhance understanding of electromagnetic radiation. *Science Activities*, 56(1), 19–26. DOI: 10.1080/00368121.2019.1638746

Sladojevic, S., Arsenovic, M., Anderla, A., Culibrk, D., & Stefanovic, D. (2016). Deep neural networks based recognition of plant diseases by leaf image classification. *Computational Intelligence and Neuroscience*, 2016, 1–11. DOI: 10.1155/2016/3289801 PMID: 27418923

Snavely, N., Seitz, S. M., & Szeliski, R. (2008). Modeling the World from Internet Photo Collections. *International Journal of Computer Vision*, 80(2), 189–210. DOI: 10.1007/s11263-007-0107-3

Snehmani, G. A., Ganju, A., Kumar, S., Srivastava, P. K., & Hari Ram, R. P. (2017). A comparative analysis of pansharpening techniques on QuickBird and WorldView-3 images. *Geocarto International*, 32(11), 1268–1284. DOI: 10.1080/10106049.2016.1206627

Soesbergen, A. van. (2016). *A review of land use change models.*

Solaimani, K., Modallaldoust, S., & Lotfi, S. (2009). Soil erosion prediction based on land use changes (A case in Neka wathershed). *American Journal of Agricultural and Biological Sciences*, 4(2), 97–104. DOI: 10.3844/ajabssp.2009.97.104

Solórzano, J. V., Mas, J. F., Gallardo-Cruz, J. A., Gao, Y., & de Oca, A. F. M. (2023). Deforestation detection using a spatio-temporal deep learning approach with synthetic aperture radar and multispectral images. *ISPRS Journal of Photogrammetry and Remote Sensing*, 199, 87–101. DOI: 10.1016/j.isprsjprs.2023.03.017

Son, N., Chen, C., & Chen, P. (2021). *A phenological object-based approach for rice crop classification using time-series Sentinel-1 synthetic aperture radar (SAR) data in Taiwan. Journal of Remote Sensing*. Retrieved from Taylor & Francis. DOI: 10.1080/01431161.2020.1862440

Sossey Alaoui F (2005). *Traitement et intégration des données satellitaires optiques et radar dans un sig en vue de l'élaboration de cartes de l'aléa lie aux instabilités de terrain dans la péninsule de Tanger (Rif septentrional, Maroc).* Universite Mohammed V, Faculté des Sciences.

Spalevic, V., Djurovic, N., Mijovic, S., Vukelic-Sutoska, M., & Curovic, M. (2013). Soil erosion intensity and runoff on the Djuricka River Basin (north of Montenegro). *The Malaysian Journal of Soil Science*, 17, 49–68.

Sultana, Q., Sultana, A., & Ara, Z. (2023). Assessment of the land use and landcover changes using remote sensing and GIS techniques. In *Water, Land, and Forest Susceptibility and Sustainability* (pp. 267-297). Elsevier. DOI: 10.1016/B978-0-323-91880-0.00022-2

Sun, A. Y., & Scanlon, B. R. (2019). How can Big Data and machine learning benefit environment and water management: A survey of methods, applications, and future directions. *Environmental Research Letters*, 14(7), 073001. DOI: 10.1088/1748-9326/ab1b7d

Sun, T., Cheng, W., Abdelkareem, M., & Al-Arifi, N. (2022). Mapping prospective areas of water resources and monitoring land use/land cover changes in an arid region using remote sensing and GIS techniques. *Water (Basel)*, 14(15), 2435. DOI: 10.3390/w14152435

Suter G (1980). Carte structurale de la chaîne rifaine au 1/500,000. *Notes et Mémoires Du Service Géologique Du Maroc*, 245.

Tamiminia, H., Salehi, B., Mahdianpari, M., Quackenbush, L., Adeli, S., & Brisco, B. (2020). Google Earth Engine for geo-big data applications: A meta-analysis and systematic review. *ISPRS Journal of Photogrammetry and Remote Sensing*, 164, 152–170. DOI: 10.1016/j.isprsjprs.2020.04.001

Tangestani, M. H. (2006). Comparison of EPM and PSIAC models in GIS for erosion and sediment yield assessment in a semi arid environment. *Journal of Asian Earth Sciences*, 27(5), 585–597. DOI: 10.1016/j.jseaes.2005.06.002

Tanvir, M. R., & Haque, A. (2023). Assessment of Development Plans for Small Urban Areas of Bangladesh: Remote Sensing and GIS-based Approach. *Applied Spatial Analysis and Policy*, 16(4), 1591–1625. DOI: 10.1007/s12061-023-09520-3

Tenhunen, H., Pahikkala, T., Nevalainen, O., Teuhola, J., Mattila, H., & Tyystjärvi, E. (2019). Automatic detection of cereal rows by means of pattern recognition techniques. *Computers and Electronics in Agriculture*, 162, 677–688. DOI: 10.1016/j.compag.2019.05.002

Tenze, L., & Canessa, E. (2024). altiro3d: Scene representation from single image and novel view synthesis. *International Journal of Information Technology : an Official Journal of Bharati Vidyapeeth's Institute of Computer Applications and Management*, 16(1), 33–42. DOI: 10.1007/s41870-023-01590-3

Tettamanzi, A. G., & Tomassini, M. (2013). *Soft computing: integrating evolutionary, neural, and fuzzy systems*. Springer Science & Business Media.

The World Bank Group. (2024). *Vietnam - Climatology*. https://climateknowledgeportal.worldbank.org/country/vietnam/climate-data-historical

The World Factbook. (2024). *Vietnam*. https://www.cia.gov/the-world-factbook/countries/vietnam/

Thomas, D. S. G., & Wiggs, G. F. S. (2021). Aeolian Processes and Dune Systems: Advances in Understanding and Challenges Ahead. *Earth-Science Reviews*, 216, 103555. DOI: 10.1016/j.earscirev.2021.103555

Tichouiti, K. (2024). NVDI-and NBR-based Comparison for Burnt Areas Characterization Using Sentinel-2 Imagery. In 2024 IEEE Mediterranean and Middle-East Geoscience and Remote Sensing Symposium (M2GARSS) (pp. 187-190). IEEE. DOI: 10.1109/M2GARSS57310.2024.10537321

Tillihal, S. B., & Shukla, A. K. (2023). River flood monitoring and management: a review. Advances in Water Resource Planning and Sustainability, 223-238. DOI: 10.1007/978-981-99-3660-1_13

Toğaçar, M. (2022). Using DarkNet models and metaheuristic optimization methods together to detect weeds growing along with seedlings. *Ecological Informatics*, 68, 101519. DOI: 10.1016/j.ecoinf.2021.101519

Tošić, R., Kapović, M., Dragićević, S., & Lovrić, N. (2013). Assessment of soil erosion potential using RUSLE and GIS: A case study of Bosnia and Herzegovina. *Fresenius Environmental Bulletin*, 22(11a), 3415–3423.

Tsegaye, L., & Bharti, R. (2021). Soil erosion and sediment yield assessment using RUSLE and GIS-based approach in Anjeb watershed, Northwest Ethiopia. *SN Applied Sciences*, 3(5), 582. DOI: 10.1007/s42452-021-04564-x

Tunçay, T., Kılıç, Ş., Dedeoğlu, M., Dengiz, O., Başkan, O., & Bayramin, İ. (2021). Assessing soil fertility index based on remote sensing and gis techniques with field validation in a semiarid agricultural ecosystem. *Journal of Arid Environments*, 190, 104525. DOI: 10.1016/j.jaridenv.2021.104525

Turner, B., & Meyer, W. (1994). Global Land-Use and Land-Cover Change: An Overview. In Meyer, W., & Turner, B. (Eds.), *Changes in Land Use and Land Cover: A Global Perspective* (pp. 3–10). Cambridge University Press.

U.S. Geological Survey. (2023a). *Landsat 5*. https://www.usgs.gov/landsat-missions/landsat-5

U.S. Geological Survey. (2023b). *Landsat 8*. https://www.usgs.gov/landsat-missions/landsat-8

Ullman, S. (1979). The Interpretation of Structure From Motion. In *The Interpretation of Visual Motion*. The MIT Press., DOI: 10.7551/mitpress/3877.003.0009

Van der Linden, S., Okujeni, A., Canters, F., Degerickx, J., Heiden, U., Hostert, P., Priem, F., Somers, B., & Thiel, F. (2018). Imaging spectroscopy of urban environments. *Surveys in Geophysics*, 40(3), 471–488. DOI: 10.1007/s10712-018-9486-y

Vanhellemont, Q., & Ruddick, K. (2018). Atmospheric correction of metre-scale optical satellite data for inland and coastal water applications. *Remote Sensing of Environment*, 216, 586–597. DOI: 10.1016/j.rse.2018.07.015

Vapnik, V. (1995). *The nature of statistical learning theory*. Springer Verlag. DOI: 10.1007/978-1-4757-2440-0

Vasileiadis, N., Tsoukala, E., & Kostopoulos, D. S. (2019). The late Miocene bovids from Platania (Drama Basin, Greece), with description of a new species of Palaeoryx. *Geobios*, 55, 57–76. DOI: 10.1016/j.geobios.2019.06.005

Veenendaal, B., Brovelli, M. A., & Li, S. (2017). Review of Web Mapping: Eras, Trends and Directions. *ISPRS International Journal of Geo-Information*, 6(10), 317. DOI: 10.3390/ijgi6100317

Victor, N., Maddikunta, P. K. R., Mary, D. R. K., Murugan, R., Chengoden, R., Gadekallu, T. R., Rakesh, N., Zhu, Y., & Paek, J. (2024). Remote Sensing for Agriculture in the Era of Industry 5.0—A Survey. *IEEE Journal of Selected Topics in Applied Earth Observations and Remote Sensing*, 17, 5920–5945. DOI: 10.1109/JSTARS.2024.3370508

Vikram, L., & Bhardwaj, M. (2023). Assessment of Urban Wastelands using GIS and IoT as Tools for Spatial Data Analysis. *Current World Environment*, 18(2), 893–903. DOI: 10.12944/CWE.18.2.35

Vioix, J.-B., Douzals, J.-P., Truchetet, F., Assémat, L., & Guillemin, J.-P. (2002). Spatial and Spectral Methods for Weed Detection and Localization. *EURASIP Journal on Advances in Signal Processing*, 2002(7), 793080. DOI: 10.1155/S1110865702204072

Wahid, K., Dinh, A., & Islam, M. (2017). Detection of potato diseases using image segmentation and multiclass support vector machine. In IEEE.

Wang, Y., Wu, B., & Li, L. (2022). *Urban Redevelopment and Traffic Congestion Management Strategies*. Springer Nature Singapore. DOI: 10.1007/978-981-19-1727-1

Wang, A., Xu, Y., Wei, X., & Cui, B. (2020). Semantic segmentation of crop and weed using an encoder-decoder network and image enhancement method under uncontrolled outdoor illumination. *IEEE Access : Practical Innovations, Open Solutions*, 8, 81724–81734. Advance online publication. DOI: 10.1109/ACCESS.2020.2991354

Wang, G., Sun, Y., & Wang, J. (2017). Automatic image-based plant disease severity estimation using deep learning. *Computational Intelligence and Neuroscience*, 2017, 1–8. DOI: 10.1155/2017/2917536 PMID: 28757863

Wang, L., Shang, Y. J., & Zhang, S. W. (2015). Plant disease recognition based on plant leaf image. *The Journal of Animal & Plant Sciences*, 25, 42–45.

Wang, L., Xue, X., Zhao, Z., & Wang, Z. (2018). The Impacts of Transportation Infrastructure on Sustainable Development: Emerging Trends and Challenges. *International Journal of Environmental Research and Public Health*, 15(6), 1172. DOI: 10.3390/ijerph15061172 PMID: 29874785

Wang, T., Hussain, A., Sun, Q., Li, S. E., & Jiahua, C. (2022). The Prediction of Urban Road Traffic Congestion by Using a Deep Stacked Long Short-Term Memory Network. *IEEE Intelligent Transportation Systems Magazine*, 14(4), 102–120. DOI: 10.1109/MITS.2021.3049383

Wang, Y., Chen, Y., & Peng, S. (2011). A GIS Framework for Changing Cropping Pattern Under Different Climate Conditions and Irrigation Availability Scenarios. *Water Resources Management*, 25(13), 3073–3090. DOI: 10.1007/s11269-011-9846-6

Wang, Y., Deng, N., Xin, B., Wang, W., Xing, W., & Lu, S. (2020). A novel three-dimensional surface reconstruction method for the complex fabrics based on the MVS. *Optics & Laser Technology*, 131, 106415. DOI: 10.1016/j.optlastec.2020.106415

Wang, Z., & Bovik, A. C. (2002). A universal image quality index. *IEEE Signal Processing Letters*, 9(3), 81–84. DOI: 10.1109/97.995823

Wellmann, T. (2020). *Remote sensing in urban planning: Contributions towards ecologically sound policies? Landscape and Urban Planning*. Elsevier., DOI: 10.1016/j.landurbplan.2020.103921

Westoby, M. J., Brasington, J., Glasser, N. F., Hambrey, M. J., & Reynolds, J. M. (2012). 'Structure-from-Motion' photogrammetry: A low-cost, effective tool for geoscience applications. *Geomorphology*, 179, 300–314. DOI: 10.1016/j.geomorph.2012.08.021

Wiatkowska, B., Słodczyk, J., & Stokowska, A. (2021). Spatial-temporal land use and land cover changes in urban areas using remote sensing images and GIS analysis: The case study of Opole, Poland. *Geosciences*, 11(8), 312. DOI: 10.3390/geosciences11080312

Wiggs, G. F. S., Livingstone, I., Thomas, D. S. G., & Bullard, J. E. (1996). Airflow and roughness characteristics over partially vegetated linear dunes in the southwest Kalahari Desert. *Earth Surface Processes and Landforms*, 21(1), 19–34. DOI: 10.1002/(SICI)1096-9837(199601)21:1<19::AID-ESP508>3.0.CO;2-P

Wischmeier, WH, Smith DD (1978). Predicting Rainfall Erosion Losses. A Guide to Conservation Planning. *The USDA Agricultural Handbook, 537.*

Wu, C. (2024). VisualSFM: A Visual Structure from Motion System. Retrieved March 27, 2024, from http://ccwu.me/vsfm/

Wu, W. (2022). Research on Urban Congestion Control Strategy Based on Smart Traffic Management System. *Journal of Humanities and Social Sciences Studies*, 4(1), 89–96. DOI: 10.32996/jhsss.2022.4.1.9

Xie, H., Zhang, Y., Zeng, X., & He, Y. (2020). Sustainable land use and management research: A scientometric review. *Landscape Ecology*, 35(11), 2381–2411. DOI: 10.1007/s10980-020-01002-y

Xie, Y. W., Yang, J. Y., Du, S. L., Zhao, J., Li, Y., & Huffman, E. C. (2012). A GIS-based fertilizer decision support system for farmers in Northeast China: A case study at Tong-le village. *Nutrient Cycling in Agroecosystems*, 93(3), 323–336. DOI: 10.1007/s10705-012-9519-8

Xie, Z., Lv, W., Huang, S., Lu, Z., Du, B., & Huang, R. (2020). Sequential Graph Neural Network for Urban Road Traffic Speed Prediction. *IEEE Access : Practical Innovations, Open Solutions*, 8, 63349–63358. DOI: 10.1109/ACCESS.2019.2915364

Xue, J., & Su, B. (2017). Significant remote sensing vegetation indices: A review of developments and applications. *Journal of Sensors*, 2017, 2017. DOI: 10.1155/2017/1353691

Xu, K., Shu, L., Xie, Q., Song, M., Zhu, Y., Cao, W., & Ni, J. (2023). Precision weed detection in wheat fields for agriculture 4.0: A survey of enabling technologies, methods, and research challenges. *Computers and Electronics in Agriculture*, 212, 108106. DOI: 10.1016/j.compag.2023.108106

Xu, Z., Zhang, Y., & Xu, X. (2016). 3D visualization for building information models based upon IFC and WebGL integration. *Multimedia Tools and Applications*, 75(24), 17421–17441. DOI: 10.1007/s11042-016-4104-9

Yadav, S. P., Singh, D., & Kumar, S. (2015). Studies on powdery mildew of rapeseed-mustard (Brassica juncea L.) caused by Erysiphe cruciferarum and its management. *Journal of Pure & Applied Microbiology*, 9(2).

Yang, C., Huang, Q., Li, Z., Liu, K., & Hu, F. (2017). Big Data and cloud computing: Innovation opportunities and challenges. *International Journal of Digital Earth*, 10(1), 13–53. DOI: 10.1080/17538947.2016.1239771

Yan, Y., Xin, Z., Bai, X., Zhan, H., Xi, J., Xie, J., & Cheng, Y. (2023). Analysis of growing season normalized difference vegetation index variation and its influencing factors on the Mongolian Plateau based on Google Earth Engine. *Plants*, 12(13), 2550. DOI: 10.3390/plants12132550 PMID: 37447111

Ying, B., Xu, Y., Zhang, S., Shi, Y., & Liu, L. (2021). Weed detection in images of carrot fields based on improved YOLO v4. *TS. Traitement du Signal*, 38(2), 341–348. DOI: 10.18280/ts.380211

Yuan, L., Huang, Y., Loraamm, R. W., Nie, C., Wang, J., & Zhang, J. (2014). Spectral analysis of winter wheat leaves for detection and differentiation of diseases and insects. *Field Crops Research*, 156, 199–207. DOI: 10.1016/j.fcr.2013.11.012

Yue, J., Tian, J., Tian, Q., Xu, K., & Xu, N. (2019). Development of soil moisture indices from differences in water absorption between shortwave-infrared bands. *ISPRS Journal of Photogrammetry and Remote Sensing*, 154, 216–230. DOI: 10.1016/j.isprsjprs.2019.06.012

Yu, J., Sharpe, S. M., Schumann, A. W., & Boyd, N. S. (2019). Deep learning for image-based weed detection in turfgrass. *European Journal of Agronomy*, 104, 78–84. DOI: 10.1016/j.eja.2019.01.004

Yun, S., Xianfeng, W., Shanwen, Z., & Chuanlei, Z. (2016). PNN based crop disease recognition with leaf image features and meteorological data. *International Journal of Agricultural and Biological Engineering*, 8(4), 61–68.

Žabota, B., Berger, F., & Kobal, M. (2023). The Potential of UAV-Acquired Photogrammetric and LiDAR-Point Clouds for Obtaining Rock Dimensions as Input Parameters for Modeling Rockfall Runout Zones. *Drones (Basel)*, 7(2), 104. DOI: 10.3390/drones7020104

Zahir, S. A. D. M., Omar, A. F., Jamlos, M. F., Azmi, M. A. M., & Muncan, J. (2022). A review of visible and near-infrared (Vis-NIR) spectroscopy application in plant stress detection. *Sensors and Actuators. A, Physical*, 338, 113468. DOI: 10.1016/j.sna.2022.113468

Zahnoun, A., & Al Karkouri, J. (2020). Estimation & Cartography the Water Erosion by Integration of the Gavrilovic "Epm" Model using a Gis in the Mediterranean Watershed: Oued Tleta (Western Rif, Morocco). *European Journal of Scientific Research*, ISSN 1450-216X / 1450-202X Vol. 155 No 3, 2020, pp.265 -278.

Zang, J., Jiao, P., Liu, S., Zhang, X., Song, G., & Yu, L. (2023). Identifying Traffic Congestion Patterns of Urban Road Network Based on Traffic Performance Index. *Sustainability (Basel)*, 15(2), 948. DOI: 10.3390/su15020948

Zeghmar, A., Marouf, N., & Mokhtari, E. (2022). Assessment of soil erosion using the GIS-based erosion potential method in the Kebir Rhumel Watershed, Northeast Algeria. *Journal of Water and Land Development*, 52. DOI: 10.24425/jwld.2022.140383

Zeng, Y., Ning, Z., Liu, P., Luo, P., Zhang, Y., & He, G. (2020). A mosaic method for multi-temporal data registration by using convolutional neural networks for forestry remote sensing applications. *Computing*, 102(3), 795–811. DOI: 10.1007/s00607-019-00716-5

Zhang, W., ... (2018). Broad-leaf weed detection in pasture. In 2018 IEEE 3rd International Conference on Image, Vision and Computing (ICIVC) (pp. 101–105). https://doi.org/DOI: 10.1109/ICIVC.2018.8492831

Zhang, J., Wang, Z., Bai, L., Song, G., Tao, J., & Chen, L. (2021, July). Deforestation Detection Based on U-Net and LSTM in Optical Satellite Remote Sensing Images. In *2021 IEEE International Geoscience and Remote Sensing Symposium IGARSS* (pp. 3753-3756). IEEE. DOI: 10.1109/IGARSS47720.2021.9554689

Zhang, J., Yu, F., Zhang, Q., Wang, M., Yu, J., & Tan, Y. (2024). Advancements of UAV and Deep Learning Technologies for Weed Management in Farmland. *Agronomy (Basel)*, 14(3), 494. DOI: 10.3390/agronomy14030494

Zhang, L., Du, B., You, J., & Tao, D. (2019). Hyperspectral image unsupervised classification by robust manifold matrix factorization. *Information Sciences*, 485, 154–169. DOI: 10.1016/j.ins.2019.02.008

Zhang, Y., Luo, X., Yang, W., & Yu, J. (2019). Fragmentation Guided Human Shape Reconstruction. *IEEE Access : Practical Innovations, Open Solutions*, 7, 45651–45661. DOI: 10.1109/ACCESS.2019.2905879

Zhao, W., Zhang, S., Wang, B., & Zhou, B. (2023). Spatio-temporal causal graph attention network for traffic flow prediction in intelligent transportation systems. *PeerJ. Computer Science*, 9, e1484. DOI: 10.7717/peerj-cs.1484 PMID: 37547406

Zhou, L., Zhang, S., Yu, J., & Chen, X. (2020). Spatial–Temporal Deep Tensor Neural Networks for Large-Scale Urban Network Speed Prediction. *IEEE Transactions on Intelligent Transportation Systems*, 21(9), 3718–3729. DOI: 10.1109/TITS.2019.2932038

Zhu, L., Suomalainen, J., Liu, J., Hyyppä, J., Kaartinen, H., & Haggren, H. (2018). A review: Remote sensing sensors. Multi-purposeful application of geospatial data, 19.

Zou, Y., & Greenberg, J. A. (2019). A spatialized classification approach for land cover mapping using hyperspatial imagery. *Remote Sensing of Environment*, 232, 111248. DOI: 10.1016/j.rse.2019.111248

About the Contributors

Mouhcine Batchi is a prominent academic figure specializing in the intricate realms of Geographic Information Systems (GIS), Remote Sensing, and human-environment interactions. Holding the esteemed position of Vice-Dean in Charge of Academic Affairs at the Faculty of Humanities and Social Sciences, University of Ibn Tofail, Dr. Batchi plays a pivotal role in shaping academic endeavors and fostering intellectual growth within the institution. Education: Dr. Batchi completed his doctoral studies in Geography at Ibn Tofail University in Kenitra, Morocco, in 2014. His doctoral dissertation titled "Agro-spatial changes and risks of environmental degradation in the coastal zone of Gharb (North-Western Morocco)" showcased his profound insights into the dynamic interplay between agricultural practices and environmental sustainability. Prior to his doctoral studies, he earned a Master's degree in Environmental Geography from Ibn Tofail University, where his thesis focused on the spatial estimation of water erosion utilizing GIS and remote sensing techniques. Appointments: In addition to his current role as Vice-Dean in Charge of Academic Affairs, Dr. Batchi has held several key positions within the academic sphere. Since 2016, he has served as a Professor in the Department of Geography at the University of Ibn Tofail, where his expertise and dedication have contributed significantly to the academic community. Furthermore, he assumed the role of Head of the Department of Geography in January 2021, further solidifying his commitment to academic leadership and excellence. Dr. Batchi's contributions extend beyond his immediate department, as he has been an active member of the Ibn Tofail University Board, serving on both the pedagogical and scientific commissions. Teaching Experience: Dr. Batchi's passion for education is evident through his extensive teaching experience at both undergraduate and master's degree levels. At the undergraduate level, he has taught a diverse array of courses ranging from Cartographic Semiology to Mediterranean Geography, equipping students with fundamental knowledge and analytical skills crucial for geographical studies. Additionally, he has supervised numerous undergraduate students, guiding them through their academic journey and fostering their research capabilities. At the master's degree level, Dr. Batchi has imparted specialized knowledge in Geomatics, Geographic Information Systems (GIS), and Remote Sensing, providing students with advanced tools and methodologies essential for cutting-edge research and practical applications. Research Projects: Dr. Batchi's scholarly pursuits extend beyond the confines of the classroom, as evidenced by his involvement in various research projects. Noteworthy among these is his participation in the realization of the Kenitra SDAU project (Urban Development Plan) from 2012 to 2014, demonstrating his practical engagement in addressing urban development challenges. Additionally, he has been an integral member of the SHS N°2 research project, focusing on spatial and environmental changes in the coastal zone of Gharb (North West Morocco) in the context of climate change. His research endeavors underscore his commitment to addressing pressing environmental issues and advancing knowledge in the field of geography. Dr. Mouhcine Batchi's contributions to academia, research, and institutional leadership exemplify his unwavering dedication to excellence and innovation in geographical studies. His multifaceted expertise and visionary leadership continue to enrich the academic community at the University of Ibn Tofail and beyond.

* * *

Sanaa Cheikh is a Ph.D. student in Geography, specializing in Urban Transportation Geography at Abdelmalek Essaâdi University, Morocco. Alongside her studies, she works in secondary education and has expertise in Geographic Information Systems (GIS) and remote sensing. She has participated in several national and international conferences, presenting research related to urban mobility and GIS applications. Her research interests focus on sustainable urban transportation, spatial analysis, and the integration of remote sensing technologies for urban planning.

Nisserine Ben Driss holds a Ph.D. in Geodynamics (specializing in Hydrogeology) from the Complutense University of Madrid. She is a Professor of Higher Education in the Department of Geography at the Faculty of Letters and Human Sciences, Abdelmalek Essaâdi University, Tétouan. Her main research areas include hydrogeology, the valorization of natural resources, the promotion of local products, and data collection and analysis using software such as Surfer, Grapher, Aquachem, Cartalinx, ArcGIS, and Aquifer Test. Dr. Ben Driss has participated in numerous national and international conferences and colloquia. She supervises research projects on a variety of topics and coordinates the undergraduate program in Space and Environment, as well as the Master's program in Space, Environment, and Sustainable Development.

Kumud Dubey holds Ph.D. (Forest Ecology). Has more than 30 years of research experience in field of Biotechnology, Environment Conservation, Biodiversity Conservation /Forest Ecology and Reclamation / Restoration / Bioremediation/ Social Forestry. Executed several research projects in field of Restoration, Forest Conservation, Forest Ecology, Forest Types, Forestry, Biodiversity and Forest Policy, funded from organizations like ICFRE, NITI Aayog, NOVOD Board, UPCAR, MoEFCC, MoWR, DST, DBT, CAMPA etc. Executed various extension & plantation programmes. Supervising Ph.D. in field of Forest Ecology, Climate Change, Conservation Social Sciences and Biotechnology. Supervising dissertations of M.Sc. and B.Sc. students. Published several research papers (more than 100) in reputed National and International journals. Published one book with LAP LAMBERT Academic Publishing, Germany. Honoured by several reputed organizations. More than 120 international and national publications and Author of One Book published by Lambert Academic Publication, Germany.

Karim Ennouri*Bacillus thuringiensis***.**

Monia Ennouri studied food science engineering at National Engineering School of Sfax, Tunisia and graduated as engineer in 1999, She then joined the University of Montpellier II (sciences et technique) for a master degree. She received her phD degree in 2006 at Faculty of Pharmacy of Montpellier and University of Sfax, She obtained in 2006 a position of an assistant professor at ISSAT Mahdia University of Monastir (Tunisia) and a position of associate professor at 2017 in the same university. Since 2013 she is an expert in sensory analysis of olive oil. Since 2020 she is an associate professor at olive tree institute of Sfax (Tunisia). She has published more than 28 research articles in food science field.

Ayoub Lahlouh holds a Ph.D. in Geography from Abdelmalek Essaâdi University, Morocco. He has extensive experience in geospatial data analysis, cartography, and Geographic Information Systems (GIS), working with both academic and municipal institutions. Dr. Lahlouh has collaborated on several projects focusing on urban planning, digital transformation, and governance, particularly within the M'diq Municipality, where he pioneered the development of specialized geoportals. His research interests include spatial analysis, remote sensing, urban mobility, and the integration of AI in geospatial systems. Dr. Lahlouh is also involved in freelance consulting, where he has provided GIS services to international clients. Currently, his work emphasizes the automation of geocoding processes and the application of machine learning for urban analysis.

Liem Nguyen graduated from Nong Lam University, Ho Chi Minh City, Vietnam, with a Bachelor's degree in Geographic Information System (GIS) in 2011. He later earned a Master's degree in Environmental and Resource Management from the same university in 2022. With a solid foundation in GIS and related fields—such as Global Positioning Systems, Remote Sensing, Surface Modeling, Modeling and Optimization, Applied Mathematics, Cartography, and Geodesy—he has developed extensive expertise in scientific research and the application of GIS in addressing natural resource and environmental issues. His skill set includes data creation, data transformation, projections, cartography, visualization, statistical analysis, GPS data collection, and visual interpretation. His research currently focuses on GIS, Remote Sensing, the Soil and Water Assessment Tool (SWAT) model, and Information Communication Technology (ICT).

Faran Masood Peerzada is a graduate of the computer systems engineering program through his bachelor's degree, as well as a master's degree in mechatronics engineering. He is now working on some projects related to remote sensing and machine learning. Faran has been engaged in programming over a long period, of which he is proficient in Python. His job is to bring about advanced technologies which in turn facilitate the solving of intricate problems in remote sensing. On the one hand, Faran applies his engineering and programming skills with true leadership to innovate in his field.

Ngan Pham holds a Bachelor's degree in Geographic Information Systems (GIS) from Nong Lam University, Ho Chi Minh City (2011) and a Master's degree in Environmental and Resource Management from the same university (2017). She has extensive experience in GIS and remote sensing applications, particularly in environmental and resource management, gained through her roles as a GIS Specialist at the Southern Sub-Institute of Forest Inventory and Planning (2011-2016) and at the Center for Youth Science and Technology Development in Ho Chi Minh City (2017-2018). Proficient in software tools such as ArcGIS, MapInfo, AutoCAD, ENVI, ERDAS, and SNAP, and skilled in programming with VBA, R, and Python, Ms. Ngan has contributed to various research projects, including mangrove forest resource evaluation and national forest inventory. She has published scientific papers on GIS applications and participated in numerous workshops on data science and advanced GIS techniques, demonstrating her commitment to professional growth and research in her field.

Zeeshan Ali Shah obtained his MS in Electrical Engineering from the National University of Sciences and Technology (NUST), Islamabad, Pakistan, and his B.Sc. in Electrical Engineering from the University of Engineering and Technology (UET),

Lahore, Pakistan. He is currently serving as a Lecturer at the University of Wah, Wah Cantt, Pakistan.

Mohamed Ali Triki is Director of Research Laboratory " Genetic Ressources of Olive Tree: Caracterisation, Valorisation and Phytosanitary Protection". He obtained his PhD in Biology in 2009 from Faculty of Science of Sfax, Tunisia and he obtained his diploma of Ingineer in Horticulture in 1992 from Superior Horticulture Institute of Chott-Mariem - Sousse, Tunisia. He works on etiology of die-back and wilting disease of olive tree caused by soilborne fungi in nurseries and in the field and biological control of soilborne fungi causing root and collar rot of olive trees by autochthones antagonistic bacteria and by plant extracts. Besides, he works on New multidisciplinary approach for reducing uncertainties in risk assessment of *Xylella fastidiosa* in Tunisia.

Kyriaki Tychola received the Diploma Degrees from the Department of Geoinformation and Surveying from the Technological Educational Institute, Serres, Greece and in Landscape Architecture from the Technological Educational Institute of Kavala, Greece in 2005 and 2013 respectively and M.Sc Degree from the Department of Geography and applied Geoinformatics in 2022 specializing in GIS and Photogrammetry. She has publications in international scientific journals in computer vision field. As a researcher, she is a member of the Machine Learning and Vision (MLV) Research Group, where she participates in research projects.

Vandita is a Ph.D. scholar of Forest Ecology & Climate Change Division, Forest Research Institute Deemed University, Dehradun, India. She holds M.Sc. degree in Environmental Sciences from University of Allahabad, India. She also holds UGC NET-JRF award, and her doctoral research is in the field of Forest Ecology. She also has research experience in the field of Atmospheric Sciences and data analysis using RS & GIS.

Index

Symbols

A

B

C

D

E

F

G

I

K

L

M

P

R

S

T

U

V

W

www.ingramcontent.com/pod-product-compliance
Lightning Source LLC
LaVergne TN
LVHW062321300926
849161LV00007B/1526